6

7

8

9

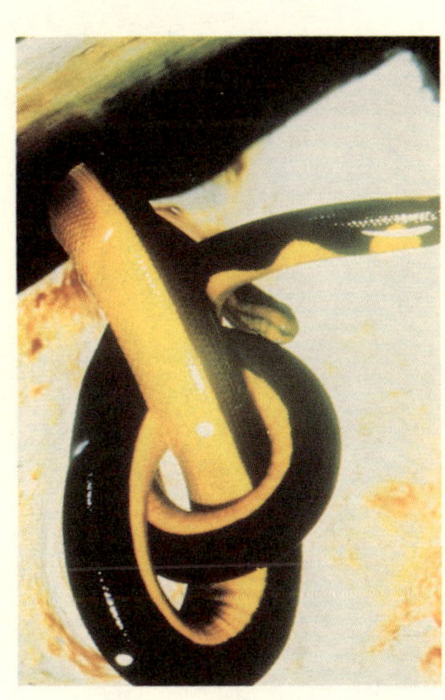

10

6. *Laticauda semifesciata*. Photo by J. Sneed.
7. *Laticauda colubrina*. Yellow-lipped sea krait. Photo by W. Farmer.
8. *Aipysurus laevis*. Olive sea snake, Great Barrier Reef. Photo by Bio-Oceanics.
9. *Pelamis platurus*. Pelagic sea snake beached at Puerto Vallarta, Mexico. Photo by G. V. Pickwell.
10. *Pelamis platurus* knotting. Pelagic sea snake. Photo by W. Farmer.

The Venomous Sea Snakes
A Comprehensive Bibliography

The Venomous Sea Snakes
A Comprehensive Bibliography

by

WENDY A. CULOTTA
University Library
California State University, Long Beach
Long Beach, California

and

GEORGE V. PICKWELL
Environmental Sciences Division
Naval Ocean Systems Center
San Diego, California

KRIEGER PUBLISHING COMPANY
MALABAR, FLORIDA
1993

Original Edition 1993

Printed and Published by
KRIEGER PUBLISHING COMPANY
KRIEGER DRIVE
MALABAR, FLORIDA 32950

Library of Congress Cataloging-in-Publication Data

Culotta, Wendy A., 1943–
  The venomous sea snakes: a comprehensive bibliography/by Wendy A. Culotta and George V. Pickwell.
      p. cm.
  Includes bibliographical references and index.
  ISBN 0-89464-469-6
  1. Sea snakes—Bibliography. I. Pickwell, George V. II. Title.
Z7996.S64C84   1991
[QL666.O645]
016.59796'09162—dc20   90-42244
                                                                CIP

10 9 8 7 6 5 4 3 2

BEYOND THE SHADOW OF THE SHIP,
I WATCHED THE WATER-SNAKES:
THEY MOVED IN TRACKS OF SHINING WHITE,
AND WHEN THEY REARED, THE ELFISH LIGHT
FELL OFF IN HOARY FLAKES.

* * * * * * * * * * * * *

O HAPPY LIVING THINGS! NO TONGUE
THEIR BEAUTY MIGHT DECLARE:
A SPRING OF LOVE GUSHED FROM MY HEART,
AND I BLESSED THEM UNAWARE . . .

from THE RIME OF THE ANCIENT MARINER
by Samuel Taylor Coleridge

Dedication

This work is respectfully dedicated to the memories of four who loved science, natural history, and herpetology and contributed much to those fields. For us they were scholars to emulate, mentors, colleagues, and friends.

Richard B. Loomis
1925–1985
Professor of Biology
California State University, Long Beach

Charles E. Shaw
1918–1971
Curator of Herpteology
San Diego Zoological Society

James P. Bacon
1939–1986
Curator of Herpetology
San Diego Zoological Society

Sheldon Campbell
1919–1985
Chairman of the Board of Trustees
San Diego Zoological Society

Contents

Foreword

Assembling the world's literature on seasnakes is a formidable task. Not only is the technical literature scattered through an amazing variety of journals, but it almost appears as though many workers, sensitive to the esoteric nature of their subject, sought out the most obscure outlets for their work. The casual and anecdotal literature is even more widely dispersed. Seasnake references touch on so many different fields of biology and chemistry that no one person is likely to have detailed familiarity with all relevant fields. Persons familiar with the taxonomic and ecological literature are seldom up-to-date on the research on biochemistry and venomology, yet both topics have received considerable attention and have a rather voluminous bibliography. All these aspects are of importance to students of this group of reptiles. The breadth of the job of comprehensively compiling the literature has daunted even the bravest of herpetologists.

Investigators of seasnakes have long recognized the need for a bibliography. Some years ago, driven by the sheer urgency for such a work, I began compiling one and found that G. O. Vigle was simultaneously engaged in the same project. We published our modest efforts as a joint paper but soon became aware that we had barely scratched the surface. Some years later I reconciled myself to the task of preparing a more complete bibliography, again to find that someone else was pursuing the same goal, this time J. C. Enderman of the Netherlands. As before, we decided to combine forces. In contacting various colleagues to solicit any obscure reference they might have, we learned of the present treatise, then nearing completion. It was with considerable relief that I finally abandoned the attempt in the knowledge that a thoroughly researched and comprehensive bibliography to the literature of seasnakes was finally on its way. No one appreciates better than I do, the years of painstaking labour that clearly went into the present work. The authors are to be congratulated for a job well done. It will be a most valuable aid to students and researchers of seasnakes. It is indeed a pleasure to join with Sherman Minton in introducing such a timely and useful publication.

Harold Heatwole
University of New England
Armidale, NWS, 2351
Australia

Preface

Tales of serpent-like sea monsters are as old as human history and pervade the mythology of peoples around the world. However, a very real group of marine snakes occurs in many parts of the tropics and subtropics where they are of interest and concern to persons working in disciplines as diverse as coral reef ecology, diving medicine, neurophysiology and aquaculture. The marine snakes represent a biologically interesting group of organisms, partly because of the morphological and physiological adaptations that have allowed them to go from a terrestrial to a marine environment and partly because of their role in marine ecosystems. They must be considered in man's growing concern with the biological resources of the sea and their utilization and conservation. Additionally, these snakes are of direct medical importance as a potential cause of injury or death to humans in the marine environment.

With one exception, the peculiar and primitive file snake *(Acrochordus granulatus)*, the existing marine snakes belong to a group of about fifty species of proteroglyphs related to the venomous land snakes of Australia and New Guinea. From a probable site of origin in this region, they have dispersed extensively through the tropical Indian and Pacific oceans from Fiji to the Persian Gulf. One species, the pelagic or yellow bellied sea snake *(Pelamis platurus)* has the widest range of any snake species with stray specimens recorded from points as remote as California, New Zealand, the Kuril Islands, and the Cape of Good Hope. No other sea snake species is so widely distributed or is found far from continental shelves, but several species are locally abundant and make up a significant part of the shallow water marine vertebrate fauna.

Some sea snakes regularly dive to depths of at least 50 m, and voluntary submergence times may be up to an hour. Most sea snakes are specialized in their choice of food. Eels and marine catfish are among the groups preyed upon. Neuroparalytic venoms evidently help with prey capture and also provide an effective defense. Two sea snake species eat only fish eggs and have a much reduced venom apparatus. Most sea snakes give birth to their young in open water, however the sea kraits *(Laticauda)* go ashore to lay eggs.

Sea snake venoms are among the most lethal of biological toxins. Their actions include blocking of nerve impulse transmission and breakdown of skeletal muscle. Sea snake bites are not uncommon among fishing peoples of southern Asia and are occasional among bathers and others engaged in aquatic recreation. Because sea snakes have small amounts of venom and a rather inefficient injecting system, many bites do not result in serious poisoning.

Some species of sea snakes can be captured in large numbers with relative ease, and their skins are a significant source of reptile leather. Commercial sea snake fisheries operate in the Philippines, Ryukyus, and northern Australia. Sea snake meat is eaten by some peoples, and the organs are believed to have medicinal value.

Information on sea snakes has been accumulating piecemeal for more than two centuries in both popular and scientific literature. Because the snakes are most plentiful in places remote from the centers of scientific activity and are animals that are not easily transported or maintained in captivity, most of the early work was descriptive and based on examination of preserved specimens. Yet it was not until 1926 that Malcolm Smith's *Monograph of the Sea-Snakes* provided a comprehensive taxonomic treatment of the group. The last half of this century has brought vast advances in scientific technology as well as expansion of scientific activity beyond the boundaries of Europe and North America. A result has been a great increase in knowledge of organisms in their environment and better understanding of their biochemistry and behavior. Yet, insofar as sea snakes are concerned, there is but one book, *The Biology of Sea Snakes*, edited by William A. Dunson (1975) that treats these reptiles in some detail [with two smaller, more recent works contributing important summaries (Mao and Chen, 1980; Heatwole, 1987)], although a large and growing number of short communications are devoted to sea snake biology and particularly to their venoms and neurotoxins. Information on incidence, clinical aspects, and treatment of sea snake bites is almost wholly confined to short reports in a wide range of medical journals.

This first comprehensive compilation of the literature on sea snakes is a highly significant undertaking. It brings together in readily available form an enormous number of references scattered through the scientific and popular literature and presents them in an organized fashion. This work provides ready access to material important for marine biologists, herpetologists, biochemists, physiologists, physicians, and others interested in these remarkable reptiles.

Sherman A. Minton, M. D.
Professor Emeritus
Department of Microbiology and Immunology
Indiana University School of Medicine
Indianapolis, Indiana

Introduction

The Science Librarian's Perspective

The Purpose of the Bibliography. The germ of the idea for this bibliography began in the summer of 1972 when I worked the summer between graduate school years at the Technical Library of what was then called the Naval Undersea Center (now Naval Ocean Systems Center). I was interested in doing a Masters project on sea snakes. The head librarian knew of one of the research scientist's (GVP) interest in, and past publications regarding these animals and introduced us. Pickwell offered to contribute the contents of his extensive collection of sea snake literature in exchange for the completed bibliographic project which I would do as part of my requirements for completing the Masters of Library and Information Science at UCLA (reference 24–2). This formed the beginning of our nearly twenty-year collaboration.

When I began my professional work at California State University, Long Beach, in 1973, online searching was just coming into heavy usage. In 1974 UCLA offered a free year of searches as part of an introductory effort to bring this service to the attention of researchers as well as librarians. This seemed an appropriate start for a potentially productive research association. The culmination is the present volume. This bibliography contains approximately 2500 references in twenty-four subject categories (chapters) and several indices including a genus-species index following each chapter.

Criteria for Inclusion. It has been our intention to collect as comprehensively as possible all works on sea snakes from the earliest literature of the ancient Greeks through the literature of 1986, with many additional references from 1987 to 1990. We concentrated on the scientific literature and rejected all references that seemed to contain only hearsay information. We also tended to reject the simple repetitive mention of sea snakes without substantive treatment often found in current text books, dictionaries, and some encyclopedias. Valuable, but generally nonscientific observations were collected in a chapter on nontechnical references.

The Entry. Each entry follows, for the most part, the *CBE Style Manual* (5th edition, Bethesda, MD: Council of Biology Editors; 1983). When a serial is cited, the journal abbreviation is listed but the full title can be found in the list of journals included in this volume. Occasionally there is a brief note indicating the original language if this is not apparent from the title. Infrequently we mentioned other editions of the reference if we had seen and reviewed these works. All names and words with an umlaut are spelled out, as Buecherl for Bücherl. All names beginning with Mac, M' or Mc are arranged in a strict alphabetical order. When a thesis (Masters) or a dissertation (Doctoral) is cited, the school granting the degree is noted. When a patent or a technical report is cited, all possible information is included. Reprints include the date and publisher responsible.

Each citation was verified, where possible, in major indexing sources. Among those used were *Zoological Record, Biological Abstracts, Index Medicus*, the Royal Society's *Catalogue of Scientific Papers*, the book catalogs from the University of California at Berkeley and Los Angeles, from Harvard University's *Catalogue of the Museum of Comparative Zoology*, from the book catalogue of the British Museum, and the *National Union Catalog*. We also used major online indexes such as OCLC, MELVYL, ORION, and online databases available through DIALOG, BRS, and STN.

Each reference in this compendium has been seen and evaluated. However, we also have a file of references that we have been unable to locate for one reason or another. Sometimes the material was too old and valuable to loan through interlibrary loan, or too large to copy. Occasionally the reference may have been incorrectly cited over and over again in the literature. We elected not to perpetuate these "ghost" citations that we were unable to locate.

References were gathered initially from computer and manual searches of the indexes named above, augmented by the holdings in Pickwell's library, and further enhanced by the references available in the bibliographies of major authors. Papers received from colleagues active in sea snake research were a further valuable source. We found that a considerable number of papers resulting from current sea snake research were not indexed in a timely manner, and often not indexed at all.

The Indexes. The author index is comprehensive for all authors whether a single author was responsible or many. Each chapter except the last has a genus-species index. This allowed us to coordinate the species discussed with the subject aspect of the paper. All species of sea snakes mentioned in the references were included, the name indexed exactly as the author published. it. Where a specific name was not mentioned, the name of the animal as the author used it was indexed, hence you may see "sea snake," "hydrophid," "Hydrophiidae," "serpent de mer," etc. The inclusion of such non-genus/species designations was our attempt to make each reference accessible by at least one other means in addition to the author index.

The list of journals cited, from abbreviated to full title, was also included as an aid for those who might have a question about the source journal cited in the entry. The list is arranged alphabetically. Below the abbreviated title is the full title as we verified it in the reference sources mentioned above.

Searching the Online Literature. Online databases of many indexing and abstracting services were searched. Among the indexes used regularly were *Biological Abstracts, Chemical Abstracts, Index Medicus, Oceanic Abstracts, Science Citation Index*. And, of course, a sometimes lengthy list of subject terms and authors known to be active in sea snake research was essential for effective use of the online databases. Nevertheless, such terms did not by themselves prevent the unwary searcher from occasionally drowning in the online literature.

Wendy A. Culotta
Long Beach, California

Introduction

The Marine Biologist's Perspective

We have attempted here to be both comprehensive and selective. This work is our effort to present the world scientific literature on the venomous sea snakes plus useful ancillary accounts from the popular press on the one hand, and yet not burden the interested investigator with unnecessarily superficial or peripheral references on the other. We list no publications here that we have not personally seen. Our definition of sea snakes includes all taxa traditionally placed in the Hydrophiidae of Malcolm Smith (1926).

We have made numerous more or less arbitrary judgments on whether to include or reject an item and our decisions may be questioned in some cases by colleagues who might have selected differently. Many references make brief mention of sea snakes or some aspect of their biology often in a comparative context. In these cases we based inclusion or rejection on our judgment of whether the brief remarks constituted a useful addition to our general fund of knowledge on the subject. A few seemingly peripheral works were included when they offered useful (even if negative) information—for example, authoritative statements regarding fossil snakes and the lack of affinities with present day sea snakes. We have included some abstracts, especially when they gave useful data or references, and also when they served to establish priority. Occasionally the abstracts reported research for which the definitive report has never appeared.

We have determined individually whether anecdotal accounts (even occasional newspaper articles) should be included here. Until perhaps the last two to three decades much of our information on sea snakes came from such sources. A number of aspects of sea snake biology, if covered at all, still rely upon such anecdotal and incidental reportings made by nonscientists, for example, the phenomenon of sea snake "rafting" or the formation of very dense aggregations of snakes at the sea surface.

The problem of inclusion or rejection of anecdotal observations became especially challenging in the case of the skin and SCUBA diving literature of the past thirty years. There has always been a great need for accurate observations of the animals in their natural habitat, but a considerable tendency on the part of divers without scientific training to exaggerate or misinterpret the behavior observed. We have for this reason included most of the interesting, but not necessarily scientific, reportings in a separate nontechnical chapter where they are available to the interested reader, but are not categorized as though we considered them scientifically authoritative. We also included in this chapter works intended largely for the layman that were written by recognized authorities in herpetology.

Papers describing work where sea snake products, such as toxins, were used as research tools were not generally included unless they also contributed useful information and details about that specific product. Published summaries of information on snakes that included sea snakes, were referenced here if they appeared useful by placing the sea snakes in a larger evolutionary, ecological or other biological context.

In the cases of some chapters such as 1, 3, or 4, for example, the inclusion or rejection of particular references was, in fact, arbitrary and reflected our opinion of where the topic or aspect most emphasized in any specific work might lie. We have thus avoided the massive cross referencing that would be required to thoroughly integrate the entire body of sea snake literature in favor of the simpler approach of subdividing the whole literature into logical subunits (chapters). Nevertheless many references are listed in two or more chapters. This is especially true of items in chapters 4 and 5. In chapter 6 we refer to evolution as that based largely on anatomical or morphological features in contrast to discussions of evolutionary affinities based on toxin molecular structure, especially primary structure and amino acid sequence homology, which was placed in chapter 16.

For us this has been a long-term collaboration begun some twenty years ago and pursued mostly as a spare time effort until we neared the final stages of checking and rechecking the many papers gathered over the years. Many of our colleagues have helped with advice, criticism and encouragement as well as supplying us with reprints of their own published work on sea snakes. Nevertheless, all errors of omission or commission remain ours alone. We would be grateful to any scientists who might care to contact us regarding major references we may have overlooked.

George V. Pickwell
San Diego, California

Acknowledgments

Many colleagues have aided us in this effort. We especially have benefited from the scholarly advice of the late Dr. Richard Loomis, California State University at Long Beach. Drs. Anthony Tu and Harold Heatwole generously provided copies of their own compiled bibliographies on aspects of sea snake venoms and sea snake biology. Drs. Sherman Minton and Harold Voris have both reviewed the entire manuscript and have made a number of helpful suggestions and comments.

The library personnel of the Naval Ocean Systems Center libraries and especially the interlibrary loan librarians during the past twenty years have proved an invaluable resource in this task. The Interlibrary Loan group at California State University Long Beach, Cathrine Lewis Ida, Sharlene LaForge, and George Ouwendijk have most recently contributed greatly to the completion of the bibliography. We have benefited from the labors of many interlibrary loan technicians and other library personnel in various institutions throughout the country and overseas.

A number of colleagues have been particularly important to us and this work. We thank Frank S. Shipp, Jr. for unflagging attention to detail in our observation and studies of living sea snakes and for his general encouragement, interest, and support during the entirety of this project; George B. Anderson for early interest and insistence that we consider production of a compilation of some sort on sea snakes; the late Dr. Gilbert Curl for encouraging one of us (GVP) as a fledgling scientist at Navy Electronics Laboratory to pursue an interest in these fascinating animals; Dr. Eric G. Barham for warm hearted and enthusiastic support for all aspects of our studies on sea snakes in the field, the laboratory and the library; and Dr. George Zug for helpful suggestions regarding possible publishers for our work.

Two colleagues from the CSULB Department of Biology who contributed their special knowledge were Dr. Ju-Shey Ho who translated the Japanese references and helped us understand the Japanese literature and Dr. Charles Galt who helped us with the translation of some of the old German articles. Lloyd Kramer, retired Associate Director of the CSULB Library, translated all the Russian articles.

At CSULB we acknowledge the administrative assistance which contributed to the completion of this project and highlight the support of Jordan Scepanski, Roman Kochan, and Gretchen Johnson, who are currently serving in the CSULB Library administration, and Dr. Peter Spyers-Duran, Helen Britton, and Lloyd Kramer who have served in past years.

How can a project of this length be completed without the support of our families? Wendy Culotta thanks her parents, Willard and Dorothy Eder, her sister Carol and Carol's husband Byron Dieckman, and her own husband Dr. M. Charles Culotta for their never-ceasing support of her efforts to complete this work.

LIST OF JOURNALS CITED

Includes Abbreviated Titles to Full Titles

Abh. Geb. Naturwiss. Nat. Hamburg
*Abhandlungen aus dem Gebiet der Naturwissenschaften hrsg. vom Naturwis-
senschaftlichen Verein in Hamburg*

Abh. K. Bayer. Akad. Wiss. Math.-Phys. Kl.
*Abhandlungen der Bayerischen Adademie der Wissenschaften. Mathematisch-
Naturwissenschaftliche Klasse. Muenchen*

Abh. Mus. Dresden
*Abhandlungen und Berichte des Koeniglichen Zoologischen und Anthropolo-
gisch-Ethnographischen Muscums zu Dresden*

Abh. Senckenb. Naturforsch. Ges.
Abhandlungen der Senckenbergischen Naturforschenden Gesellschaft

Abstr. Upssal. Diss. Sci.
Abstracts of Uppsala Dissertations in Science

Abstr. West. Soc. Nat. Annu. Meeting
Abstracts of the Western Society of Naturalists Annual Meeting

Acarologia

Acta Anat. Nippon
Acta Anatomica Nipponica

Acta Endocrinol.
Acta Endocrinologica

Acta Herpetol. Jpn.
Acta Herpetologica Japonica

Acta Herpetol. Sin.
    Acta Herpetologica Sinica

Acta Histochem.
    Acta Histochemica

Acta Med. Univ. Kagoshima.
    Acta Medica Universitatis Kagoshimaensis

Acta Trop.
    Acta Tropica

Acta Zool. Sin.
    Acta Zoologica Sinica

Actes Soc. Linn. Bordeaux
    Actes de la Societe Linneenne de Bordeaux

Actual. Pharmacol.
    Actualities Pharmacologiques

Adv. Anat. Embryol. Cell Biol.
    Advances in Anatomy, Embryology and Cell Biology

Adv. Behav. Biol.
    Advances in Behavioral Biology

Adv. Cytopharmacol.
    Advances in Cytopharmacology

Adv. Enzymol.
    Advances in Enzymology

Adv. Neurol. Sci.
    Advances in Neurological Sciences (see also: Shinkei Kenkyu No Shinpo)

Adv. Physiol. Sci.
    Advances in Physiological Sciences

Afr. Wildl.
    African Wildlife

Africana

Agric. Food Chem.
  Agricultural and Food Chemistry

Akad. Wissenschaften, Vienna. Math.-Naturw. Situngsb.
  Akademie der Wissenschaften, Vienna. Mathematische-Naturwissenschaftliche Klasse. Sitzungsberichte

Am. J. Physiol.
  American Journal of Physiology

Am. J. Trop. Med.
  American Journal of Tropical Medicine

Am. J. Trop. Med. Hyg.
  American Journal of Tropical Medicine and Hygiene

Am. Midl. Nat.
  American Midland Naturalist

Am. Month. Mag. Crit. Rev.
  American Monthly Magazine and Critical Review

Am. Mus. Nat. Hist. Central Asiatic Expedition
  American Museum of Natural History Central Asiatic Expedition

Am. Mus. Novitates
  American Museum Novitates

Am. Nat.
  American Naturalist

Am. Sci.
  American Scientist

Am. Zool.
  American Zoologist

Amoeba (Tokyo)

Amphib.-Reptilia
  Amphibia-Reptilia

An. Inst. Biol. Univ. Nac. Auton. Mex.
  Anales del Instituto de Biologia Universidad Nacional Autonoma de Mexico

Anaesthesiol. Intens. Care Med.
   Anaesthesiology and Intensive Care Medicine

Anal. Soc. Esp. Hist. Nat.
   Anales de la Sociedad Espanola de Historia Natural. Madrid

Anim. Kingdom
   Animal Kindgom

Anim. Kingdom Rept.
   Animal Kingdom Reports

Ann. Biol.
   Annee Biologique, Paris

Ann. Carnegie Mus.
   Annals of the Carnegie Museum

Ann. Clin. Lab. Sci.
   Annals of Clinical and Laboratory Science

Ann. Emerg. Med.
   Annals of Emergency Medicine

Ann. Hyg. Med. Colon.
   Annales d'Hygiene et de Medecine Coloniales

Ann. Inst. Oceanogr.
   Annales de l'Institut Oceanographique

Ann. Inst. Pasteur (Paris)
   Annales de l'Institut Pasteur. Paris

Ann. Mag. Nat. Hist.
   Annals and Magazine of Natural History

Ann. Mus. Civ. Genova
   Annali del Museo Civico di Storia Naturale di Genova

Ann. Mus. Nat. Hist. Nat. Paris
   Annales de la Museum National d'Histoire Naturelle. Paris

Ann. Parasitol. Hum. Comp.
   Annales de Parasitologie Humaine et Comparee

Ann. Philos.
  Annals of Philosophy

Ann. Queensl. Mus.
  Annals of the Queensland Museum

Ann. Rep. Sado Mar. Biol. Stn Niigata Univ.
  Annual Report of the Sado Marine Biological Station Niigata University

Ann. S. A. Mus.
  Annals of the South African Museum

Ann. Sci. Nat.
  Annales des Sciences Naturelles

Ann. Sci. Nat. Zool. Biol. Anim.
  Annales des Sciences Naturelles. Zoologie et Biologie Animale

Ann. Soc. Belg. Med. Trop.
  Annales de la Societe Belge de Medicine Tropicale

Ann. Transvaal Mus.
  Annals of the Transvaal Museum

Ann. Zool. (Agra)
  Annals of Zoology. Agra

Annexe Bull. Gen. Instr. Pub.
  Annexe au Bulletin General de l'Instruction Publique

Annot. Zool. Jpn.
  Annotationes Zoologicae Japonenses

Annu. Rep. Inst. Med. Res. (Kuala Lumpur)
  Annual Report of the Institute for Medical Research. Kuala Lumpur

Annu. Rep. Sado Mar. Biol. Stn Niigata Univ.
  Annual Report of the Sado Marine Biological Station Niigata University

Annu. Rep. Sankyo Res. Lab.
  Annual Report of Sankyo Research Laboratories

Annu. Rep. U.S. Natl Mus.
  Annual Report of the U.S. National Museum

Annu. Rev. Biochem.
  Annual Review of Biochemistry

Annu. Rev. Pharmacol.
  Annual Review of Pharmacology

Aquarien Terrarien

Aquasphere

Arch. Biochem. Biophys.
  Archives of Biochemistry and Biophysics

Arch. Mus. Hist. Nat. Lyon
  Archives du Museum d'Histoire Naturelle de Lyon

Arch. Mus. Nat. Hist. Nat.
  Archives du Museum National d'Histoire Naturelle

Arch. Nat. (Berlin)
  Archive fuer Naturgeschichte. Berlin

Arch. Neerl. Zool.
  Archives Neerlandaises de Zoologie

Arch. Schiffs- Tropen- Hyg.
  Archiv fuer Schiffs- und Tropen-Hygiene

Arch. Sci. Med. Colon.
  Archivio Italiano di Scienze Mediche Coloniali

Argus (Melbourne)

Asia Pac. J. Pharmacol.
  Asia Pacific Journal of Pharmacology

Asiat. Res.
  Asiatick Researches

Atoll Res. Bull.
  Atoll Research Bulletin

Atti Accad. Veneto-Trent.-Istriana
  Atti dell'Accademia Scientifica Veneto-Trentino-Istriana

Atti Soc. Nat. Mat. Modena
  Atti della Societa dei Naturalisti e Matematici di Modena

Auk

Aus der Natur (Berlin)

Aust. Fam. Physician
  Australian Family Physician

Aust. J. Exp. Biol. Med. Sci.
  Australian Journal of Experimental Biology and Medical Science

Aust. J. Herpetol.
  Australian Journal of Herpetology

Aust. J. Mar. Freshwater Res.
  Australian Journal of Marine and Freshwater Research

Aust. J. Zool.
  Australian Journal of Zoology

Aust. Med. Gaz.
  Australian Medical Gazette

Aust. Mus. Mag.
  Australian Museum Magazine

Aust. Nat.
  Australian Naturalist

Aust. Nat. Hist.
  Australian Natural History

Aust. Terr.
  Australian Territories

Aust. Wildl. Heritage
  Australia's Wildlife Heritage

Australas. J. Pharm. Sci. Suppl.
  Australasian Journal of Pharmacy. Science Supplement

Australas. Nurses J.
  Australiasian Nurses Journal

Bangladesh J. Zool.
  Bangladesh Journal of Zoology

Basteria

Batavia Nat. Tidjsch.
  Naturkundig Tijdschrift voor Nederlandsch-Indie

BBC Wildl.
  BBC Wildlife

Beitr. Palaeont. Geol. Oest. Ung.
  Beitraege zur Palaeontologie von Geologie Oesterreich-Ungarns u. des Orients

Ber. Offenbacher Vereins Naturkd.
  Bericht des Offenbacher Vereins fuer Naturkunde

Ber. Senckenb. Nat. Ges. Frankfurt a. Main
  Bericht der Senckenbergischen Naturforschenden Gesellschaft in Frankfurt am Main

Bihang K. Svenska VetenskAkad. Handl.
  Bihang till K. Svenska Vetenskapasakademiens Handlingar

Bijdr. Dierkd.
  Bijdragen tot de Dierkunde

Biochem. Arch.
  Biochemical Archives

Biochem. Biophys. Res. Commun.
  Biochemical and Biophysical Research Communications

Biochem. J.
  Biochemical Journal

Biochem. Pharmacol.
  Biochemical Pharmacology

Biochem. Soc. Trans.
  Biochemical Society, London. Transactions

Biochemistry

Biochim. Biophys. Acta
  Biochimica et Biophysica Acta

Biochimie (Paris)
  Biochimie. Paris

Biol. Action
  Biology in Action

Biol. Bull. (Woods Hole)
  Biological Bulletin. Woods Hole

Biol. J. Linn. Soc.
  Biological Journal of the Linnean Society

Biol. Jaarb. (Ghent)
  Biologisch Jaarboek (Ghent)

Biol. Morya (Vladivost.)
  Biologiya Morya. Vladivostok

Biol. Rev. Camb. Philos. Soc.
  Biological Reviews of the Cambridge Philosophical Society

Biologia

Biologist

Biomed. News
  Biomedical News: Newspaper for the Life Sciences

Bioorg. Khim.
  Bioorganicheskaya Khimiya

Biophys. J.
  Biophysical Journal

Biopolymers

Biosci. Rep.
   Bioscience Reports

Biotropica

Blaett. Aquar.-u. Terr. (Stuttgart)
   Blaetter fuer Aquarien.- und Terrarienkunde. Stuttgart

Bol. Mus. Nac. Hist. Nat. Chile
   Boletin Museo Nacional de Historia Natural, Chile

Boll. Mus. Ist. Biol. Univ. Genova
   Bollettino dei Musei e degli Istituti Biologici dell'Universita di Genova

Br. Herpetol. Soc. Newsl.
   British Herpetological Society Newsletter

Br. J. Herpetol.
   British Journal of Herpetology

Br. J. Pharmacol.
   British Journal of Pharmacology

Br. Med. J.
   British Medical Journal

Brain Res.
   Brain Research

Bull. Acad. R. Belgique. Class Sciences
   Bulletin de l'Academie R. de Belgique. Classe des Sciences

Bull. Am. Mus. Nat. Hist.
   Bulletin of the American Museum of Natural History

Bull. Antivenin Inst. Am.
   Bulletin of the Antivenin Institute of America

Bull. Bingham Ocean. Coll.
   Bulletin of the Bingham Oceanographic Collection

Bull. Biogeogr. Soc. Jpn
   Bulletin of the Biogeographical Society of Japan

Bull. Biol. Soc. Wash.
   Bulletin of the Biological Society of Washington

Bull. Br. Mus. (Nat. Hist.) Zool.
   Bulletin of the British Museum (Natural History) Zoology

Bull. Br. Ornithol. Club
   Bulletin of the British Ornithologists' Club

Bull. Chic. Herpetol. Soc.
   Bulletin of the Chicago Herpetological Society

Bull. Dept. Biol. Yenching Univ.
   Bulletin of the Department of Biology. Yenching University

Bull. Emirates Nat. Hist. Group (Abu Dhabi)
   Bulletin of the Emirates Natural History Group (Abu Dhabi)

Bull. Essex Inst.
   Bulletin of the Essex Institute

Bull. Fan Meml Inst. Biol. (Peiping)
   Bulletin of the Fan Memorial Institute of Biology. Peiping

Bull. Field Mus. Nat. Hist.
   Bulletin of the Field Museum of Natural History

Bull. Gen. Instr. Pub.
   Bulletin General de l'Instruction Publique

Bull. Haffkine Inst.
   Bulletin of the Haffkine Institute

Bull. Harv. Univ. Mus. Comp. Zool.
   Bulletin of the Harvard University Museum of Comparative Zoology (see also
   Bull. Mus. Comp. Zool.)

Bull. Inst. Pasteur
   Bulletin de l'Institut Pasteur

Bull. Kyoto Gakugei Univ. Ser. B Math. Nat. Sci.
   *Bulletin of the Kyoto Gakugei University Series B Mathematics and Natural
   Science*

Bull. Madras Gov. Mus. (Nat. Hist. Sect.)
   Bulletin of the Madras Government Museum (Natural History Section)

Bull. Mar. Sci.
   Bulletin of Marine Science

Bull. Mar. Sci. Gulf Caribb.
   Bulletin of Marine Science of the Gulf and Caribbean

Bull. Md Herpetol. Soc.
   Bulletin of the Maryland Herpetological Society

Bull. Mus. Comp. Zool.
   Bulletin of the Museum of Comparative Zoology (see also Bull. Harv. Univ. Mus. Comp. Zool.)

Bull. Mus. Natl Hist. Nat.
   Bulletin du Museum National d'Histoire Naturelle

Bull. Mus. R. Hist. Nat. Belg.
   Bulletin du Musee R. d'Histoire Naturelle de Belgique. Bruxelles

Bull. N.Y. Zool. Soc.
   Bulletin of the New York Zoological Society

Bull. Osaka Mus. Nat. Hist.
   Bulletin of the Osaka Museum of Natural History

Bull. Post-Grad. Comm. Med. Univ. Sydney
   Bulletin of the Post Graduate Committee in Medicine, University of Sydney

Bull. Raffles Mus.
   Bulletin of the Raffles Museum

Bull. Sado Mus.
   Bulletin of the Sado Museum

Bull Sci. Nat.
   Bulletin des Sciences Naturelles et de Geologies

Bull. Sci. Nat. Geol.
   Bulletin Universel des Sciences et de l'Industrie

Bull. Soc. Cent. Aquic. Peche
  Bulletin de la Societe Centrale d'Aquiculture et de Peche

Bull. Soc. Fribourg. Sci. Nat.
  Bulletin de la Societe Fribourgeoise des Sciences Naturelles

Bull. Soc. Pathol. Exot.
  Bulletin de la Societe de Pathologie Exotique

Bull. Soc. Philomath. Paris
  Bulletin de la Societe Philomathique de Paris

Bull. Stn Biol. Arcachon
  Bulletin de la Station Biologique d'Arcachon

Bull. U.S. Natl Mus.
  Bulletin of the United States National Museum

Bull. W. H. O.
  Bulletin of the World Health Organization

Bull. West. Aust. Fish. Dept. Fauna
  Bulletin of the Western Australian Fishers Department. Fauna

Bull. Zool. Nomencl.
  Bulletin of Zoological Nomenclature

Cah. Pac.
  Cahiers du Pacifique

Calif. Fish Game
  California Fish and Game

Caldasia

Can. J. Zool.
  Canadian Journal of Zoology

Cell. Mol. Biol.
  Cellular and Molecular Biology

Ceylon Med. J.
  Ceylon Medical Journal

Chem. Engr. News
    Chemical & Engineering News

Chem. Zool.
    Chemical Zoology

China J. (Shanghai)
    China Journal. Shanghai

Chromosoma (Berl.)
    Chromosoma. Berlin

Chronmy Przyr. Ojczysta
    Chronmy Przyrode Ojczysta

Clin. Med.
    Clinical Medicine

Clin. Nephrol.
    Clinical Nephrology

Clin. Pharmacol. Therap.
    Clinical Pharmacology and Therapeutics

Clin. Toxicol.
    Clinical Toxicology

Collect. Culture
    Collection and Culture

Colloq. Int. Cent. Natl Rech. Sci.
    Colloques Internationaux du Centre National de la Recherche Scientifique

Commun. Behav. Biol. Part A Orig. Artic.
    Communications in Behavioral Biology Part A Origional Articles

Commun. Mus. Nac. Buenos Aires
    Communicaciones del Museo Nacional de Historia Natural de Buenos Aires

Comp. Biochem. Physiol.
    Comparative Biochemistry and Physiology

Comp. Biochem. Physiol. A Comp. Physiol.
    Comparative Biochemistry and Physiology A Comparative Physiology

Comp. Biochem. Physiol. B Comp. Biochem.
  Comparative Biochemistry and Physiology B Comparative Biochemistry

Consultant (Phila.)
  Consultant. Philadelphia

Contemp. Neurol. Ser.
  Contemporary Neurology Series

Contrib. Life Sci. Div. R. Ont. Mus.
  Contributions. Life Science Division, Royal Ontario Musuem. Toronto

Copeia

Country Life

C. R. Acad. Sci. Paris
  Compte Rendu Hebodomadaire des Seances de l'Academie des Sciences

Cranbrook Inst. Sci. Bull.
  Cranbrook Institute of Science Bulletin

CRC Crit. Rev. Biochem.
  CRC Critical Reviews in Biochemistry

Crit. Rev. Biochem.
  Critical Reviews in Biochemistry

Crustaceana

Curr. Sci.
  Current Science

Dana-Rep. Carlsberg Found.
  Dana-Report Carlsberg Foundation

Denkschr. Math. Nat. Kl. Akad. Wiss. Wien.
  Denkschriften der Akademie der Wissenschaften, Mathematisch-Naturwissenschaftliche Klasse. Wien

Desalination

Dev. Biochem.
  Developmental Biology

Discovery (Lond.)
  Discovery. London

Diss. Abstr. Int. B Sci. Eng.
  Dissertation Abstracts International B Sciences and Engineering

Diver Int.
  Diver International

Dobutsugaku Zasshi
  Dobutsugaku Zasshi (Zoological Society of Japan)

Drum Croaker
  Drum and Croaker

Dtsch. Arch. Physiol.
  Deutsches Archiv fuer die Physiologie

Ecology

Edinb. Med. J.
  Edinburgh Medical Journal

Ekologiya

Emerg. Med. Clin. North Am.
  Emergency Medicine Clinics of North America

Emu

Eur. J. Biochem.
  European Journal of Biochemistry

Eur. J. Pharmacol.
  European Journal of Pharmacology

Everyday Sci.
  Everyday Science

Evolution

Excur. Reconn.
  Excursions et Reconnaissances

Exp. Brain Res.
  Experimental Brain Research

Experientia (Basel)

Explor. J.
  Explorers Journal

FASEB (Fed. Am. Soc. Exp. Biol.) J.
  FASEB (Federation of American Societies for Experimental Biology) Journal

Fathom

Fauna

FEBS (Fed. Eur. Biochem. Soc.) Lett.
  FEBS (Federation of European Biochemical Societies) Letters

Fed. Proc.
  Federation Proceedings

Field (Lond.)
  Field. London

Field Mus. Nat. Hist. Publ. Zool. Ser.
  Field Museum of Natural History Publications, Zoological Series

Fieldiana Zool.
  Fieldiana Zoology

Filos. Let. Quito Univer. Cent. Ecuador
  Filosofía y Letras. Quito, Universidad Central del Ecuador

Fla. Sci.
  Florida Scientist

Fld Stream
  Field and Stream

Flora (Quito)

Folia Pharmacol. Jpn.
    Folia Pharmacologica Japonica

Folletos Divulg. Cient. Inst. Biol.
    Folletos de Divulgacion Cientifica. Instituto de Biologia. Chapultepec

Fortschr. Naturw. Forsch.
    Forteschritte der Naturwissenschaftlichen Forschung

Frontiers
    Frontiers, a Magazine of Natural History

Fujian Shida Xuebao

Garcia de Orta (Lisb.)
    Garcia de Orta. Lisbon

Gaz. Hosp. Civil. Mil.
    Gazette des Hospitaux Civils et Militaires

Gazelle Dubai Nat. Hist. Group Newsl.
    Gazelle, the Dubai Natural History Group Newsletter

Gen. Comp. Endocrinol.
    General and Comparative Endocrinology

Geneesk. Tijdschr. Ned.-Indie
    Geneeskundig Tijdschrift voor Nederlandsch-Indie

Geo
    Geo, Australasia's Geographical Magazine

Geol. Soc. Am. Mem.
    Geological Society of America Memoir

Great Basin Nat.
    Great Basin Naturalist

Gunma J. Med. Sci.
    Gunma Journal of Medical Science

Guy's Hosp. Rep.
    Guy's Hospital Reports

Handb. Zool.
    Handbuch der Zoologie. Berlin

Helminthologica (Bratisl.)
    Helminthologia. Bratislava

Helv. Chim. Acta
    Helvetica Chimica Acta

Herpetol. Rev.
    Herpetological Review

Herpetofauna

Herpetologica

High Polymers (Kobun Shi)

Hong Kong Univ. Fish. J.
    Hong Kong University Fisheries Journal

Hongkong Nat.
    Hongkong Naturalist

Honolulu Star Bull.
    Honolulu Star Bulletin

Illawarra Nat. Hist. Soc.
    Illawarra Natural History Society. Wallongong, N.S.W.

Illus. Weekly India
    Illustrated Weekly of India

Ind. Med. Surg.
    Industrial Medicine and Surgery

Indian Ann. Mag. Nat. Sci.
    Indian Annals and Magazine of Natural Science

Indian J. Biochem. Biophys.
    Indian Journal of Biochemistry and Biophysics

Indian J. Med. Res.
   Indian Journal of Medical Research

Indian J. Zootomy
   Indian Journal of Zootomy

Indian Med. Gaz.
   Indian Medical Gazette

Indian Sci. Cong. Assoc. Proc.
   Indian Science Congress Association Proceedings

Indones. J. Nat. Sci.
   Indonesian Journal for Natural Science

Indust. Med. Surg.
   Industrial Medicine and Surgery

Inst. Pasteur Viet-Nam Rapp. Annu. Fonct. Tech.
   Institut Pasteur du Viet-Nam Rapport Annuel sur le Fonctionnement Technique

Int. J. Parasitol.
   International Journal for Parasitology

Int. J. Pept. Protein Res.
   International Journal of Peptide and Protein Research

Int. Wildl.
   International Wildlife

Int. Zoo Yearb.
   International Zoo Yearbook

Inter-Am. Trop. Tuna Comm. Bull.
   Inter-American Tropical Tuna Commission Bulletin

Isis von Oken

Izv. Akad. Nauk Gruz. S. S. R. Ser. Biol.
   Izvestiya Akademii Nauk Gruzinskoi S. S. R. Seriya Biologicheskaya

J. Acad. Nat. Sci. Phil.
Journal of the Academy of Natural Sciences of Philadelphia

J. Am. Med. Assoc.
Journal of the American Medical Association

J. Anat.
Journal of Anatomy

J. Anim. Ecol.
Journal of Animal Ecology

J. Anim. Morphol. Physiol.
Journal of Animal Morphology and Physiology

J. Anat. Physiol.
Journal of Anatomy and Physiology

J. Anthropol. Inst. London
Journal of the Anthropological Institute, London

J. Asiat. Soc. Beng.
Journal of the Asiatic Society of Bengal. Calcutta

J. Biochem. (Tokyo)
Journal of Biochemistry. Tokyo

J. Biol. Chem.
Journal of Biological Chemistry

J. Bombay Nat. Hist. Soc.
Journal of the Bombay Natural History Society

J. Chin. Biochem. Soc.
Journal of the Chinese Biochemical Society

J. Col.- Wyo. Acad. Sci.
Journal of the Colorado-Wyoming Academy of Science

J. Coll. Sci. Imp. Univ. Tokyo
Journal of the College of Science, Imperial University of Tokyo

J. Comp. Physiol.
Journal of Comparative Physiology

J. Comp. Physiol. B Biochem. Syst. Environ. Physiol.
Journal of Comparative Physiology B Biochemical Systemic Environmental
Physiology

J. Electron Microsc.
Journal of Electron Microscopy

J. Exp. Biol.
Journal of Experimental Biology

J. Exp. Zool.
Journal of Experimental Zoology

J. Fac. Sci. Hokkaido Univ. Zool.
Journal of the Faculty of Science, Hokkaido University. Zoology

J. Fed. Malay States Mus.
Journal of the Federated Malay States Museums. Kuala Lumpur

J. Fish. Res. Board Can.
Journal of the Fisheries Research Board of Canada

J. Formosan Med. Assoc.
Journal of the Formosan Medical Association

J. Helminthol.
Journal of Helminthology

J. Herpetol.
Journal of Herpetology

J. Histochem. Cytochem.
Journal of Histochemistry and Cytochemistry

J. Linn. Soc. Lond. Zool.
Journal of the Linnean Society of London. Zoology

J. Malay. Branch R. Asiat. Soc.
Journal of the Malayan Branch Royal Asiatic Society. Singapore

J. Mar. Biol. Assoc. India
Journal of the Marine Biological Association of India

J. Med. Entomol.
    Journal of Medical Entomology

J. Mol. Biol.
    Journal of Molecular Biology

J. Mol. Evol.
    Journal of Molecular Evolution

J. Morphol.
    Journal of Morphology

J. N. China Br. R. Asiat. Soc.
    Journal of the North China Branch of the Royal Asiatic Society. Shanghai

J. Nat. Hist. Soc. Siam (Bangkok)
    Journal of the Natural History Society of Siam. Bangkok

J. Nat. Hist. Soc. Siam Suppl.
    Journal of the Natural History Society of Siam. Natural History Supplement

J. No. Ohio Assoc. Herpetol.
    Journal of the Northern Ohio Association of Herpetology

J. Parasitol.
    Journal of Parasitology

J. Path.
    Journal of Pathology

J. Pharm. Pharmacol.
    Journal of Pharmacy and Pharmacology

J. Pharmacol. Exp. Ther.
    Journal of Pharmacology and Experimental Therapeutics

J. Phys.
    Journal de Physique

J. Phys. Chem.
    Journal of Physical Chemistry

J. Physiol. (Lond.)
    Journal of Physiology (London)

J. Proc. Asiat. Soc. Beng.
  Journal and Proceedings of the Asiatic Society of Bengal

J. Protein Chem.
  Journal of Protein Chemistry

J. R. Soc. West. Aust.
  Journal of the Royal Society of Western Australia

J. Saudi Arabian Nat. Hist. Soc.
  Journal of the Saudi Arabian Natural History Society

J. Sci. Math. Phys. Nat. (Lisboa)
  Jornal de Sciencias Mathematicas, Fiscicas et Naturais (Lisboa)

J. Soc. Bibliogr. Nat. Hist.
  Journal of the Society for the Bibliography of Natural History

J. South Pac. Underwater Med. Soc.
  Journal of the South Pacific Underwater Medicine Society

J. Straits Br. R. Asiat. Soc.
  Journal of the Straits Branch of the Royal Asiatic Society. Singapore

J. Theor. Biol.
  Journal of Theoretical Biology

J. Toxicol. Toxin Rev.
  Journal of Toxicology Toxin Reviews

J. Trans. Victoria Inst.
  Journal of the Transactions of the Victoria Institute. London

J. Trop. Med. Hyg.
  Journal of Tropical Medicine and Hygiene

J. Univ. Bombay
  Journal of the University of Bombay

J. Zool. (Lond.)
  Journal of Zoology. London

J. Zool. Soc. India
  Journal of the Zoological Society of India

Jahrb. Hamb. Wiss. Anst.
  Jahrbuch der Hamburgischen Wissenschaftlichen Anstalten

Jahrb. Nassau. Naturk.
  Jahrbuecher des Nassauischer verein fuer Naturkunde

Jena. Z. Med. Naturwiss.
  Jenaische Zeitschrift fuer Medizin und Naturwissenschaft

Jena. Z. Naturwiss.
  Jenaische Zeitschrift fuer Naturwissenschaft

Jpn. J. Bacteriol.
  Japanese Journal of Bacteriology

Jpn. J. Exp. Med.
  Japanese Journal of Experimental Medicine

Jpn. J. Herpetol.
  Japanese Journal of Herpetology

Jpn. J. Leg. Med.
  Japanese Journal of Legal Medicine (Nippon Hoigaku Zasshi)

Jpn. J. Pharmacol.
  Japanese Journal of Pharmacology

Jpn. J. Sanit. Zool.
  Japanese Journal of Sanitary Zoology (Eisei Dobutsu)

Jpn. J. Zool.
  Japanese Journal of Zoology

Jpn. Med. Res. Found. Publ.
  Japan Medical Research Foundation Publication

K. Akad. Wet. Amsterdam. Afd. Nat. Proc.
  Kungliga Akademie van Wetenschappen, Amsterdam. Aftdeeling Natu-urkunde. Proceedings

K. Svenska Vetensk.-Akad. Handl.
  Kungliga Svenska Vetenskapsakademiens Handlingar

Kagaku, Zokan (Kyoto)

Kagoshimaken Hakubutsu Chosajhe
   Reports on Natural History of the Kagoshima Prefecture

Kon. Inst. Tropen Amsterdam
   Koninklijk Instituut voor de Tropen. Amsterdam

Lacerta

Lancet

Land and Water

Life Sci.
   Life Sciences

Limnol. Oceanogr.
   Limnology and Oceanography

Lingnan Sci. J.
   Lignan Science Journal

Loris

Macromolecules

Madras Fish. Bull.
   Madras Fisheries Bulletin

Madras Quart. J. Med. Sci.
   Madras Quarterly Journal of Medical Science

Malay. Nat. J.
   Malayan Nature Journal

Malays. J. Sci.
   Malayasian Journal of Science

Managua

Mar. Biol. (Berl.)
  Marine Biology. Berlin

Mar. Life; Occas. Pap.
  Marine Life; Occasional Papers

Mar. Observer
  Marine Observer

Mar. Res. Indones.
  Marine Research in Indonesia

Med. J. Aust.
  Medical Journal of Australia

Med. J. Kagoshima Univ.
  Medical Journal of Kagoshima University

Med. J. Malaya
  Medical Journal of Malaya

Med. J. Malays.
  Medical Journal of Malaysia

Med. Rec. Aust.
  Medical Record of Australia

Med. Reporter
  Medical Reporter

Med. Rev. Mex.
  Medicina Revista Mexicana

Meded. Dienst Volksgezond. Ned.-Indie
  Mededeelungen van den Dienst der Volksgezondheid in Nederlandsche-Indie

Mem. Acad. Imp. Sci. St. Petersbourg
  Memoires de l'Academie Imperiale des Sciences de St. Petersbourg

Mem. Acad. Imp. Sci. St. Petersbourg. Cl. Phys.-Math.
  Memoires de l'Academie Imperiale des Sciences de St. Petersbourg. Classe de Sciences Physico-Mathematiques

Mem. Acad. Sci. Paris
  Memoires de l'Academic des Sciences, Paris

Mem. Asiat. Soc. Bengal
  Memoirs of the Asiatic Society of Bengal

Mem. Coll. Sci. Kyoto Imp. Univ. Series B
  Memoirs of the College of Science, Kyoto Imperial University. Series B

Mem. Fac. Educ. Kumamoto Univ.
  Memoirs of the Faculty of Education, Kumamoto University (Kumamoto Daigaku Kyoikugakubu Kiyo, Shizen Kagaku)

Mem. Hong Kong Nat. Hist. Soc.
  Memoirs of the Hong Kong Natural History Society

Mem. Indian Mus.
  Memoirs of the Indian Museum

Mem. Inst. Butantan (Sao Paulo)
  Memorias do Instituto Butantan. Sao Paulo

Mem. Inst. Scient. Madagascar
  Memoires de l'Institut Scientifique de Madagascar

Mem. Mus. Comp. Zool. Harv.
  Memoirs of the Museum of Comparative Zoology of Harvard College

Mem. Mus. Hist. Nat. Belg.
  Memoires du Musee Royal d'Histoire Naturelle de Belgique

Mem. Mus. Natl Hist. Nat. Ser. A Zool.
  Memoirs du Museum National d'Histoire Naturelle Serie A Zoologie

Mem. Nat. Cult. Res. San-in Reg.
  Memoirs of Natural and Cultural Researches of the San-in Region

Mem. Queensl. Mus.
  Memoirs of the Queensland Museum

Mem. Soc. Endocrinol.
  Memoires of the Society for Endocrinology

Mem. Soc. Linn. Normandie
Memoires de la Societe Linneenne de Normandie

Mem. Soc. Nat. Sci. Nat. Math. Cherbourg
Memoires de la Societe Nationale de Sciences Naturelles et Mathematiques de Cherbourg

Mem. Soc. Zool. Fr.
Memoires Societe Zoologique de France

Mem. Wernerian Nat. Hist. Soc.
Memoirs of the Wernerian Natural History Society, Edinburgh

Memorie R. Accad. Sci. Torino
Memorie della Accademia delle Scienze di Torino

Mil. Med.
Military Medicine

Milw. Public Mus. Contrib. Biol. Geol.
Milwaukee Public Museum Contributions in Biology and Geology

Misc. Publ. Mus. Zool. Univ. Mich.
Miscellaneous Publications Museum of Zoology University of Michigan

Misc. Rep. Res. Inst. Nat. Resour. (Tokyo)
Miscellaneous Reports of the Research Institute for Natural Resources (Tokyo)

Misc. Zool. Sumatr.
Miscellanea Zoologica Sumatrana

Mitt. Deut. Ges. Nat. Voelkerk. Ostasiens
Mitteilungen der Deutschen Gesellschaft fuer Natur- u. Voelkerkunde Ostasiens im Tokyo

Mitt. Geogr. Ges. Nat. Mus. Luebeck
Mitteilungen der Geographischen Gesellschaft und des Naturhistorischen Museums in Luebeck

Mitt. Naturh. Mus. Hamburg
Mitteilungen aus dem Naturhistorischen Museum in Hamburg

Mitt. Zool. Samml. Mus. Naturk. Berl.
  Mitteilungen aus der Zoologischen Sammlung des Museums fuer Naturkunde in Berlin

Mol. Biol.
  Molecular Biology (English translation of *Molekulyarnaya Biologiya (Moscow))*

Mol. Pharmacol.
  Molecular Pharmacology

Monatsber. Deut. Akad. Wiss. Berl.
  Monatsbericht der Deutschen Akademie der Wissenschaften zu Berlin

Monatsber. K. Preuss. Akad. Wiss. Berl.
  Monatsberichte der Koeniglichen Preussischen Akademie der Wissenschaften zu Berlin

Monatsschr. Ornithol. Vivarienkd. Ausg. B Aquarien Terrarien
  Monatsschrift fuer Ornithologie und Vivarienkunde Ausgabe B Aquarien Terrarien

Monit. Zool. Ital.
  Monitore Zoologico Italiano

Mus. Nac. Hist. Nat. Not. Mens. (Santiago)
  Museo Nacional de Historia Natural Noticiario Mensual. Santiago

N. S. W. Med. Gaz.
  New South Wales Medical Gazette. Sydney

Nat. Geogr. Mag.
  National Geographic Magazine

Nat. Hist.
  Natural History

Nat. Hist. Bull. Siam Soc.
  Natural History Bulletin of the Siam Society

Nat. Hist. Rennell Isl. Br. Solomon Isl.
  Natural History of Rennell Island British Solomon Islands

Nat. New Biol.
  Nature New Biology

Nat. Sci.
  Natural Science

Nat. Volk (Frankf.)
  Natur und Volk (Frankfurt)

Natur Mus. (Arhus)
  Natur og Museum. Arhus

Nature (Lond.)
  Nature. London

Nature (Paris)

Naturens Verd.
  Naturens Verden

Naturw. Beob.
  Naturwissenschaftlicher Beobachter

Naturw. Rundsch.
  Naturwissenschaftliche Rundschau

Natuurw. Tijdschr. Ned.-Indie
  Natuurwetenschappelijk Tijdschrift voor Nederlandsch-Indie

Natuurwet. Tijdschr.
  Natuurwetenschappelijk Tijdschrift

Navorscher
  Navorscher. Amsterdam

Ned. Tijdschr. Geneeskd.
  Nederlands Tijdschrift voor Geneeskunde

Neurosci. Lett.
  Neuroscience Letters

New Sci.
  New Scientist

New York Times

Newsl. Ahmadi Nat. Hist. Field Study Group
    Newsletter of the Ahmadi Natural History Field Study Group

Newsl. Tucson Herpetol. Soc.
    Newsletter of the Tucson Herpetological Society

Newsl. Zool. Surv. India
    Newsletter of the Zoological Survey of India (Calcutta)

Nippon Herpetol. J.
    Nippon Herpetological Journal

Nord. Med.
    Nordisk Medicin

North Queensl. Nat.
    North Queensland Naturalist

Northland Holiday (New Zealand)

Not. Galapagos
    Noticias de Galapagos

Notes Leyden Mus.
    Notes from the Leyden Museum

Nouv. Annal. Mus. Hist. Nat.
    Nouvelles Annales du Museum d'Histoire Naturelle

Nouv. Arch. Mus. Hist. Nat.
    Nouvelles Archives du Museum d'Histoire Naturelle. Paris

Nova Acta Acad. Caesar. Leop. Carol.
    *Nova Acta Academia Caesareae Leopoldino-Carolinae Germanicae Naturae
    Curiosorum*

Nova Acta Phys.-Med. Acad. Natur. Curios.
    *Nova Acta Physico-Medica Academiae Caesareae Leopoldino Carolinae Na-
    turae Curiosorum*

Nucleus (Calcutta)

Nuovi Annal. Sci. Natu.
  Nuovi Annali delle Scienze Naturali

Occas. Pap. Dept. Geog. Univ. Papua, New Guinea
  Occasional Papers of the Department of Geography. University of Papua, New Guinea

Occas. Pap. Mus. Zool. Univ. Mich.
  Occasional Papers of the Museum of Zoology University of Michigan

Oceans Mag.
  Oceans Magazine

Oecologia (Berl.)
  Oecologia. Berlin

Of Sea and Shore

Omni

Opusc. Zool.
  Opuscula Zoologica

Orange Co. Daily Pilot
  Orange County Daily Pilot

Orient. Insects
  Oriental Insects

Outside

Pac. Discovery
  Pacific Discovery

Palaeontographica

Paleont. Zh.
  Paleontologicheskii Zhurnal

Pap. Mich. Acad. Sci. Arts Lett.
  Papers of the Michigan Academy of Science Arts and Letters

Pap. Proc. R. Soc. Tasm.
   Papers and Proceedings of the Royal Society of Tasmania

Papua New Guinea Med. J.
   Papua New Guinea Medical Journal

Papua New Guinea. Wildl. Br. Dept. Nat. Res. Wildl. Publ.
   Papua New Guinea. Wildlife Branch. Department of Natural Resources. Wildlife Publications

Paradise
   Paradise (In Flight with Air Niugini)

Parasitology

Paris. Bull. Soc. Centr. Aquiculture
   Paris. Bulletin de la Societe Centrale d'Aquiculture et de Peche

Pathology

Pharm. J. N. Z.
   Pharmaceutical Journal of New Zealand

Pharmacol. Rev.
   Pharmacological Reviews

Pharmacol. Ther.
   Pharmacology & Therapeutics

Philipp. J. Sci.
   Philippine Journal of Science

Philipp. J. Sci. Sect. A.
   Philippine Journal of Science. Section A

Philipp. J. Sci. Sect. D. Gen. Biol., Ethnol., Anthropol.
   Philippine Journal of Science. Section D: General Biology, Ethnology, Anthropology

Philipp. Sci.
   Philippine Scientist

Philos. Mag.
   Philosophical Magazine

Philos. Trans. R. Soc.
Philosophical Transactions of the Royal Society. London

Physiol. Zool.
Physiological Zoology

Physiologist

Practitioner

Priroda (Mosc.)
Priroda. Moscow

Proc. Acad. Nat. Sci. Phil.
Proceedings of the Academy of Natural Sciences of Philadelphia

Proc. Am. Philos. Soc.
Proceedings of the American Philosophical Society

Proc. Asiat. Soc. Beng.
Proceedings of the Asiatic Society of Bengal

Proc. Aust. Physiol. Pharmacol. Soc.
Proceedings of the Australian Physiological and Pharmacological Society

Proc. Biol. Soc. Wash.
Proceedings of the Biological Society of Washington

Proc. Boston Soc. Nat. Hist.
Proceedings of the Boston Society of Natural History

Proc. Calif. Acad. Sci.
Proceedings of the California Academy of Sciences

Proc. Helminthol. Soc. Wash.
Proceedings of the Helminthological Society of Washington

Proc. Indian Acad. Sci. Chem. Sci.
Proceedings of the Indian Academy of Sciences Chemical Sciences

Proc. Indian Acad. Sci. Sect. B
Proceedings of the Indian Academy of Sciences Section B

Proc. Indiana Acad. Sci.
  Proceedings of the Indiana Academy of Science

Proc. Int. Conf. Raman Spectrosc.
  Proceedings of the International Conference on Raman Spectroscopy

Proc. Int. Congr. Biochem.
  Proceedings of the International Congress of Biochemistry

Proc. Int. Congr. Pharmacol.
  Proceedings of the International Congress of Pharmacology

Proc. Int. Congr. Zool.
  Proceedings of the International Congress of Zoology

Proc. Int. Coral Reef Symp.
  Proceedings of the International Coral Reef Symposium

Proc. K. Ned. Akad. Wet.
  Proceedings of the K. Nederlandse Akademie van Wetenschappen. Amsterdam

Proc. K. Ned. Akad. Wet. Ser. C Biol. Med. Sci.
  Proceedings of the Koninklijke Nederlandse Akademie van Wetenschappen Series C Biological and Medical Sciences

Proc. Linn. Soc. Lond.
  Proceedings of the Linnean Society of London

Proc. Linn. Soc. N.S.W.
  Proceedings of the Linnean Society of New South Wales

Proc. Nat. Acad. Sci. U.S.A.
  Proceedings of the National Academy of Sciences of the United States of America

Proc. New Engl. Zool. Club
  Proceedings of the New England Zoological Club

Proc. Pacif. Sci. Congr.
  Proceedings of the Pacific Science Congress

Proc. Phys. Soc. Edinb.
  Proceedings of the Royal Physical Society of Edinburgh

Proc. Physiol. Soc. (London)
Proceedings of the Physiological Society, London

Proc. R. Soc.
Proceedings of the Royal Society. London

Proc. R. Soc. Med.
Proceedings of the Royal Society of Medicine

Proc. U.S. Natl Mus.
Proceedings of the United States National Museum

Proc. West. Pharmacol. Soc.
Proceedings of the Western Pharmacology Society

Proc. Zool. Soc. Lond.
Proceedings of the Zoological Society of London

Protein Nucleic Acid Enzyme (Tokyo)
see also *Tanpakushitsu Kakusan Koso*

Publ. Ext. Cult. Mus. Argent. Cienc. Nat.
Publicaciones de Extension Cultural y Didactica. Museo Argentino de Ciencias Naturales "Bernardino Rivadavia"

Publ. Field Mus. Nat. Hist. Zool. Ser.
Publications of the Field Museum of Natural History. Chicago Zoological Series

Publ. Herpetol.
Publications in Herpetology

Publ. Mus. Hist. Nat. 'Javier Prado' Ser. A. Zool.
Publicaciones del Museo de Historia Natural 'Javier Prado' Series A Zoologia

Publ. Seto Mar. Biol. Lab.
Publications of Seto Marine Biological Laboratory

Publ. Univ. Calif. Los Ang. Biol. Sci.
Publications of the University of California at Los Angeles in Biological Sciences

Q. J. Exp. Physiol.
   Quarterly Journal of Experimental Physiology

Q. J. Taiwan Mus. (Taipei)
   Quarterly Journal of the Taiwan Museum. Taipei

Quart. Rev. Biol.
   Quarterly Review of Biology

Rec. Aust. Mus.
   Records of the Australian Museum

Rec. Dom. Mus. (Wellington)
   Records of the Dominion Museum. Wellington

Rec. S. Aust. Mus.
   Records of the South Australian Museum. Adelaide

Rec. West. Aust. Mus.
   Records of the Western Australian Museum

Rec. Zool. Surv. India
   Records of the Zoological Survey of India

Rec. Zool. Surv. Pakistan
   Records of the Zoological Survey of Pakistan

Rep. Br. Assoc. Adv. Sci.
   Reports of the British Association for the Advancement of Science, London

Res. Commun. Chem. Pathol. Pharmacol.
   Research Communications in Chemical Pathology and Pharmacology

Res. Rep. Smithson. Inst.
   Research Reports of the Smithsonian Institution

Resen. Cient. R. Soc. Esp. Hist. Nat.
   Resenas Cientificas de la Real Sociedad Espanola de Historia Natural. Madrid

Respir. Physiol.
   Respiration Physiology

Rev. Acad. Colomb. Cienc. Exactas Fis. Nat.
  Revista de la Academia Colombiana de Ciencias Exactas Fisicas y Naturales

Rev. Biol. Mar.
  Revista de Biologia Marina

Rev. Biol. Trop.
  Revista de Biologia Tropical

Rev. Costarric. Cienc. Med.
  Revista Costarricense de Ciencias Medicias

Rev. Fac. Nac. Agron. Medellin
  Revista de la Facultad Nacional de Agronomia Medellin

Rev. Gen. Sci. Pures Appl.
  Revue Generale des Sciences Pures et Appliques

Rev. Inst. Def. Cafe Costa Rica
  Revista del Instituto de Defensa del Cafe de Costa Rica

Rev. Mex. Cienc. Med. Biol.
  Revista Mexicana de Ciencias Medicas y Biologicas

Rev. Pharm.
  Revue Pharmaceutique

Rev. Suisse Zool.
  Revue Suisse de Zoologie

Revue Mag. Zool.
  Revue et Magasin de Zoologie Pure et Appliquee

Revue Scient.
  Revue Scientifique

Ric. Sci.
  Ricerca Scientifica

Sabah Soc. J.
  Sabah Society Journal

Saito Ho-on Kai Mus. Nat. Hist. Res. Bull.
  Saito Ho-on Kai Museum of Natural History Research Bulletin

San Diego Evening Tribune

San Diego Union

Sar. Gaz.
  Sarawak Gazette

Sar. Mus. J.
  Sarawak Museum Journal

Schr. Ges. Koenigsb.
  Schriften der Physikalisch-Oekonomischen Gesellschaft zu Koenigsberg in Preussen

Sci. Am.
  Scientific American

Sci. Cult.
  Science and Culture

Sci. Mem. Med. Offrs. Army India
  Scientific Memoirs by Medical Officers of the Army of India

Sci. News
  Science News

Sci. Rep.
  Science Reporter

Science (Wash. D.C.)
  Science. Washington D.C.

Scott. Med. Surg. J.
  Scottish Medical and Surgical Journal

Sea Front.
  Sea Frontiers

Sea Secrets

Senckenb. Biol.
  Senckenbergiana Biologica

Sentinel (Pacific Beach: CA)

Shengwu Huaxue Zazhi

Shinkei Kenkyu No Shinpo
  (see also: Adv. Neurol. Sci.)

Singapore Nat.
  Singapore Naturalist

Sitz. Deut. Akad. Wiss. Berlin
  Sitzungsberichte der Deutschen Akademie der Wissenschaften zu Berlin

Sitz. Deut. Akad. Wiss. Wien Math. Naturw. Kl.
  Sitzungsberichte der Akademie der Wissenschaften in Wien. Mathematisch-Naturwissenschaftliche Klasse

Sitz. Ges. Nat. Freunde Berlin
  Sitzungsberichte der Gesellschaft Naturforschender Freunde zu Berlin

Skin Diver

Skindiving Aust.
  Skindiving in Australia and New Zealand

Smithson. Herpetol. Inf. Serv.
  Smithsonian Herpetological Information Service

Smithson. Inst. Annu. Rep.
  Smithsonian Institution Annual Report

Smithson. Inst. Res. Rep.
  Smithsonian Institution Research Reports

Smithson. Misc. Collect.
  Smithsonian Miscellaneous Collection

Smithson. Scient. Ser.
  Smithsonian Scientific Series

Smithsonian

Snake

Soc. Geogr. Hanoi
  Societe de Geographie de Hanoi

South Pac. J. Nat. Sci.
  South Pacific Journal of Natural Science

Southeast Asian J. Trop. Med. Public Health
  Southeast Asian Journal of Tropical Medicine and Public Health

Sov. J. Bioorg. Chem.
  *Soviet Journal of Bioorganic Chemistry (*English translation of *Biooganiches-
  kaya Khimiya)*

Sov. J. Ecol.
  *Soviet Journal of Ecology (*English translation of *Ekologiya)*

Sov. J. Mar. Biol.
  *Soviet Journal of Marine Biology (*English translation of *Biologiya Morya)*

Spolia Zeylan.
  Spolia Zeylanica

Sr. Sch.
  Senior Scholastic

Stud. Inst. Med. Res. Kuala Lumpur
  Studies from the Institute for Medical Research. Kuala Lumpur

Stuttg. Beitr. Naturkd.
  Stuttgarter Beitrage zur Naturkunde

Stuttg. Beitr. Naturkd. Ser. A (Biol.)
  Stuttgarter Beitrage zur Naturkunde Serie A (Biologie)

Sunday Advertiser (Honolulu)

Sydney Morning Herald

Sydney Univ. Med. J.
  Sydney University Medical Journal

Symp. Zool. Soc. Lond.
Symposium of the Zoological Society of London

Syst. Zool.
Systematic Zoology

Taiwan Sotokufu Kenkyujo Hokoku
Annual Report of the Institute of Science, Government of Formosa

Tanpakushitsu Kakusan Koso
see also *Protein Nucleic Acid Enzyme (Tokyo)*

Tap Chi Sinh Vat Hoc

Taprobanian

Tasman. J. Nat. Sci. Agric. Stat.
Tasmanian Journal of Natural Science, Agriculture, Statistics, etc.

Tauchen

Terre et la Vie

Tethys

Thromb. Haemostasis
Thrombosis and Haemostasis

Thromb. Res.
Thrombosis Research

Tico Times (San Jose, Costa Rica)

Tier (Stutt.)
Tier. Stuttgart

Tohoku J. Exp. Med.
Tohoku Journal of Experimental Medicine

Toxicon

Tr. Zool. Inst. Akad. Nauk S. S. S. R.
Trudy Zoologicheskogo Instituta, Akademiya Nauk S. S. S. R.

Trans. Am. Philos. Soc.
  Transactions of the American Philosophical Society

Trans. Kans. Acad. Sci.
  Transactions of the Kansas Academy of Science

Trans. Nat. Hist. Soc. Taiwan
  Transactions of the Natural History Society of Taiwan

Trans. Phil. Soc. N.S.W.
  Transactions of the Philosophical Society of New South Wales. Sydney

Trans. Proc. N. Z. Inst.
  Transactions and Proceedings of the New Zealand Institute

Trans. Proc. Palaeontol. Soc. Jpn. New Ser.
  Transactions and Proceedings of the Palaeontological Society of Japan. New Series

Trans. R. Soc. N. Z.
  Transactions of the Royal Society of New Zealand

Trans. R. Soc. S. Aust.
  Transactions of the Royal Society of South Australia

Trans. R. Soc. Trop. Med. Hyg.
  Transactions of the Royal Society of Tropical Medicine and Hygiene

Trans. San Diego Soc. Nat. Hist.
  Transactions of the San Diego Society of Natural History

Trans. Sci. Soc. China
  Transactions of the Science Society of China. Shanghai

Trans. Zool. Soc. Lond.
  Transactions of the Zoological Society of London

Trends Biochem. Sci.
  Trends in Biochemical Sciences

Treubia

Trop. Dis. Bull.
  Tropical Diseases Bulletin

Trop. Natuur (Weltevreden)
  Tropische Natuur. Weltevreden

Tuatara

U.S. Fish Wildl. Ser. Fish. Bull.
  U.S. Fish and Wildlife Service Fishery Bulletin

U.S. Fish Wildl. Ser. Res. Rep.
  U.S. Fish and Wildlife Service Research Report

U.S. Fish Wildl. Ser. Sp. Sci. Rep.-Fish.
  U.S. Fish and Wildlife Service Special Scientific Report Fisheries

U.S. Fish Wildl. Ser. Sp. Sci. Rep.-Wildl.
  U.S. Fish and Wildlife Service Special Scientific Report Wildlife

Undersea Technol.
  Undersea Technology

Univ. Kans. Mus. Nat. Hist. Misc. Publ.
  University of Kansas Museum of Natural History Miscellaneous Publications

Univ. Kans. Publs Mus. Nat. Hist.
  University of Kansas Publications of the Museum of Natural History

Univ. Kans. Sci. Bull.
  University of Kansas Science Bulletin

Veld & Vlei

Verh. Naturforsch. Ges. Basel
  Verhandlungen der Naturforschenden Gesellschaft in Basel

Verh. Zool.-Bot. Ges. Wien
  Verhandlungen der Zoologisch-Botanischen Gesellschaft in Wien

Vet. Med. Small Anim. Clin.
  Veterinary Medicine and Small Animal Clinician

Victorian Nat.
  Victorian Naturalist

Vidensk. Medd. Dan. Naturhist. Foren.
  Videnskabelige Meddelelser fra Dansk Naturhistorisk Forening

Walkabout (Melbourne)

Wash. Post
  Washington Post

Wasmann Collector

Wasmann J. Biol.
  Wasmann Journal of Biology

West. Aust. Nat.
  Western Australian Naturalist

West. Aust. Shell Collect.
  Western Australia Shell Collector

West. Aust. Fish. Dept. Fauna Bull.
  Western Australia. Fisheries Department. Fauna Bulletin

West. Boatman
  Western Boatman

Wiad. Parazytol.
  Wiadomosci Parazytologiczne

Wildlife (Lond.)
  Wildlife. London

Yamashina Chorui Kenkyujo, Tokyo

Yearb. Herpetol.
  Yearbook of Herptetology

Z. Morphol. Oekol. Tiere
  Zeitschrift fuer Morphologie und Oekologie der Tiere

Z. Parasitenkd.
  Zeitschrift fuer Parasitenkunde

Z. Tropenmed. Parasit.
   Zeitschrift fuer Tropenmedizen und Parasitologie

Ziva

Zoo Life (Lond.)
   Zoo Life. London

Zoogeographica

Zool. Anz.
   Zoologischer Anzeiger

Zool. Bidrag Uppsala
   Zoologiska Bidrag fran Uppsala

Zool. Garten
   Zoologische Garten

Zool. J. Linn. Soc.
   Zoological Journal of the Linnean Society

Zool. Jahrb. Abt. Allg. Zool. Physiol. Tiere
   Zoologische Jahrbuecher Abteilung fuer Allgemeine Zoologie und Physiologie der Tiere

Zool. Jahrb. Abt. Anat. Ontog. Tiere
   Zoologische Jahrbuecher Abteilung fuer Anatomie und Ontogenie der Tiere

Zool. Jahrb. Abt. Syst. Geogr. Biol. Tiere
   Zoologische Jahrbuecher Abteilung fuer Systematik Oekologie und Geographie der Tiere

Zool. Jahrb. Abt. Syst. Oekol. Geogr. Tiere
   Zoologische Jahrbuecher Abteilung fuer Systematik Oekologie und Geographie der Tiere

Zool. Mag. (Tokyo)
   Zoological Magazine. Tokyo

Zool. Meded. (Leiden)
   Zoologische Mededeelingen. Leiden

Zool. Meded. Rijksmus. Nat. Hist. Leiden
  Zoologische Mededeelingen. Rijksmuseum van Natuurlijke Historie. Leiden

Zool. Miscell.
  Zoological Miscellaney

Zool. Res. (China)
  Zoological Research (China)

Zool. Sci. (Tokyo)
  Zoological Science (Tokyo)

Zool. Verh. (Leiden)
  Zoologische Verhandelingen. Leiden

Zool. Zh.
  Zoologicheskii Zhurnal

Zoologica (N.Y.)
  Zoologica. New York

Zoologica (Stuttg.)
  Zoologica. Stuttgart

Zoonooz (San Diego)

Chapter 1

PRE-LINNEAN AND OTHER PRE-NINETEENTH CENTURY REFERENCES

1-1
Aelian. ca 200–225 A.D.

On the characteristics of animals. Cambridge, MA: Harvard University Press; 1959. Volume 1.

1-2
Anonymous. ca 60 A.D.

The Periplus of the Erythraean Sea: travel and trade in the Indian Ocean by a merchant of the first century. Translated from the Greek by William H. Schoff. 2nd ed. New Delhi: Oriental Books Reprint Corp; 1974.

1-3
Aristotle. ca 350 B.C.

History of Animals. Barnes, J., ed. *The complete works of Aristotle.* Princeton, NJ: Princeton University Press; 1984. Book 2.

1-4
Cook, J. 1784.

A voyage to the Pacific Ocean. Undertaken, by command of His Majesty, for making discoveries in the northern hemisphere. To determine the position and extent of the west side of North America; its distance from Asia; and the practicability of a northern passage to Europe. Performed under the direction of Captains Cook, Clarke, and Gore, and His Majesty's ships the Resolution and Discovery. In the years 1776, 1777, 1778, 1779 and 1780. London: G. Nicol and T. Cadell. Volume 1.

1-5
Cooke, E. 1712.
A voyage to the South Sea, and round the world, perform'd in the years 1708, 1709, 1710, and 1711, by the ships Duke and Dutchess of Bristol. London: B. Lintot and R. Gosling. New York: De Capo; 1969.

1-6
Dampier, W. 1717.
A new voyage round the world. London: James Knapton. Dampier, W. *A collection of voyages.* London: Grant Richards; 1906. Volume 2.

1-7
Fernandez de Oviedo y Valdez, G. 1519.
Historia general y natural de las Indias. Edicion y estudio preliminar de Juan Perez de Tudela Bueso. Biblioteca de Autores Espanoles. Madrid: Ediciones Atlas; 1959. Volumes 2, 3, and 4.

1-8
Forster, G. 1777.
A voyage round the world in His Brittanic Majesty's sloop, Resolution, commanded by Capt. James Cook, during the years 1772, 3, 4, and 5. London: B. White, J. Rosson, P. Elmsly, and G. Robinson. Berlin: Akademie-Verlag; 1968.

1-9
Forster, J. R. 1778.
Observations made during a voyage round the world on physical geography, natural history, and ethic philosophy. Especially on 1. The earth and its strata, 2. Water and the ocean, 3. The atmosphere, 4. The changes of the globe, 5. Organic bodies, and 6. The human species. London: G. Robinson.

1-10
Forster, J. R. 1844.
Descriptiones animalium quae in itinere ad maris australis terras per annos 1772, 1773, et 1774 suscepto collegit observavit et delineavit Joanned Reinoldus Forster. Berolini: Ex Officina Academica.

1-11
Gray, E. W. 1789.
Observations on the class of animals called, by Linnaeus, Amphibia; particularly on the means of distinguishing those serpents which are venomous, from those which are not so. *Philos. Trans. R. Soc.* 79:21–36.

1-11.5
Gray, E. W. 1790.

Observations sur la classe des animaux, nominee amphibia par Linnaeus, & en particulier, sur les moyens de distinguer les serpens venimeux de ceux qui ne le sont pas. *J. Phys.* 37(2):321–331 (Translation of Gray, E. W. 1789)

1-12
Herbert, T. 1665.

Some years travels into divers parts of Africa and Asia the Great describing more particularly the empires of Persia and Industan: interwoven with such remarkable occurrences as hapned in those parts during these later times. As also, many other rich and famous kingdoms in the oriental India, with the isles adjacent. Severally relating their religion, language, customs and habit. 3rd impression. London: Andrew Crook.

1-13
Herbert, T. 1677.

Some years travels into divers parts of Africa and Asia the Great describing more particularly the empires of Persia and Industan: interwoven with such remarkable occurrences as hapned in those parts during these later times. As also, many other rich and famous kingdoms in the oriental India, with the isles adjacent. Severally relating their religion, language, customs and habit. 4th impression. London: R. Scot, T. Basset, J. Wright, and R. Chiswell.

1-14
Klein, J. T. 1753.

Tentamen herpetologiae. Leidae: Eliam Luzac.

1-14.5
Leske, N. G. 1784.

Anfangsgrunde der naturgeschichte. Leipzig: Siegfried Lebrecht Crusius. Volume 1.

1-15
Linschoten, J. H. van. 1598.

The voyage of John Huyghen van Linschoten to the East Indies. From the Old English translation of 1598. London: Hakulyt Society; 1885. Volume 1.

1-16
Pennant, T. 1798.

The view of Hindoostan. London: Volumes 1 and 2.

1-17
Philips, J. 1744.
An authentic journal of the late expedition under the command of Commodore Anson. London: J. Robinson.

1-18
Pliny The Elder. ca 77 A.D.
The natural history of Pliny. Translated by John Bostock and H. T. Riley. London: Henry G. Bohn; 1855–1857. Book 29.

1-19
Raveneau de Lussan. 1693.
Raveneau de Lussan, buccaneer of the Spanish main and early French filibuster of the Pacific. A translation into English of his Journal of a voyage into the south seas in 1684 and the following years with the Filibusters. Paris: Jean Baptiste Coignard. Cleveland: Arthur H. Clark; 1930.

1-19.5
Rumphius, G. E. 1705.
D'amboinsche rariteitkamer, behelzende eene beschryvinge van allerhande zoo weeke als harde schaalvisschen, te weeten raare krabben, kreeften, en diergelyke zeedieren als mede allerhande hoorntjes en schulpen, die men in d'amboinsche zee vindt: daar beneven zommige mineraalen, gesteenten en soorten van aarde, die in d'amboinsche, en zommige omleggende eilanden gevonden worden. Amsterdam: Francois Halma.

1-20
Seba, A. 1753.
Locupletissimi rerum naturalium thesauri accurata descriptio et iconibus artificiosissimis expressio, per universam physices historiam. Amstelaedami. Volume 2.

1-21
Topsell, E. 1658.
The history of four-footed beasts and serpents and insects. London: G. Sawbridge. New York: Da Capo; 1967. Volume 2.

1-22
Valmont de Bomare, J. C. 1775.
Dictionnaire raissone universel d'histoire naturelle; contenant l'histoire des animaux, des vegetaux et des mineraux et celle des corps celestes, des meteores, & des autres principaux phenomenes de la nature, avec l'histoire et la description des drogues simples tirees des trois regnes; et le detail de leurs usages dans la medicine, dans l'economie domestique & champetre, & dans les arts & metiers; plus, une table

concordante des noms latins, & le renvoi aus objets mentionnes dans cet ouvrage.
Nouvelle edition, revue & considerablement augmentee. Paris: Chez Brunet. Volume
8.

1-23
Vosmaer, A. 1774.
  *Beschryving van twee. Platstaart slangen. Natuurkundige beschryving eener
uitm untende verzameling van zeldsaame gedierten, bestaande in Oost- en Westin-
dische viervoetige dieren, vogelen en slangen, weleer leevend voorhanden geweest
zynde, buiten den Haag, op het kleine Loo van Z. D. H. den Prins van Oranje-
Nassau.* Amsterdam: J. B. Elwe.

Genus-Species Index
for Chapter 1

Chapter 2

GENERAL REFERENCES—TECHNICAL

2-.05
Alcala, A. C. 1986.
  Guide to Philippine flora and fauna. Manila: Natural Resources Management Center, Ministry of Natural Resources, and University of the Philippines. Volume 10.

2-1
Angel, F. 1950.
  Vie et moeurs des serpents. Paris: Payot.

2-2
Bellairs, A. 1970.
  The life of reptiles. New York: Universe Books. Volumes 1 and 2.

2-3
Bergman, R. A. M. 1942.
  Enhydrina schistosa. Natuurwet. Tijdschr. Ned.-Indie 102(1):9–12. (In Dutch with English abstract)

2-4
Bergman, R. A. M. 1954.
  Thalassophis anomalus Schmidt. Amsterdam: Koninklijk Instituut voor de Tropen. (In French)

2-4.5
Bolanos, R. 1982.
  Serpientes venenosas de Centro America: distribucion, caracteristicas y patrones cariologicos. *Mem. Inst. Butantan (Sao Paulo)* 46:275–291

2-4.7
Bolanos, R. 1984.
Serpientes venenos y ofidismo en Centroamerica. San Jose, Costa Rica: Editorial Universidad de Costa Rica.

2-5
Boquet, P. 1948.
Venins de serpents et antivenins. Paris: Editions Medicales Flammarion.

2-6
Boulenger, G. A. 1892.
Marine snakes. *Nat. Sci.* 1(1):44–49.

2-7
Bourret, R. 1935.
Les serpents marins de l'Indochine francaise. Hanoi: Institute Oceanographique de l'Indochine. No. 25:1–69.

2-8
Branch, W. R. 1979.
The venomous snakes of Southern Africa, Part 2. Elapidae and Hydrophidae. *Snake* 11:199–225.

2-9
Branch, W. R. 1981.
The venomous snakes of Southern Africa, Part 2. Elapidae and Hydrophidae. *Bull. Md Herpetol. Soc.* 17(1):1–47.

2-9.5
Calmette, A.; Bruyant, L. 1914.
Die tropische intoxikationskrankheiten. Mense, C. *Handbuch der tropenkrankheiten*. 2nd ed. Leipzig: J. A. Barth. Vol. 1, pp. 617–678.

2-10
Cogger, H. G. 1975.
Reptiles and amphibians of Australia. Sydney: A. H. & A. W. Reed.

2-11
Cogger, H. G. 1975.
Sea snakes of Australia and New Guinea. Dunson, W. A., ed. *The biology of sea snakes*. Baltimore: University Park Press. Pp. 59–139.

2-12
Cogger, H. G. 1979.
   Reptiles and amphibians of Australia. Rev. ed. Sydney: A. H. & A. W. Reed.

2-12.5
Cogger, H. G. 1983.
   Reptiles and amphibians of Australia. 3rd ed. Sanibel, FL: Ralph Curtis Books.

2-13
Cope, E. D. 1900.
   The crocodilians, lizards, and snakes of North America. *Annu. Rep. U.S. Natl Mus.* 1898:155–1270.

2-14
Darlington, P. J. 1957.
   Zoogeography: the geographical distribution of animals. New York: Wiley.

2-15
Dumeril, A. M. C.; Bibron, G. 1844.
   Erpetologie generale ou histoire naturelle complets des reptiles. Paris: Librairie Encyclopedique de Roret. Volume 6.

2-16
Dumeril, A. M. C.; Bibron, G.; Dumeril, A. 1854.
   Erpetologie generale ou histoire naturelle complete des reptiles. Paris: Librarie Encyclopedique de Roret. Volume 7, part 2.

2-17
Dunson, W. A. 1975.
   Adaptations of sea snakes. Dunson, W. A., ed. *Biology of sea snakes*. Baltimore: University Park Press. Pp. 3–19.

2-18
Dunson, W. A., editor. 1975.
   The biology of sea snakes. Baltimore: University Park Press.

2-19
Fayrer, J. 1871.
   The thanatophidia of India. *Indian Med. Gaz.* 6(1):1–3.

2-20

Fayrer, J. 1872.

The thanatophidia of India, being a description of the venomous snakes of the Indian Peninsula, with an account of the influence of their poison on life; and a series of experiments. London: Churchill.

2-21

Fayrer, J. 1874.

The thanatophidia of India, being a description of the venomous snakes of the Indian Peninsula with an account of the influence of their poison on life; and a series of experiments. 2nd ed. rev'd and enlarged. London: J. and A. Churchill.

2-22

Fischer, J. G. 1856.

Die familie der seeschlangen systematisch beschrieben. 2nd ed. Hamburg: Druck von J. A. Meissner. (also published in *Abh. Geb. Naturwiss. Nat. Hamburg* 3:1–78)

2-23

Fischer, J. G. 1888.

Herpetologische mitteilungen. Ueber zwei von Liukiu-Insel Okinawa stammende schlangen. 1. *Platurus colubrinus* Schn. und dessen giftigkeit. *Jahrb. Hamb. Wiss. Anst.* 5:18–20.

2-24

Fish, C. J.; Cobb, M. C. 1954.

Noxious marine animals of the central and western Pacific ocean. *U.S. Fish. Wildl. Ser. Res. Rep.* No. 36:1–45.

2-25

Girard, C. 1858.

Herpetology. *United States Exploring Expedition during the years 1838, 1839, 1840, 1841, 1842. Under the command of Charles Wilkes, U.S.N.* Philadelphia: J. B. Lippincott. Volume 20.

2-25.5

Guenther, A. 1858.

On the geographical distribution of reptiles. *Proc. Zool. Soc. Lond.* 1858:373–397.

2-26

Guenther, A. C. L. G. 1864.

The reptiles of British India. London: Printed for the Ray Society by Robert Hardwicke.

2-26.5
Guinea, M. L. 1986.
   Aspects of the biology of the common Fijian sea snake <u>Laticauda colubrina</u>
(Schneider). Suva, Fiji: University of the South Pacific. Thesis. 206 pp.

2-27
Halstead, B. W. 1970.
   Poisonous and venomous marine animals of the world. Washington, DC: Government Printing Office. Volume 3.

2-28
Halstead, B. W. 1978.
   Poisonous and venomous marine animals of the world. Rev. ed. Princeton: Darwin Press.

2-28.5
Halstead, B. W. 1988.
   Poisonous and venomous marine animals of the world. 2nd rev. ed. Princeton: Darwin Press.

2-29
Heatwole, H. 1978.
   Adapatations of marine snakes. *Am. Sci.* 66(5):594–604.

2-29.5
Heatwole, H. 1987.
   Sea snakes. Kensington, N.S.W.: New South Wales University Press.

2-30
Hoffmann, C. K. 1890.
   Reptilien III. Schlangen und entwicklungsgeschichte der Reptilien. Bronn, H. G., ed. *Klassen und ordnungen des thier-reichs.* Leipzig: C. F. Winter'sche Verlagshandlung. Vol. 6, pt. 3, pp. 1401–1873.

2-30.5
Hoffstetter, R. 1955.
   Squamates de type moderne. Piviteau, J., ed. *Traite de paleontologie.* Paris: Masson et Cie. Vol. 5, pp. 605–662.

2-31
Institut Pasteur du Viet-Nam. 1960.
   Serpents marins venimeux (Hydrophiidae) et leurs venins. *Inst. Pasteur Viet-Nam Rapp. Annu. Fonct. Tech.* 1960:52–61.

2-31.2
Kinghorn, J. R. 1951.
   Reptiles. Macinnes, I. G., ed. *Australian fisheries: a handbook prepared for the second meeting of the Indo-Pacific Council, Sydney, April 1950.* Sydney: Halstead Press.

2-31.5
Knauer, F. 1887.
   Handwoerterbuch der zoologie. Stuttgart: Ferdinand Enke.

2-31.7
Kopstein, F. 1926.
   Die giftigen tiere von Niederlaendisch-ost-indien. *Natuur. Tijdschr. Ned.-Indie* 86(2):123–146.

2-32
Kraus, R.; Werner, F. 1931.
   Giftschlangen und die serumbehandlng der schlangenbisse. Jena: Gustav Fischer.

2-33
Krefft, G. 1869.
   The snakes of Australia; an illustrated and descriptive catalogue of all the known species. Sydney: Thomas Richards.

2-33.5
Krefft, G. 1962.
   Seeschlangen. Pax, F., ed. *Meersprodukte, ein handwoerterbuch der marinen rohstoffe.* Berlin: Gebrueder Borntraeger. Pp. 361–362.

2-34
Kropach, C. 1975.
   The yellow-bellied sea snake, *Pelamis*, in the Eastern Pacific. Dunson, W. A. ed. *The biology of sea snakes.* Baltimore: University Park Press. Pp. 185–213.

2-35
Kuntz, R. E. 1963.
   Snakes of Taiwan. *Q. J. Taiwan Mus. (Taipei)* 16(1–2):1–79. (Reprinted as U.S. Naval Medical Research Unit Publication No. 2)

2-35.5
Lichtenstein, H. 1818.
   Das zoologische museum der Universitat zu Berlin. Berlin: F. Dummler.

2-36
Luedicke, M. 1962.
   Ordnung der klasse reptilia: serpentes. *Handb. Zool.* 7(1,5):1–298.

2-37
Mao, S. H.; Chen, B. Y. 1980.
   Sea snakes of Taiwan: a natural history of sea snakes. Taipei, Taiwan: National Science Council.

2-38
Minton, S. A. 1966.
   A contribution to the herpetology of West Pakistan. *Bull. Am. Mus. Nat. Hist.* 134(2):27–184.

2-39
Nicholson, E. 1874.
   Indian snakes. An elementary treatise on ophiology with a descriptive catalogue of the snakes found in India and the adjoining countries. Madras: Higginbotham.

2-40
Peters, W. 1872.
   Ueber den *Hydrus fasciatus* Schneider und einige andere seeschlangen. *Monatsber. Deut. Akad. Wiss. Berl.* 1872:848–861.

2-41
Picado Twight, C. 1931.
   Serpientes venenosas de Costa Rica sus venenos seroterapia anti-ofidica. San Jose, Costa Rica: Sauter, Arias & Co. (Reissued 1976 by Editorial Universidad de Costa Rica)

2-42
Pickwell, G. V.; Culotta, W. A. 1980.
   Pelamis and *Pelamis platurus.* Zweifel, R. G., ed. *Catalogue of American amphibians and reptiles.* New York: Society for the Study of Amphibians and Reptiles. No. 255:1–4.

2-43
Saint Girons, H. 1972.
   Les serpents du Cambodge. *Mem. Mus. Natl Hist. Nat. Ser. A. Zool.* 74:1–170.

2-43.5
Schlegel, H. 1827.
   Erpetologische nachtrichten. *Isis von Oken* 20(3):281–294.

2-44

Schlegel, H. 1844.

Abbildungen neuer oder unvollstandig bekannter Amphibien, nach der Natur oder dem Leben entworfen. Dusseldorf: Arnz & Comp.

2-44.5

Schlegel, H. 1858.

Handleiding tot de beoefening der dierkunde. Breda: Koninklijke Militaire Akademie. Volume 2.

2-45

Schmidt, P. 1852.

Beitrage zur ferneren kenntniss der Meerschlangen. *Abh. Geb. Naturwiss. Nat. Hamburg* 2(2):No. 3:71–86.

2-46

Shaw, G. 1802.

General zoology or systematic natural history. London: G. Kearsley. Volume 3, part 2.

2-47

Shuntov, V. P. 1965.

Morskie zmei (Hydrophiidae) Tonkinskogo (Severo-V'etnamskogo) Zaliva. Sea snakes (Hydrophiidae) of the Gulf of Tonkin (Northern Viet-Nam). (Trans-240 by the U.S. Naval Oceanographic Office. of *Zool. Zh.* 41(8):1203–1209, 1962)

2-48

Smith, H. M. 1943.

Comments on G. Jan's papers on venomous serpents and the Coronellidae. *Trans. Kans. Acad. Sci.* 46:241–242.

2-49

Smith, M. A. 1920.

On sea snakes from the coasts of the Malay Peninsula, Siam and Cochin-China. *J. Fed. Malay States Mus.* 10(1):1–63.

2-50

Smith, M. 1926.

Monograph of the sea-snakes (Hydrophiidae). London: British Museum (Natural History).

2-51
Smith, M. A. 1943.
The fauna of British India, Ceylon and Burma, including the whole of the Indo-Chinese sub-region. London: Taylor and Francis. Volume 3.

2-52
Stephenson, J. 1838.
Medical zoology, and mineralogy; or illustrations and descriptions of the animals and minerals employed in medicine, and of the preparations derived from them: including also an account of animal and mineral poisons. London: John Churchill.

2-53
Swainson, W. 1839.
The natural history of fishes, amphibians, & reptiles, or monocardian animals. London: Longman, Orme, Brown, Green & Longmans.

2-54
Terentiev, P. V. 1961.
Herpetologie. Moskau. (In Russian)

2-55
Terent'ev, P. V. 1965.
Herpetology: a manual on amphibians and reptiles. Jerusalem: Israel Program for Scientific Translations. TT65-50015.

2-56
Tu, A. T. 1976.
Investigation of the sea snake, *Pelamis platurus* (Reptilia, Serpentes, Hydrophiidae), on the Pacific coast of Costa Rica, Central America. *J. Herpetol.* 10(1):13–18.

2-57
Tweedie, M. W. F. 1983.
The snakes of Malaya. 3rd ed. Singapore: Singapore National Printers.

2-60
Viaud Grand Marais, A. 1881.
Serpents venimeux. Dechambre, A., ed. *Dictionnaire encyclopedique des sciences medicales.* Paris: P. Asselin. Vol. 9, pp. 387–417.

2-61
Volsoe, H. 1939.
The sea snakes of the Iranian Gulf and the Gulf of Oman. With a summary of the biology of the sea snakes. Jessen, K., ed. *Danish Scientific Investigations in Iran.* Copenhagen: Ejnar Munksgaard. Vol. 1, pp. 9–45.

2-62
Wall, F. 1906.
A descriptive list of the sea snakes (Hydrophiidae) in the Indian Museum, Calcutta. *Mem. Asiat. Soc. Bengal* 1906:277–299.

2-63
Wall, F. 1909.
A monograph of the sea-snakes (Hydrophiinae). *Mem. Asiat. Soc. Bengal* 2(8):169–251.

2-64
Wall, F. 1919.
A popular treatise on the common Indian snakes. *J. Bombay Nat. Hist. Soc.* 26:430–437.

2-65
Wall, F. 1919.
A popular treatise on the common Indian snakes, 28. *Enhydrina valakadyn* (Boie) (vel *schistosa* (Daudin)). *J. Bombay Nat. Hist. Soc.* 26:803–810.

2-66
Wall, F. 1921
Ophidia taprobanica or the snakes of Ceylon. Colombo, Ceylon: H. R. Cottle, Government Printer.

2-67
Werner, F. 1900.
Die reptilien- und batrachierfauna des Bismarck-Archipels. *Mitt. Zool. Samml. Mus. Naturk. Berl.* 1(4):1–132.

2-68
Wright, A. H.; Wright, A. A. 1957.
Handbook of snakes of the United States and Canada. Ithaca, NY: Comstock.

Genus-Species Index
for Chapter 2

Chapter 3

GENERAL NATURAL HISTORY AND BEHAVIOR OBSERVATIONS

3-1
Abel, O. 1924.
Die eroberungszuege der wirbeltiere in die Meere der Vorzeit. Jena: Gustav Fischer.

3-2
Aiyar, T. V. R. 1906.
Notes on some sea-snakes caught at Madras. *J. Asiat. Soc. Beng.* 2(3):69–72.

3-4
Bennett, G. 1862.
Observations on the snakes of New South Wales. *Sydney Morning Herald,* Jun. 21:8.

3-5
Bennett, G. 1862.
The snakes of New South Wales. *Med. Rec. Aust.* 2(8):92–93.

3-6
Bergman, R. A. M. 1941.
Het vervellen van zeeslangen en van *Acrochordus. Trop. Natuur. (Weltevreden)* 30(9):145–148.

3-7
Boettger, O. 1890.
Protokoll-Auszuge uber die wissenschaftlichen sitzungen wahrend 1889–90. *Ber. Senckenb. Nat. Ges. Frankfurt a. Main* 1890:70–74.

3-7.5
Brehm, A. E. 1869.
Illustrites thierleben. Ein allgemeine kunde des thierreichs. Hildburghausen: Bibliographisches Institut.

3-8
Brehm, A. E. 1878.
Brehms thierleben. Leipzig: Bibliographischen Instituts. Volume 7.

3-9
Brehm, A. E. 1892.
Brehms tierleben. 3rd ed. Leipzig: Bibliographisches Institut. Volume 7.

3-10
Brongersma, L. D. 1958.
The animal world of Netherlands New Guinea. Groningen: J. B. Wolters.

3-11
Burns, G. W. 1984.
Aspects of population movements and reproductive biology of Aipysurus laevis, the olive sea snake. Armidale, New South Wales: University of New England. 171 pp. (Dissertation)

3-11.5
Campbell, J. A.; Lamar, W. W. 1989.
The venomous reptiles of Latin America. Ithaca, NY: Comstock Publishing Associates.

3-12
Cantor, T. 1838.
Observations on marine serpents. *Proc. Zool. Soc. Lond.* 1838(4):80.

3-13
Cantor, T. 1841.
Observations upon pelagic serpents. *Trans. Zool. Soc. Lond.* 2(4):303–313.

3-14
Caras, R. 1974.
Venomous animals of the world. Englewood Cliffs, NJ: Prentice Hall.

3-15
Carrington, R. 1960.
A biography of the sea: the story of the world ocean, its animal and plant populations, and its influence on human history. New York: Basic Books.

3-15.3
Cloitre, J. 1905.
Note sur le serpent corail en Nouvelle-Caledonie. *Ann. Hyg. Med. Colon.* 8:131–134.

3-15.7
Cogger, H.; Heatwole, H.; Ishikawa, Y.; McCoy, M.; Tamiya, N.; Teruuchi, T. 1987.
The status and natural history of the Rennell Island sea krait, *Laticauda crockeri* (Serpentes: Laticaudidae). *J. Herpetol.* 21(4):255–266.

3-16
DeCosta, N. 1980.
Gray's sea snake. *Loris* 15(3):155–156, 195.

3-16.5
Deraniyagala, P. E. P. 1932.
Herpetological notes. Reptiles and fish associates. *Spolia Zeylan.* 17(1):44–45.

3-16.7
Desmarest, E. 1857.
Reptiles et poissons. Chenu, J. C. *Encyclopedie d'histoire naturelle ou traite complet de cette science.* Paris: E. Girard et A. Boitte. Volume 6.

3-18
Finn, F. 1929.
Sterndale's Mammalia of India. A new and abridged edition, thoroughly revised and with an appendix on the Reptilia. Calcutta: Thacker, Spink & Co.

3-19
Gans, C. 1979.
Momentarily excessive construction as the basis for protoadaptation. *Evolution* 33(1):227–233.

3-19.5
Graham, J. B.; Lowell, W. R.; Rubinoff, I.; Motta, J. 1987.
Surface and subsurface swimming of the sea snake *Pelamis platurus. J. Exp. Biol.* 127:27–44.

3-20
Gray, M. E. 1930.
Notes on exhibit of living fishes. *Aust. Nat.* 8(4):87–88.

3-21

Guinea, M. L. 1981.

Sea snakes of Fiji. *Proc. 4th Int. Coral Reef Symp., Manila* 2:581–585.

3-22

Heatwole, H.; Minton, S. A.; Taylor, R.; Taylor, V. 1978.

Underwater observations on sea snake behaviour. *Rec. Aust. Mus.* 31(18):737–761.

3-22.5

Heatwole, H. 1987.

Sea snakes. Kensington, N.S.W.: New South Wales University Press.

3-23

Herre, A. W. C. T. 1942.

Notes on Philippine sea-snakes. *Copeia* 1942(1)1:7–9.

3-24

Hopley, C. C. 1882.

Snakes: curiosities and wonders of serpent life. New York: E. P. Dutton.

3-25

J. J. A. 1879.

Animal life in New Caledonia. *Land and Water*, Mar. 29:266–267.

3-26

Klemmer, K. 1966.

Observations on the biology of sea snakes—Hydrophiidae—with remarks on their systematics. *Mem. Inst. Butantan (Sao Paulo)* 33(1);101–103.

3-27

Kropach, C. 1972.

A field study of the sea snake, <u>Pelamis platurus</u> (Linnaeus) in the Gulf of Panama. New York: City University of New York. 200 pp. (Dissertation)

3-29

MacPherson, J. 1933.

Freshwater snakes and sea snakes. *Australas. Nurses J.* 31(4):71–74.

3-30

MacPherson, J. 1933.

Freshwater snakes and sea snakes. *Australas. Nurses J.* 31(5):100–103.

3-31

Mahadevan, S.; Nayar, K. N. 1965.

Underwater ecological observations in the Gulf of Mannar, off Tuticorin. I. Association between a fish (*Gnathanodon*) and a sea-snake. *J. Mar. Biol. Assoc. India* 7(1):197–199.

3-32

Mason, F. 1860.

Burmah, its people and natural productions, or notes on the nations, fauna, flora, and minerals of Tenasserim, Pegu and Burmah, with systematic catalogues of the known mammals, birds, fish, reptiles, insects, mollusks, crustaceans, annalids, radiates, plants and minerals, with vernacular names. 2nd ed. New York: Phinney, Blakeman & Mason.

3-33

Myers, G. S. 1945.

Nocturnal observations on sea-snakes in Bahia Honda, Panama. *Herpetologica* 3(1):22–23.

3-34

Pernetta, J. C. 1977.

Observations on the habits and morphology of the sea snake *Laticauda colubrina* (Schneider) in Fiji. *Can. J. Zool.* 55(10):1612–1619.

3-35

Pickwell, G. V. 1971.

Knotting and coiling behavior in the pelagic sea snake *Pelamis platurus* (L.). *Copeia* 1971(2):348–350.

3-36

Pimento, R. J. 1972.

Some notes on the sea snake *Laticauda colubrina* (Schneider). *J. Bombay Nat. Hist. Soc.* 69(1):191–192.

3-37

Pringle, E. H. 1881.

Modern snake lore. *Field (Lond.)* Sept. 3:352.

3-38

Read, K. R. H. 1970.

Palauan lobsters. *Aquasphere* 4(3):9–13.

3-38.5
Schnee, S. 1905.
   Ein neuer fall von mimikry? *Aus der Natur (Berlin)* 1(11):337–339.

3-39
Shaw, C. E. 1957.
   Sea serpents *do* exist! *Zoonooz (San Diego)* 30(4):10–12.

3-40
Stone, F. H. S. 1913.
   An unusually large sea-snake (*Distira brugmansi*). *J. Bombay Nat. Hist. Soc.* 22:403–404.

3-41
Van der Meer Mohr, J. C. 1930.
   Notes on the fauna of Pulau Berhala. *Treubia* 12(3–4):277–298.

3-42
Voris, H.; Voris, H. 1975.
   Sea snakes: a field report. *Bull. Field Mus. Nat. Hist.* 46(7):3–4, 21.

3-43
Voris, H. K. 1985.
   Population size estimates for a marine snake (*Enhydrina schistosa*) in Malaya. *Copeia* 1985(4):955–961.

3-44
Wolff, T. 1970.
   Lake Tegano on Rennell Island, the former lagoon of a raised atoll. *Nat. Hist. Rennell Isl. Br. Solomon Isl.* 6:7–29.

Genus-Species Index
for Chapter 3

Chapter 4

TAXONOMY AND SYSTEMATICS, SPECIES ACCOUNTS,CATALOGS OF COLLECTIONS

4-.05
Acharji, M. N.; Mukherjee, A. K. 1966.
   Report on a collection of snakes from lower Bengal (Reptilia: Ophidia). *J. Zool. Soc. India* 16(1–2):76–81.

4-1
Acton, H. W.; Knowles, R. 1921.
   Snakes and snake poisoning. Byam, W.; Archibald, R. G., eds. *The practice of medicine in the tropics.* London: Henry Frowde and Hodder & Stouhton. Vol. 1, pp. 683–762.

4-2
Agassiz, L. 1846.
   Nomenclator zoologicus. Soloduri: Sumtibus et Typis Jent et Gassmann. (In Latin)

4-3
Anderson, J. 1871.
   A list of the reptilian accession to the Indian Museum, Calcutta, from 1865 to 1870, with a description of some new species. *J. Asiat. Soc. Beng.* 40(1):12–39.

4-4
Anderson, J. 1871.
   On some Indian reptiles. *Proc. Zool. Soc. Lond.* 1871:149–211.

4-5
Anderson, J. 1872.
   On some Persian, Himalayan, and other reptiles. *Proc. Zool. Soc. Lond.* 1872:371–404.

4-6
Anderson, J. 1889.
Report on the mammals, reptiles, and batrachians, chiefly from the Mergui Archipelago, collected for the Trustees of the Indian Museum. *J. Linn. Soc. Lond. Zool.* 21:331–350.

4-7
Andersson, L. G. 1899.
Catalogue of Linnean type-specimens of snakes in the Royal Museum in Stockholm. *Bihang K. Svenska VetensAkad. Handl.* Ser. 4, 24(pt. 4, no. 6):1–35.

4-8
Andersson, L. G. 1916.
Notes on the reptiles and batrachians in the Zoological Museum at Gothenburg with an account of some new species. Goteborg: Wald. Zachrissons Boktryckeri.

4-9
Annandale, N. 1905.
Additions to the collection of oriental snakes in the Indian Museum. Part 2. Specimens from the Andamans and Nicobars. *J. Asiat. Soc. Beng.* N.S. 1(7):173–176.

4-10
Annandale, N. 1915.
Fauna of the Chilka Lake: reptiles and batrachians. *Mem. Indian Mus.* 5(2):167–174.

4-10.3
Anonymous. 1878.
Geschenke an das naturhistorische museum in den jahren 1873 bis 1877. *Verh. Naturforsch. Ges. Basel* 6:740–757.

4-10.5
Anonymous. 1921.
Contributions to the museum. *J. Bombay Nat. Hist. Soc.* 27(4): 970–973.

4-11
Anonymous. 1963.
Opinion 651: *Anilius* Oken, 1816 (Reptilia); validation under the plenary powers. *Bull. Zool. Nomencl.* 20(2):111–113.

4-12
Anonymous. 1982.
Opinion 1201: Elapidae Boie, 1827 (Reptilia, Serpentes): ruling to stabilize nomenclature of taxa in this family (and others). *Bull. Zool. Nomencl.* 39(1):21–26.

4-13
Barbour, T. 1912.
A contribution to the zoogeography of the East Indian Islands. *Mem. Mus. Comp. Zool. Harv.* 44(1):1–203.

4-14
Barbour, T.; Loveridge, A. 1929.
Typical reptiles and amphibians in the Museum of Comparative Zoology. *Bull. Mus. Comp. Zool.* 69(10):205–360.

4-15
Bavay, A. 1869.
Catalogue des reptiles de la Nouvelle-Caledonie et description d'especes nouvelles. *Mem. Soc. Linn. Normandie* 15(5):1–37.

4-16
Berg, C. 1901.
Herpetological notes. *Commun. Mus. Nac. Buenos Aires* 1(8):289–291.

4-17
Blainville, H. D. de. 1816.
Prodrome d'une nouvelle distribution systematique du regne animal. *Bull. Soc. Philomath. Paris* Ser. 3, 2:105–124.

4-17.5
Blainville, H. D. de. 1835.
Description de quelques especes de reptiles de la Californie, precedee de l'analyse d'un systeme general d'erpetologie et d'amphibiologie. *Nouv. Annal. Mus. Hist. Nat.* 4:233-296.

4-18
Blanford, W. T. 1879.
Notes on reptilia. *J. Asiat. Soc. Beng.* 48(3):127–132.

4-19
Blanford, W. T. 1881.
On a collection of Persian reptiles recently added to the British Museum. *Proc. Zool. Soc. Lond.* 1881:671–682.

4-20

Bleeker, P. 1858.

Bestuursvergadering. Gehouden den 12^n November 1857, ten huize van den heer steenstra toussaint. *Natuurw. Tijdschr. Ned.-Indie* 16:42–59.

4-21

Boettger, O. 1888.

Materialien zur herpetologischen fauna von China II. *Ber. Offenbacher Vereins Naturkd.* 26–28:53–191.

4-22

Boettger, O. 1889.

Herpetologische miscellen. *Ber. Senckenb. Nat. Ges. Frankfurt a. Main.* 1888–1889 2:267–316.

4-23

Boettger, O. 1892.

Listen von kriechtieren und lurchen aus dem tropischen Asien u. aus Papuasien. *Ber. Offenbacher Vereins Naturkd.* 29–32:65–164.

4-24

Boettger, O. 1895.

Neue froesche und schlangen von den Liukiu-Inseln. *Ber. Offenbacher Vereins Naturkd.* 33–36:101–117.

4-25

Boettger, O. 1898.

Katalog der reptilien-sammlung im Museum der Senckenbergischen Natur-forschenden Gesellschaft in Frankfurt am Main. Frankfurt am Main: Druck von Gebruder Knauer. Volume 2.

4-26

Bogert, C. M. 1940.

Herpetological results of the Vernay Angola Expedition. With notes on African reptiles in other collections. Part I. Snakes, including an arrangement of African Colubridae. *Bull. Am. Mus. Nat. Hist.* 77(1):1–107.

4-27

Boie, F. 1826.

Bemerkungen ueber Merrem's Versuch eines systems der amphibien. *Isis von Oken* 1826:cols:508–566.

4-28
Boie, F. 1826.
Generaluebersicht der familien und gattungen der ophidier. *Isis von Oken* 19(10):cols.981–982.

4-28.5
Boie, H. 1926.
Notice sur l'erpetologie de l'ile de Java. *Bull. Sci. Nat. Geol.* 9:233–240.

4-29
Bonaparte, C. L. 1831.
Saggio di una distribuzione metodica degli animali vertebrati. Roma.

4-30
Bonaparte, C. L. 1832
Versuch einer methodischen eintheilung der wirbelthiere. *Isis von Oken* 1832 cols:283–320.

4-31
Bonaparte, C. L. 1838.
Amphibiorum. Tabula analytica. *Nuovi Annal. Sci. Natu.* 1:391–393.

4-32
Bonaparte, C. L. 1839.
Amphibia europaea ad systema nostrum vertebratorum ordinata. *Memorie R. Accad. Sci. Torino* Ser. 2, 2:385-456.

4-33
Bonnaterre, P. J. 1789.
Ophiologie. *Tableau encyclopedique et methodique des trois regnes de la nature.* Paris: Panckoucke. Volume 1.

4-34
Bosc, L. A. G. 1803.
Hydrophis. *Nouveau dictionnaire d'histoire naturelle.* Paris: Deterville. Vol. 15, pp. 491–492.

4-34.5
Bosc, L. A. G. 1817.
Hydre, Hydrus. *Nouveau dictionnaire d'histoire naturelle.* Paris: Chez Deterville. Vol. 15, pp. 453 454.

4-35
Boulenger, G. A. 1887.
A list of the reptiles and batrachians obtained near Muscat, Arabia, and presented to the British Museum by Surgeon-Major A. S. G. Jayakar. *Ann. Mag. Nat. Hist.* Ser. 5, 20:407–408.

4-36
Boulenger, G. A. 1890.
Reptilia and batrachia. Blanford, W., ed. *The fauna of British India, including Ceylon and Burma.* London: Taylor and Francis. Part 1, section 3.

4-37
Boulenger, G. A. 1896.
Catalogue of snakes in the British Museum (Natural History). London: British Museum. Volume 3. New York: Hafner Pub. Co; 1961.

4-38
Boulenger, G. A. 1898.
Description of a new sea-snake from Borneo. *Proc. Zool. Soc. Lond.* 1898:106–107.

4-39
Boulenger, G. A. 1898.
On a little-known sea-snake from the South Pacific. Willey, A. *Zoological results based on material from New Britain, New Guinea, Loyalty Islands and elsewhere, collected during the years 1895, 1896 and 1897.* Cambridge: University Press. Vol. 1, pp. 57–58.

4-40
Boulenger, G. A. 1899.
A new sea-snake of the genus *Distira,* from Kurrachee. *J. Bombay Nat. Hist. Soc.* 12:642.

4-41
Boulenger, G. A. 1900.
Descriptions of new reptiles and batrachians from Borneo. *Proc. Zool. Soc. Lond.* 1900:182–187.

4-42
Boulenger, G. A. 1900
Descriptions of new reptiles from Perak, Malay Peninsula. *Ann. Mag. Nat. Hist.* Ser. 7, 5:306–308.

4-43

Boulenger, G. A. 1903.

Description of a new sea-snake from Rangoon. *J. Bombay Nat. Hist. Soc.* 14:719.

4-44

Boulenger, G. A. 1903.

Report on the batrachians and reptiles. Annandale, N.; Robinson, H. C. *Fasciculi Malayenses. Anthropological and sociological results of an expedition to Perak and the Siamese Malay States, 1901-1902, undertaken by Nelson Annandale and Herbert C. Robinson under the auspices of the University of Edinburgh and University College of Liverpool.* London: Published for the University Press of Liverpool by Longmans, Green & Co. Vol. 1 (Zoology), pt. 1, pp. 131–176.

4-45

Boulenger, G. A. 1908.

Note on the ophidian genus *Emydocephalus. Ann. Mag. Nat. Hist.* Ser. 8, 1:231.

4-46

Boulenger, G. A. 1912.

Reptilia and batrachia. Robinson, H. C., ed. *A vertebrate fauna of the Malay Peninsula from the Isthmus of Kra to Singapore including the adjacent islands.* London: Taylor and Francis.

4-47

Bourret, R. 1935.

Comment determiner un serpent d'Indochine. *Bull. Gen. Instr. Pub.* No. 3:1–28.

4-48

Bourret, R. 1936.

Les serpents de l'Indochine. Toulouse: Henri Basuyau. Volumes 1 and 2.

4-49

Brongersma, L. D. 1931.

Reptilia. Resultats scientifiques du voyage aux Indes orientales Neerlandaisis de. ll.aa.rr., Le Prince et Princesse Leopold de Belgique. *Mem. Mus. Hist. Nat. Belg.* 5(2):3–39.

4-50

Brongersma, L. D. 1933.

Herpetological note V. The herpetological fauna of Pulu Weh. *Zool. Meded. Rijksmus. Nat. Hist. Leiden* 16(1–2):1–29.

4-51
Brongersma, L. D. 1934.

Contributions to Indo-Australian herpetology. *Zool. Meded. Rijksmus. Nat. Hist. Leiden.* 17(3–4):161–251.

4-52
Brongersma, L. D. 1947.

Zoological notes from Port Dickson, I. Amphibians and reptiles. *Zool. Meded. Rijksmus. Nat. Hist. Leiden.* 27:300-308.

4-53
Brongersma, L. D. 1956.

Notes on New Guinean reptiles and amphibians. V. *Proc. K. Ned. Akad. Wet. Ser.C Biol. Med. Sci.* 59:599–610.

4-54
Burger, W. L.; Natsuno, T. 1974.

A new genus for the Arafura smooth seasnake and redefinitions of other seasnake genera. *Snake* 6(2):61–75.

4-55
Burt, C. E.; Burt, M. D. 1932.

Herpetological results of the Whitney South Sea Expedition. VI. Pacific island amphibians and reptiles in the collection of the American Museum of Natural History. *Bull. Am. Mus. Nat. Hist.* 38(2):461–597.

4-56
Cantor, T. 1847.

Catalogue of reptiles inhabiting the Malayan Peninsula and islands. *J. Asiat. Soc. Beng.* 16:1025–1078.

4-57
Cantor, T. 1886.

Catalogue of reptiles inhabiting the Malayan Peninsula and islands. Miscellaneous papers relating to Indo-China. London: Truebner Co. Volume 2, pp. 112-257. (Reprint of Cantor, T. 1847. *J. Asiat. Soc. Beng.* 16:1025–1078)

4-58
Cardew, A. G. 1897.

A rough key to the identification of Indian ophidia. *J. Bombay Nat. Hist. Soc.* 10:585–596.

4-59
Cloquet, H. 1830.
Encyclopedie methodique: system anatomique. Paris: Chez Mme Veuve Agasse. Volume 4.

4-59.3
Cochran, D. M. 1961.
Type specimens of reptiles and amphibians in the U.S. National Museum. *Bull. U.S. Natl Mus.* 220:1–291.

4-60
Cogger, H. G. 1983.
Zoological catalogue of Australia. Canberra: Australian Government Publishing Service. Volume 1.

4-60.5
Cogger, H. G. 1985.
Australian proteroglyphous snakes—an historical overview. Grigg, G.; Shine, R.; Ehrmann, H., eds. *Biology of Australian frogs and reptiles.* Chipping Norton, N.S.W.: Surrey Beatty & Sons. Pp. 143–154.

4-61
Constable, J. D. 1949.
Reptiles from the Indian Peninsula in the Museum of Comparative Zoology. *Bull. Mus. Comp. Zool.* 103(2):59–160.

4-62
Cope, E. D. 1859.
Catalogue of the venomous serpents in the Museum of the Academy of Natural Sciences in Philadelphia, with notes on the families, genera and species. *Proc. Acad. Nat. Sci. Phil.* 1859:332–347.

4-62.5
Cope, E. D. 1860
Supplement to "A catalog of the venomous serpents in the museum of the Academy," etc. *Proc. Acad. Nat. Sci. Phil.* 1860:72–74.

4-63
Cope, E. D. 1864.
On the characters of the higher groups of Reptilia Squamata—and especially of the Diploglossa. *Proc. Acad. Nat. Sci. Phil.* 16:224–231.

4-64
Cope, E. D. 1886.
   An analytical table of the genera of snakes. *Proc. Am. Philos. Soc.* 23:479–499.

4-65
Cope, E. D. 1887.
   Catalogue of batrachians and reptiles of Central America and Mexico. *Bull. U.S. Natl Mus.* 32:1–98.

4-66
Cope, E. D. 1895.
   The classification of the ophidia. *Trans. Am. Philos. Soc.* Ser. 2, 18:185–219.

4-67
Corkill, N. L. 1932.
   The snakes of Iraq. *J. Bombay Nat. Hist. Soc.* 35(3–4):552–572.

4-68
Covacevich, J. 1971.
   Amphibian and reptile type-specimens in the Queensland Museum. *Mem. Queensl. Mus.* 16(1):49–67.

4-69
Cuvier, G. 1831.
   Animal kingdom arranged in conformity with its organization by the Baron Cuvier, member of the Institute of France, etc. etc, etc, with additional descriptions of all the species hitherto named, and of many not before noticed, by Edward Griffith and others. London: Printed for Whittaker, Treacher and Co. Volume 9. ("A new edition" was published in 1859 in London by A. Fullarton and Co.)

4-70
Cuvier, G. 1836.
   Le regne animal, distribue d'apres son organisation, pour servir de base a l'anatomie comparee. 3rd ed. Bruxelles: L. Hauman. 3 volumes.

4-71
Dao, T. V. 1981
   Keys to the species of Vietnam snakes, Pt. 1. *Tap Chi Sinh Vat Hoc* 3(4):1–6. (In Vietnamese with English abstract)

4-72
Dao, T. V. 1981.
List of the Vietnamese snakes at present known in Vietnam. *Tap Chi Sinh Vat Hoc* 3(1):6–10. (In Vietnamese with English abstract)

4-73
Daudin, F. M. 1803.
Histoire naturelle, generale et particuliere des reptiles. Paris: F. Dufart. Volume 7.

4-74
Deraniyagala, P. E. P. 1955.
A colored atlas of some vertebrates from Ceylon. Colombo: Government Press. Volume 3.

4-75
Deraniyagala, P. 1960.
Some southern temperate zone snakes, birds and whales that enter the Ceylon area. *Spolia Zeylan.* 29(1):79–85.

4-76
Deraniyagala, P. E. P. 1974.
The history of marine herpetology of the Indian and Pacific Oceans. *J. Mar. Biol. Assoc. India* 16(3):793–806.

4-77
DeSilva, P. H. D. H. 1980.
Snake fauna of Sri Lanka with special reference to skull, dentition and venom in snakes. *Spolia Zeylan.* 34(1–2):1–472.

4-77.5
Desmarest, E. 1857.
Reptiles et poissons. Chenu, J. C. *Encyclopedie d'histoire naturelle ou traite complet de cette science.* Paris: E. Girard et A. Boitte. Volume 6.

4-78
DeVis, C. W. 1905.
Reptilia. *Ann. Queensl. Mus.* 6:46–52.

4-79
Donndorff, J. A. 1793.
Handbuch der thiergeschichte. Leipzig.

90 VENOMOUS SEA SNAKES

4-80
Donndorff, J. A. 1798.
Zoologische beytrage zur XIII ausgabe des Linneischen Natursystems. Leipzig.
Volume 3.

4-81
Donoso-Barros, R. 1966.
Reptiles de Chile. Santiago: Universidad de Chile.

4-82
Dowling, H. G. 1959.
Classification of the serpentes: a critical review. *Copeia* 1959(1):38–52.

4-83
Dowling, H. G. 1959.
Hemipenes and other characters in colubrid classification. *Herpetologica* 23(2):138–142.

4-84
Dowling, H. G. 1974.
A provisional classification of snakes. *Yearb. Herpetol.* 8:167–170.

4-85
Dowling, H. G.; Duellman, W. E. 1974.
Systematic herpetology: a synopsis of families and higher categories. *Publ. Herpetol.* 7:1–240.

4-86
Dumeril, A. M. C. 1806.
Zoologie analytique, ou methode naturelle de classification des animaux, rendue plus facile a l'aide de tableauz synoptiques. Paris: Allais.

4-87
Dumeril, A. A. 1853.
Prodrome de la classification des reptiles ophidiens. *Mem. Acad. Sci. Paris* 23:399–536

4-88
Dumeril, A.; Bocourt, F. 1870.
Etudes sur les reptiles et les batraciens. Part 3 of *Mission scientifique au Mexique et dans l'Amerique Central. Recherches Zoologiques*. Paris: Imprimerie Imperiale.

4-89
Duvernoy, G. L. 1836.
   Reptiles. Cuvier, G. *Regne animal distribue d'apres sur organisation*. Paris: Fortin, Masson et Cie. Volume 3.

4-90
Emelianov, A. A. 1929.
   Snakes of the far eastern district. Vladivostok. (In Russian with English abstract)

4-91
Ewart, J. 1878.
   The poisonous snakes of India. For the use of the officials and others residing in the Indian empire. London: Churchill.

4-92
Fayrer, J. 1871.
   The thanatophidia of India. *Indian Med. Gaz.* 6(1):1–3.

4-92.5
Fayrer, J. 1871.
   The thanatophidia of India. *Indian Med. Gaz.* Feb. 1871:21–24.

4-93
Fayrer, J. 1872.
   The thanatophidia of India, being a description of the venomous snakes of the Indian Peninsula, with an account of the influence of their poison on life; and a series of experiments. London: Churchill.

4-94
Fayrer, J. 1874.
   The thanatophidia of India, being a description of the venomous snakes of the Indian Peninsula, with an account of the influence of their poison on life; and a series of experiments. 2nd ed. rev'd and enl'd. London: Churchill.

4-95
Fischer, J. G. 1856.
   Die familie der seeschlangen systematisch beschrieben. 2nd ed. Hamburg: Druck von J. A. Meissner. (Also published in *Abh. Geb. Naturwiss. Nat. Hamburg* 3:1–78)

4-96
Fischer, J. G. 1888.
    Ueber zwei von Liukiu-Insel Okinawa stammende, schlangen. 1. *Platurus colubrinus* Schn. und dessen giftigkeit. *Jahrb. Hamb. Wiss. Anst.* 5:18–20.

4-97
Fitzinger, L. I. 1826.
    Neue classification der reptilien nach ihren naturlichen verwandtschaften. Wien: Verlage von G. Heubner.

4-97.5
Fitzinger, L. J. 1827.
    Ueber die Hydrien oder wasserschlangen. *Isis von Oken* 20(9):731–741.

4-98
Fitzinger, L. 1843.
    Systema reptilium. Vindobonae: Apud Braumueller et Seidel Bibliopolas. Volume 1.

4-99
FitzSimons, V. F. M. 1962.
    Snakes of Southern Africa. London: MacDonald.

4-100
Fleming, J. 1822.
    The philosophy of zoology; or a general view of the structure, functions, and classifications of animals. Edinburgh: Constable. Volume 2.

4-101
Gans, C. 1978.
    Reptilian venoms: some evolutionary considerations. Gans, C.; Gans, K. A., eds. *Biology of the reptilia.* New York: Academic. Vol. 8, pp. 1–106.

4-102
Garman, S. 1881.
    New and little-known reptiles and fishes in the Museum collections. *Bull. Mus. Comp. Zool.* 8(3):85–93.

4-103
Garman, S. 1892.
    On reptiles collected by Dr. Geo. Baur near Guayaquil, Ecuador. *Bull. Essex Inst.* 24:88–95.

4-103.5
Gervais, P. 1847.

Pelamide. *Dictionnaire universel d'histoire naturelle.* Paris: Chez l'Editeur Mm. Renard, Martinet et Cie. Vol. 9, pp. 544.

4-104
Girard, C. 1858.

Herpetology. *United States Exploring Expedition during the years 1838, 1839, 1840, 1841, 1842. Under the command of Charles Wilkes, U.S.N.* Philadelphia: J. B. Lippincott. Volume 20.

4-104.5
Goldfuss, G. 1820.

Handbuch der zoologie. Neurnberg: Johann Leonhard Schrag. Volume 2.

4-105
Golvan, Y. J. 1962.

Repertoire systematique des noms de genres de vertebres (suite). *Ann. Parasitol. Hum. Comp.* 37:870–997.

4-106
Gray, J. E. 1825.

A synopsis of the genera of reptiles and amphibia, with a description of some new species. *Ann. Philos.* N.S. 10:193–217.

4-107
Gray, J. E. 1829.

Synopsis generum reptilium et amphibiorum. *Isis von Oken* 22(2):187–206.

4-107.5
Gray, J. E. 1831.

A synopsis of the species of the class reptilia. Bound with Griffith, E. *The animal kingdom arranged in conformity with its organization, by the Baron Cuvier with additional descriptions of all the species hitherto named, and of many not before noticed.* London: Whittaker, Treacher. Volume 9.

4-108
Gray, J. E. 1837.

Revised arrangement of the Ophidians. *Proc. Zool. Soc. Lond.* 5:135.

4-109
Gray, J. E. 1842.

Monographic synopsis of the water snakes, or the family Hydridae. *Zool. Miscell.* 1842:59–68.

4-110
Gray, J. E. 1843.
   Description of two new species of reptiles from the collection made during the voyages of the H. M. S. Sulphur. *Ann. Mag. Nat. Hist.* 11:46.

4-111
Gray, J. E. 1846.
   Descriptions of some new Australian reptiles. Stokes, J. L. *Discoveries in Australia with an account of the coasts and rivers explored and surveyed during the voyage of the H. M. S. Beagle in the years 1837-38-39-40-41-42-43.* London: T. and W. Boone. Vol. 1, pp. 498-504. Adelaide: Libraries Board of South Australia; 1969.

4-112
Gray, J. E. 1846.
   A new genus of sea-snake from Port Essington. *Ann. Mag. Nat. Hist.* 18:284.

4-113
Gray, J. E. 1847.
   Description of a new genus of snakes. Jukes, J. B. *Narrative of the surveying voyage of H. M. S. Fly, commanded by Captain F. P. Blackwood, R. N. in Torres Strait, New Guinea, and other islands of the Eastern Archipelago, during the years 1842–1846: together with an excursion into the interior of the eastern part of Java.* London: T. & W. Boone. Vol. 2, pp. 332–334.

4-114
Gray, J. E. 1849.
   Catalogue of the specimens of snakes in the collection of the British Museum. London: Printed by order of the Trustees.

4-115
Gray, J. E. 1858.
   Appendix (2.) to the Catalogue of the specimens of snakes in the collection of the British Museum. First Part (Crotalidae, Viperidae, Hydridae and Boidae). Guenther, A. *Catalogue of the colubrine snakes in the collection of the British Museum.* London: The Trustees. Pp. 265–281.

4-116
Guenther, A. C. L. G. 1864.
   The reptiles of British India. London: Printed for the Ray Society by Robert Hardwicke.

4-117
Guenther, A. 1872.
   On the reptiles and amphibians of Borneo. *Proc. Zool. Soc. Lond.* 1872:586–600.

4-118
Guenther, A. 1872.
   Seventh account of new species of snakes in the collection of the British Museum. *Ann. Mag. Nat. Hist.* Ser. 4, 9:13–35.

4-119
Guenther, A. 1874.
   A contribution to the fauna of Savage Island. *Proc. Zool. Soc. Lond.* 1874(20):295–297.

4-120
Guerin-Meneville, F. E. 1829.
   Iconographie du regne animal de G. Couvier. Paris: J. B. Bailliere. Volume 3.

4-121
Hallowell, E. 1860.
   Report upon the reptilia of the North Pacific Exploring Expedition, under command of Capt. John Rogers, U.S.N. *Proc. Acad. Nat. Sci. Phil.* 1860:480–510.

4-122
Haly, A. 1886.
   First report on the collection of snakes in the Colombo Museum. Colombo: Government Printer.

4-123
Haly, A. 1891.
   Report on the collections of reptilia & batrachia in the Colombo Museum. Colombo.

4-124
Hanitsch, R. 1912.
   List of the birds, reptiles and amphibians in the Raffles Museum, Singapore. Singapore: The Museum.

4-124.5
Harding, K. A.; Welch, K. R. G. 1980.
   Venomous snakes of the world. New York: Pergamon.

4-125
Hermann, J. 1783.
   Tabula affinitatum animalium olim academico specimine edita nunc uberiore commentario illustrata cum annotationibus. Argentorati: Imprensis Joh. Georgii Treuttel.

4-126
Hermann, J. 1804.
   Observationes zoologicae quibus novae complures, aliaeque animalium species describuntur et illustrantur opus posthumum edidit Fridericus Ludovicus Hammer. Parisiis: Argentorati apud Amandum Koenig.

4-126.3
Hoeven, J. van der. 1856.
   Handbuch der zoologie. Leipzig: Leopold Voss. Volume 2.

4-126.5
Hoffstetter, R. 1946.
   Remarques sur la classification des ophidiens et particulierement des Boidae des Mascareignes (Bolyerinae *subfam. nov.*). *Bull. Mus. Natl Hist. Nat.* Ser. 2, 18(1):132–135.

4-127
Holtzinger Tenever, H. 1920.
   Herpetologische mitteilungen aus dem museum fuer Naturkunde in Oldenburg, Gr. *Arch. Nat. (Berlin)* Abt. A. 85(11):81–98.

4-128
In Den Bosch, H. A. J. 1985.
   Snakes of Sulawesi: checklist, key and additional biogeographical remarks. *Zool. Verh. (Leiden)* No. 217:3–50.

4-128.5
Jacquinot, H.; Guichenot, A. 1855.
   Reptiles et poissons. Dumont-D'Urville, M. J. *Voyage au Pole Sud et dans l'oceanie sur les corvettes l'Astrolabe et la Zelee; execute par ordre du roi pendant les annees 1837–1838–1839–1840: Zoologie.* Paris: Guide et J. Baudry. Volume 3.

4-129
Jan, G. 1857.
   Cenni sul Museo Civico di Milano ed indice sistematico dei rettili ed anfibi esposti nel medesimo. Milano: Luigi Giacomo Pirola.

4-130
Jan, G. 1859.
Plan d'une iconographie descriptive des ophidiens et description sommaire de nouvelles especes de serpents. *Revue Mag. Zool.* 1859:148–152.

4-131
Jan, G. 1863.
Elenco sistematico degli ofidi descritti e disegnati per l'iconografia generale. Milano.

4-132
Jan G.; Sordelli, F. 1872.
Iconographie generale des ophidiens. Paris: J.-B. Balliere. Ninth through forty-first livraisons. (Atlas)

4-132.5
Jouan, H. 1863.
Notes sur quelques animaux observes a la Nouvelle-Caledonie, pendant les annees 1861 et 1862. *Mem. Soc. Nat. Sci. Nat. Math. Cherbourg* 9:88–127.

4-133
Kano, T. 1932.
Formosan native names for different animals. 2. Yami-tribe Kotosho. *Amoeba (Tokyo)* 2:82-91. (In Japanese)

4-133.5
Kaup, J. J. 1836.
Das thierreich in seinen hauptformen systematisch beschreiben. Darmstadt: J. P. Diehl. Volume 2, part 2.

4-134
Kharin, V. E. 1981.
A review of sea snakes of the genus *Aipysurus* (Serpentes, Hydrophiidae). *Zool. Zh.* 60(2):257–264. (In Russian with English abstract)

4-135
Kharin, V. E. 1983.
A new species of the genus *Hydrophis* sensu lato (Serpentes, Hydrophiidae) from the North Australian shelf. *Zool. Zh.* 62(11):1751–1753. (In Russian with English abstract)

4-136
Kharin, V. E. 1984.
   The first record of three species of sea snakes from Vietnam with a note on rare variety of *Praescutata viperina*. *Biol. Morya (Vladivost.)* No. 2:26–30. (In Russian)

4-137
Kharin, V. E. 1984.
   A review of sea snakes of the group *Hydrophis* sensu lato (Serpentes, Hydrophiidae). 3. The genus *Leioselasma*. *Zool. Zh.* 63(10):1535–1546. (In Russian with English abstract)

4-138
Kharin, V. E. 1984.
   Revision of sea snakes of subfamily Laticaudinae Cope 1879 sensu lato (Serpentes, Hydrophiidae). *Tr. Zool. Inst. Akad. Nauk S.S.S.R.* 124:128–139. (In Russian with English abstract)

4-139
Kharin, V. E. 1984.
   Sea snakes of the genus *Hydrophis* sensu lato (Serpentes, Hydrophiidae). On taxonomic status of the New Guinea *H. obscurus*. *Zool. Zh.* 63(4):630–632. (In Russian with English abstract)

4-140
Kharin, V. E. 1985.
   A new species of sea snakes of the genus *Enhydrina* (Serpentes, Hydrophiidae) from waters of New Guinea. *Zool. Zh.* 64(5):785–787. (In Russian with English abstract)

4-140.5
Kiewiet de Jong, G. W. 1916.
   Met een verhandeling over de verschijnselen en over de behandeling van giftigen slangenbeet. Leiden: E. J. Brill.

4-141
Kinghorn, J. R. 1926.
   A new genus of sea snake and other reptiles collected by H.M.A.S. 'Geranium' during a cruise to North Australia. *Proc. Zool. Soc. Lond.* 1926:71–74.

4-142

Klemmer, K. 1963.

Liste der rezenten giftschalangen: Elapidae, Hydropheidae, Viperidae und Crotalidae. *Die Giftschlangen der Erde.* Marburg: N. G. Elwert Universitats und Verlags-Buchhandlung. Pp. 255–464.

4-142.5

Kopstein, F. 1937.

Schlangen van Enggano. *Treubia* 16(2):239–244.

4-143

Kopstein, F. 1938.

Ein beitrag zur eierkunde und zur fortpflanzung der Malaiischen reptilien. *Bull. Raffles Mus.* 14:81–167.

4-144

Krefft, G. 1869.

Descriptions of new Australian snakes. *Proc. Zool. Soc. Lond.* 1869:321–322.

4-145

Krefft, G. 1869.

The snakes of Australia: an illustrated and descriptive catalogue of all the known species. Sydney: Thomas Richards Government Printer.

4-145.5

Kuhn, O. 1961.

Die familien der rezenten und fossilen amphibien und reptilien. Bamberg: Meisenbach.

4-146

Kuhn, O. 1966.

Die reptilien: system und stammesgeschichte. Muenchen: Verlag Oeben.

4-147

Kuhn, O. 1967.

Amphibien und reptilien. Katalog der subfamilien und hoheren taxa mit nachweis des ersten auftretens. Stuttgart: Gustav Fischer.

4-148

LaCepede, B. 1800.

Herrn de La Cepede's Naturgeschichte der amphibien oder der eyerlegenden vierfussigen thiere und der schlangen. Eine fortsetzung von Buffon's Naturges-

chichte. *Aus dem franzosischen uebersetzt und mit Anmerkungen und zusatzen versehen von Johann Mathour Bechstein.* Weimar: Comptoir's. Volume 3.

4-149
LaCepede, B. 1804.
    Memoire sur plusieurs animaux de la Nouvelle-Hollande dont la description n'a pas encore ete publiee. *Ann. Mus. Nat. Hist. Nat. Paris* 4:184–211.

4-150
LaCepede, B. 1830.
    Histoire naturelle des serpents. Paris: Pillot. Volumes 3, 4, 7.

4-151
Laidlaw, F. F. 1901.
    List of a collection of snakes, crocodiles, and chelonians from the Malay Peninsula, made by members of the "Skeat Expedition", 1899-1900. *Proc. Zool. Soc. Lond.* 2:575–586.

4-152
Lamarck, J. B. 1830.
    Philosophie zoologique. Paris: J. B. Bailliere. Volume 1. New York: Hafner; 1960.

4-153
Lamarck, J. B. 1914.
    Zoological philosophy, translated by Hugh Elliot. London: Macmillan. (A translation of the 1809 edition of *Philosophie zoologique*)

4-153.2
Lampe, E. 1902.
    Catalog der reptilien- und amphibien-sammlung (schlangen; frosch-, schwanz- und schleichenlurch) des naturhistorischen museums zu Wiesbaden. *Jahrb. Nassau. Naturk.* 55:1–66.

4-153.4
Lampe, E. 1911.
    Erster nachtrag zum katalog der reptilien- und amphibien-sammlung des naturhistorischen museums der stadt Wiesbaden. *Jahrb. Nassau. Naturk.* 64:137–236.

4-153.6
Lampe, E. 1911.
    Reptilien und amphibien aus Deutsch-Neuguinea. *Jahrb. Nassau. Naturk.* 66:80–86.

4-153.8
Latreille, P. A. 1825.
Familles naturelles du regne animal, exposees succinctement et dans un ordre analytique, avec l'indication de leurs genres. Paris: J. B. Bailliere.

4-154
Laurenti, J. N. 1768.
Specimen medicum exhibens synopsin reptilium emendatam cum experimentis circa venena et antidota reptilium austriacorum, quod authoritate et consensu. Viennae: typis Joan. Thomae, nob. de Trattnern. Amsterdam: A. Asher and Co.; 1966.

4-154.5
Lesson, R. P. 1830.
Zoologie. Duperrey, L. I. *Voyage autour du monde, execute par ordre du roi, sur la corvette de la majeste, La Coquille, pendant les annees 1822, 1823, 1824 et 1825.* Paris: Arthus Bertrand.

4-155
Lesson, R. P. 1834.
Reptiles. Belanger, C. *Voyage aux Indes-Orientales: Zoologie.* Paris: Arthus Bertrand.

4-156
Liat, L. B.; Balasingam, E. 1969.
A collection of sea snakes from Port Swettenham, Selangor and Tumpat, Kelantan. *J. Fed. Malay States Mus.* 14:123–126.

4-157
Lichtenstein, H. 1856.
Nomenclator reptilium et amphibiorum musei zoologici Berolinensis; namen-verzeichniss der in der zoologischen sammlung der koeniglichen Universitat zu Berlin aufgestellten arten von reptilien und amphibien nach ihren ordnungen, familien und gattungen. Berlin.

4-157.3
Lidth de Jeude, T. W. van. 1898.
Catalogue osteologique des poissons, reptiles et amphibies. Leide: E. J. Brill.

4-157.5
Lidth de Jeude, T. W. van. 1911.
Reptilien (schlangen). *Nova Guinea. Resultats de l'expedition scientifique Neerlandaise a la Nouvelle-Guinee en 1903.* Leide: E. J. Brill. Vol. 5, pt. 4, pp. 519–530.

4-158
Linne, C. 1754.
   Museum Sae Rae Mtis Adolphi Friderici regis Svecorum, Gothorum, Vandalorumque. Holmiae. (In Latin and Swedish)

4-159
Linne, C. 1758.
   Systema naturae. 10th ed. Holmiae: Laurentii Salvii. London: British Museum (Natural History); 1956. Volume 1 (In Latin)

4-160
Linne, C. 1766.
   Systema naturae. 12th ed. Holmiae: Laurentii Salvii. Ohio: Ohio Herpetological Society; 1963. Volume 1.

4-161
Linne, C. 1788.
   Systema naturae. 13th ed. "Cura" Jo. Frid. Gmelin. Lipsiae: Imprensis Georg. Emanuel Beer. Volume 1.

4-162
Loennberg, E. 1896.
   Linnean type-specimens of birds, reptiles, batrachians and fishes in the Zoological Museum of the R. University of Upsala. *Bihang K. Svenska VetensAkad. Handl.* 22(4,1):1–45.

4-163
Loennberg, E.; Andersson, L. G. 1913.
   Results of Dr. E. Mjoebergs Swedish Scientific Expeditions to Australia 1910–13. III. Reptiles. *K. Svenska Vetensk.-Akad. Handl.* N. F. 52(3):1–17.

4-164
London, Eng. (City). Royal College of Surgeons of England. Museum. 1859.
   Descriptive catalogue of the specimens of natural history in spirit contained in the Museum of the Royal College of Surgeons of England. London: Richard Taylor and William Francis.

4-165
Longman, H. A. 1918.
   Notes on some Queensland and Papuan reptiles. *Mem. Queensl. Mus.* 6:37–44.

4-166
Luetken, C. F. 1862.
   Nogle nye krybdyr og padder. *Vidensk. Medd. Dan. Naturhist. Foren.* 1862:292–311.

4-167
Mack, G.; Gunn, S. B. 1953.
   De Vis' types of Australian snakes. *Mem. Queensl. Mus.* 13:58-70.

4-168
Maki, M. 1931.
   A monograph of the snakes of Japan. Tokyo: Dai-Ichi Shobo.

4-169
Mao, S. H.; Chen, B. Y. 1975.
   Disteira stokesi (Gray): an addition to the sea snake fauna of Taiwan. *Herpetologica* 31(4):456–458.

4-170
Mao, S. H.; Dessauer, H. C.; Chen, B. Y. 1978.
   Fingerprint correspondence of hemoglobins and the relationships of sea snakes. *Comp. Biochem. Physiol. B Comp. Biochem.* 59(4):353-361.

4-171
Mao, S. H.; Chen, B. Y. 1980.
   Sea snakes of Taiwan: a natural history of sea snakes. Taipei, Taiwan: National Science Council.

4-172
Mao, S. H.; Chen, B. Y.; Yin, F. Y.; Guo, Y. W. 1983.
   Immunotaxonomic relationships of sea snakes to terrestrial elapids. *Comp. Biochem. Physiol. A Comp. Physiol.* 74(4):869–872.

4-173
McCoy, M. 1980.
   Reptiles of the Solomon Islands. Hong Kong: Sheck Wah Tong Printing Press Limited. (*WAU Ecology Institute Handbook* No. 7)

4-174
McDowell, S. B. 1967.
   Aspidomorphus, a genus of New Guinea snakes of the family Elapidae, with notes on related genera. *J. Zool. (Lond.)* 151:497–543.

4-175
McDowell, S. B. 1969.

Notes on the Australian sea-snake *Ephalophis greyi* M. Smith (Serpentes: Elapidae, Hydrophiinae) and the origin and classification of sea-snakes. *Zool. J. Linn. Soc.* 48:333–349.

4-176
McDowell, S. B. 1972.

The genera of sea-snakes of the *Hydrophis* group (Serpentes: Elapidae). *Trans. Zool. Soc. Lond.* 32:189–247.

4-177
McDowell, S. B. 1974.

Additional notes on the rare and primitive sea-snake, *Ephalophis greyi*. *J. Herpetol.* 8(2):123–128.

4-178
Mengden, G. A. 1975.

The taxonomy of Australian elapid snakes: a review. *Rec. Aust. Mus.* 35(5):195–222.

4-179
Merrem, B. 1820.

Tentamen systematis amphibiorum: versuch eines systems der amphibien. Marburg: J. C. Kreiger. (In German and Latin)

4-179.5
Meyer, A. B. 1887.

Verzeichniss der von mir in den jahren 1870–1873 im Ostindischen Archipel gesammelten reptilien und batrachier. *Abh. Mus. Dresden* 2:1–16.

4-180
Minton, S. A. 1966.

A contribution to the herpetology of West Pakistan. *Bull. Am. Mus. Nat. Hist.* 134(2):27–184.

4-181
Mittleman, M. B. 1947.

Geographic variation in the sea snake, *Hydrophis ornatus* (Gray). *Proc. Biol. Soc. Wash.* 60:1–6.

4-182
Mueller, F. 1878.
    Katalog der im Museum und Universitatskabinet zu Basel aufgestellten amphibien und reptilien nebst Anmerkungen. *Verh. Naturforsch. Ges. Basel* 6:559–709.

4-183
Mueller, F. 1882.
    Zweiter nachtrag zum katalog der herpetologischen sammlung des Basler Museums. *Verh. Naturforsch. Ges. Basel* 7:166–174.

4-184
Mueller, F. 1885.
    Vierter nachtrag zum katalog der herpetologischen sammlung des Basler Museums. *Verh. Naturforsch. Ges. Basel* 7:668–717.

4-185
Mueller, F. 1890.
    Fuenfter nachtrag zum katalog der herpetologischen sammlung des Basler Museums. *Verh. Naturforsch. Ges. Basel* 8:249–296.

4-186
Mueller, F. 1892.
    Siebenter nachtrag zum katalog der herpetologischen sammlung des Basler Museums. *Verh. Naturforsch. Ges. Basel* 10:195–215.

4-187
Murray, J. A. 1884.
    The vertebrate zoology of Sind. A systematic account, with descriptions of all the known species of mammals, birds, and reptiles inhabiting the province; observations on their habits, &c., tables of their geographical distribution in Persia, Beloochistan, and Afghanistan; Punjab, North-West provinces, and the peninsula of India generally. London: Richardson & Co.

4-188
Murray, J. A. 1886.
    The reptiles of Sind. London: Richardson & Co.

4-189
Murray, J. A. 1887.
    Description of three new species of *Hydrophis* from the Bombay Harbour and the Mekran Coast. *J. Bombay Nat. Hist. Soc.* 2:32–35. (Also published in *Indian Ann. Mag. Nat. Sci.* 1(1):20–22)

4-190
Myers, G. S. 1947.
Murray's *Reptiles of Sind*, with a note on three forgotten descriptions of Indian sea-snakes, published therein. *Herpetologica* 3:167–168.

4-191
Nicholson, E. 1874.
Indian snakes. An elementary treatise on ophiology with a descriptive catalogue of the snakes found in India and the adjoining countries. Madras: Higginbotham.

4-192
Nicholson, E. 1893.
Indian snakes. An elementary treatise on ophiology with a descriptive catalogue of the snakes found in India and the adjoining countries. 2nd ed. Madras: Higginbotham.

4-193
Nikolskii, A. M. 1905.
Herpetologia Rossica. St. Petersbourg. (also in *Mem. Acad. Imp. Sci. St. Petersbourg. Cl. Phys.-Math.* Ser. 8, 17(1):1–517) (In Russian)

4-194
Nikolskii, A. M. 1916.
Faune de la Russie et des pays limitrophes. Petrograd. Volume 2. (In Russian) (Translated 1964 by the Israel Program for Scientific Translations: *Fauna of Russia and adjacent countries*)

4-195
Ocaranza, F. 1930.
Sistematica de los animales ponzonosos de la America Latina y accion biologica de sus venenos. *Med. Rev. Mex.* 10(121):357–374.

4-196
Ofuchi, M. 1934.
Pelamydrus platurus (Linnaeus) caught off Kinkazan Pt. *Dobutsugaku Zasshi* 46(554):537–539. (In Japanese)

4-197
Ogilby, J. D. 1898.
Ophidia and Pices, Pt. 2 of Thomas Steele's Contributions to a knowledge of the fauna of British New Guinea. *Proc. Linn. Soc. N.S.W.* 23:357–369.

4-197.5
Okada, Y. 1953.
Reptiles. Uchida, K. *Illustrated encyclopedia of the fauna of Japan (exclusive of insects)*. Rev. ed. Tokyo: Hokuryukan. Pp. 223–271. (In Japanese)

4-198
Oken, L. 1816.
Oken's Lehrbuch der Naturgeschichte. Leipzig: C. H. Reclam. Volume 3.

4-199
Oken, L. 1817.
Cuvier's und Oken's Zoologien neben einander gestellt. *Isis von Oken* 8(144):cols.1145–1186.

4-199.5
Oken, L. 1836.
Allgemeine naturgeschichte fuer alle stande. Stuttgart: Hoffmann. Volume 6.

4-199.6
Oppel, M. 1810.
Memoire sur la classification des reptiles: Ordre II. Reptiles a escailles. Section II. Ophidicns. *Ann. Mus. Nat. Hist. Nat. Paris* 16:254–295.

4-199.7
Oppel, M. 1810.
Suite du Ier memoire sur la classification des reptiles: Ord. II. Squammata mihi. Sect. II. Ophidii. Ord. III. Ophidii. *Ann. Mus. Nat. Hist. Nat. Paris* 16:376–393.

4-200
Oppel, M. 1811.
Die ordnungen, familien, und gattungen der reptilien als prodrom einer naturgeschichte derselben. Muenchen: In Commission bey J. Lindauer.

4-200.5
Oshima, M. 1916.
Key to the snakes of Formosa and the Riu-Kius. *Trans. Nat. Hist. Soc. Taiwan* 6(23):1–12. (In Japanese)

4-201
Oshima, S. 1944.
Explanations of poisonous snakes in the greater East Asia coprosperity sphere. Tokyo: Hokuryu-kan. (In Japanese)

4-202
Ouwens, P. A. 1916.
   De voornaamste giftslangen van Nederlandsch Oost-Indie. Leiden: E. J. Brill.

4-203
Peters, W. 1872.
   Ueber den *Hydrus fasciatus* Schneider und einige andere seeschlangen. *Monatsber. Deut. Akad. Wiss. Berl.* 1872:848–861.

4-204
Peters, W. 1876.
   Ueber die von S. M. S. Gazelle mitgebrachten amphibien. *Monatsber. Deut. Akad. Wiss. Berl.* 1876:528–535.

4-205
Peters, W. 1877.
   Herpetologische notizen 1. Ueber die von spix in Brasilien gesammelten eidechsen des Koeniglichen naturalien-kabinets zu Muenchen. *Monatsber. Deut. Akad. Wiss. Berl.* 1877:407–423.

4-206
Peters, W.; Doria, G. 1878.
   Catalogo dei rettili e dei batraci raccolti da O. Beccari, L. M. D'Albertis e A. A. Bruijn nella sotto-regione austro-malese. *Ann. Mus. Civ. Genova* 13:323–450.

4-207
Phillipps, W. J. 1941.
   Recent occurrence of *Pelamis platurus* Linn. in New Zealand. *Trans. R. Soc. N. Z.* 71(1):23–24.

4-208
Phipson, H. M. 1886.
   Catalogue of snakes in the Society's collection. *J. Bombay Nat. Hist. Soc.* 1:84–86.

4-209
Phipson, H. M. 1888.
   Catalogue of the snakes in the Society's collection. *J. Bombay Nat. Hist. Soc.* 3:49–53.

4-210

Pickwell, G. V.; Culotta, W. A. 1980.

Pelamis and *Pelamis platurus*. Zweifel, R. G., ed. *Catalogue of American amphibians and reptiles*. New York: Society for the Study of Amphibians and Reptiles. No. 255:1–4.

4-211

Ping, C. 1926.

A sea snake from Yenting *Disteira* sp. *Trans. Sci. Soc. China* 4:45–47.

4-212

Pope, C. H. 1935.

The reptiles of China. New York: American Museum of Natural History. (Also published as *Central asiatic expeditions of the American Museum of Natural History*, New York, Vol. 10)

4-213

Rafinesque, C. S. 1815.

Analyse de la nature ou tableau de l'univers et des corps organises. Palermo.

4-214

Rai, M. 1969.

Sur deux serpents recoltes en Iran *Hydrophis spiralis* Gray et *Hydrophis ornatus* Guenther. *Actes Soc. Linn. Bordeaux* Ser.A 106(4):3 pages. (Reprinted in *Bull. Stn Biol. Arcachon* (N. S.) No. 22:4pp, 1969)

4-214.5

Rasmussen, A. R. 1989.

An analysis of *Hydrophis ornatus* (Gray), *H. lamberti* Smith, and *H. inornatus* (Gray) (Hydrophiidae, Serpentes) based on samples from various localities, with remarks on feeding and breeding biology of *H. ornatus*. *Amphib.-Reptilia* 10:397–417.

4-215

Ritgen, F. A. 1828.

Versuch einer natuerlichen eintheilung der amphibien. *Nova Acta Phys.-Med. Acad. Natur. Curios.* 14(1):245–284.

4-216

Romer, A. S. 1956.

Osteology of the reptiles. Chicago: University of Chicago Press.

4-217

Romer, J. D. 1970.

Revised annotated checklist with keys to the snakes of Hong Kong. *Mem. Hong Kong Nat. Hist. Soc.* No. 8:1–22.

4-218

Rooij, N. de. 1917.

The reptiles of the Indo-Australian Archipelago. Leiden: E. J. Brill.

4-219

Rosen, N. 1905.

List of the snakes in the Zoological Museums of Lund and Malmo, with descriptions of new species and a new genus. *Ann. Mag. Nat. Hist.* Ser. 7, 15:168–181.

4-219.5

Rosenberg, P. 1987.

Common name index: poisonous animals, plants and bacteria. *Toxicon* 25(8):799–801.

4-220

Roux, J. 1910

Reptilien und amphibien den Aru- und Kei-Inseln. *Abh. Senckenb. Naturforsch. Ges.* 33(3):211–247.

4-221

Roux, J. 1913.

Les reptiles de la Nouvelle-Caledonie et des iles Loyalty. Sarasin, F.; Roux, J. *Nova Caledonia.* Wiesbaden: C. W. Kreidels Verlag. Vol. 1A, pt. 1, pp. 79–160.

4-221.6

Rueppel, E. 1845.

Verzeichniss der in dem Museum der Senckenbergischen naturforschenden gesellscharft aufgestellten sammlungen. Amphibien. *Museum Senckenbergianum.* Frankfurt am Main. Vol. 3, pt. 3, pp. 293–318.

4-222

Russell, P. 1796.

An account of Indian serpents, collected on the coast of Coromandel; containing descriptions and drawings of each species; together with experiments and remarks on their several poisons. London: George Nicol.

4-223
Russell, P. 1801.
   A continuation of An account of Indian serpents, containing descriptions and figures, from specimens and drawings, transmitted from various parts of India. London: G. and W. Nicol.

4-224
Saint Girons, H. 1972.
   Les serpents du Cambodge. *Mem. Mus. Natl Hist. Nat. Ser.A Zool.* 74:1–170.

4-225
Sauvage, M. H. E. 1877.
   Sur quelques ophidiens d'especes nouvelles ou peu connues de la collection du Museum. *Bull. Soc. Philomath. Paris* 7(1):107–115.

4-226
Savage, J. M. 1961.
   Anilius Oken, 1816 (Reptilia, Squamata): proposed validation under the plenary powers. *Bull. Zool. Nomencl.* 18(3):181–183.

4-226.3
Schenkel, E. 1902.
   Achter nachtrag zum katalog der herpetologischen sammlung des Basler museums. *Verh. Naturforsch. Ges. Basel* 13(1):142–199.

4-226.5
Schlegel, H. 1827.
   Erpetologische nachtrichten. *Isis von Oken* 29(3):281–294.

4-227
Schmidt, P. 1846.
   Beschreibung zweier neuen Reptilien, aus dem naturhistorischen Museum zu Hamburg. *Abh. Geb. Naturwiss. Nat. Hamburg* 1(5):166–172.

4-228
Schneider, J. G. 1799.
   Historiae amphibiorum, naturalis et literariae. Jena: Friederici Frommanni. Amsterdam: A. Asher; 1968.

4-228.5
Schulze, F. E.; Kuekenthal, W.; Heider, K., eds. 1926–1935.
   Nomenclator animalium generum et subgenerum. Berlin: Preussischen Akademie der Wissenschaften. 5 volumes.

4-229
Sclater, W. L. 1891.
    List of snakes in the Indian Museum. Calcutta: Trustees of the Indian Museum.

4-230
Sclater, W. L. 1892.
    Notes on the collection of snakes in the Indian Museum with descriptions of several new species. *J. Asiat. Soc. Beng.* 60(2):230–250.

4-231
Scortecci, G. 1939.
    Gli ofidi velenosi dell'Africa Italiana. Milano: Dell'Istituto Sieroterapico Milanese.

4-231.5
Siebold, P. F. 1824.
    De historiae naturalis in Japonia statu, nec non de augmento emolumentisque in decursu perscrutationum exspectandis dissertatio, cui accedunt spicilegia faunae japonicae. Bataviae.

4-232
Siebold, P. F. 1826.
    De historiae naturalis in Japonia statu, nec non de augmento emolumentisque in decursu perscrutationum exspectandis dissertatio, cui accedunt spicilegia faunae japonicae. Wirceburgi: Ex officina literaria C. P. Bonitas.

4-233
Slevin, J. R. 1934.
    The Templeton Crocker Expedition to western polynesian and melanesian islands, 1933. No. 15. Notes on the reptiles and amphibians, with the description of a new species of sea-snake. *Proc. Calif. Acad. Sci.* Ser. 4, 21(15):183–188.

4-234
Slevin, J. R.; Leviton, A. E. 1956.
    Holotype specimens of reptiles and amphibians in the collection of the California Academy of Sciences. *Proc. Calif. Acad. Sci.* Ser. 4, 28(14):529–560.

4-235
Smedley, N.; Kloss, C. B. 1926.
    An abnormal, or unnamed, sea-snake. *J. Malay. Branch R. Asiat. Soc.* 4(1):163–164.

4-236
Smith, H. M. 1969.
The nomenclature of certain taxa of higher categories in snakes. *J. Herpetol.* 3(1–2):19–25.

4-237
Smith, H. M.; Taylor, E. H. 1945.
An annotated checklist and key to the snakes of Mexico. *Bull. U.S. Natl Mus.* No. 187:1–239.

4-238
Smith, H. M.; Taylor, E. H. 1950.
Type localities of Mexican reptiles and amphibians. *Univ. Kans. Sci. Bull.* 33(2):313–380.

4-239
Smith, H. M.; Smith, R. B.; Sawin, H. L. 1977.
A summary of snake classification (Reptilia, Serpentes). *J. Herpetol.* 11(2):115–121.

4-240
Smith, L. A. 1974.
The sea snakes of Western Australia (Serpentes: Elapidae, Hydrophiinae) with a description of a new subspecies. *Rec. West Aust. Mus.* 3(2):93–110.

4-240.5
Smith, L. A. 1986.
A new species of *Hydrophis* (Serpentes: Hydrophiidae) from north-west Australian waters. *Rec. West. Aust. Mus.* 13(1):151–153.

4-241
Smith, M. A. 1916.
Note on a rare sea snake (*Thalassophis anomalus*) from the coast of Siam. *J. Nat. Hist. Soc. Siam Suppl.* 2(2):176–177.

4-242
Smith, M. A. 1917.
Preliminary diagnoses of four new sea snakes. *J. Nat. Hist. Soc. Siam (Bangkok)* 2(4):340–342.

4-243
Smith, M. A. 1919.
Remarks on Col. Wall's identification of *Hydrophis cyanocinctus*. *J. Bombay Nat. Hist. Soc.* 26:682–683.

4-244
Smith, M. A. 1920.
    On sea snakes from the coasts of the Malay Peninsula, Siam and Cochin-China. *J. Fed. Malay States Mus.* 10(1):1–63.

4-245
Smith, M. 1926.
    Monograph of the sea-snakes (Hydrophiidae). London: British Museum (Natural History).

4-246
Smith, M. A. 1930.
    The reptilia and amphibia of the Malay Peninsula from the isthmus of Kra to Singapore including the adjacent islands. *Bull. Raffles Mus.* No. 3:1–149. (A supplement to Boulenger, G. A. *Reptilia and batrachia*, 1912.)

4-247
Smith, M. A. 1931.
    Description of a new genus of sea-snake from the coast of Australia, with a note on the structures providing for complete closure of the mouth in aquatic snakes. *Proc. Zool. Soc. Lond.* 1931:397–398.

4-248
Smith, M. 1935.
    The sea snakes (Hydrophiidae). *Dana-Rep. Carlsberg Found.* No. 8:1–6.

4-249
Smith, M. A. 1943.
    The fauna of British India, Ceylon and Burma, including the whole of the Indo-Chinese sub-region. London: Taylor and Francis. Volume 3.

4-250
Sonnini, C. S.; Latreille, P. A. 1801.
    Histoire naturelle des reptiles. Paris: Deterville. Volume 4.

4-251
Stanley, A. 1916.
    Museum acquisitions from June 1, 1915, to May 31, 1916. *J. N. China Br. R. Asiat. Soc.* 47:xiii–xv.

4-252
Stanley, A. 1917.
    Museum acquisitions from June 1, 1916 to May 31, 1917. *J. N. China Br. R. Asiat. Soc.* 48:xii–xiii.

4-253

Steindachner, F. 1869

Reptilien. *Reise der osterreichischen Fregatte Novara um die erde in den jahen 1857, 1858, 1859*. Wein: Kaiserlich-Koeniglichen Hof- und Staatsdruckerei. Volume 3, pp. 1–98.

4-254

Steindachner, F. 1914.

Bericht ueber die von Hans Sauter auf Formosa gesammelten schlangenarten. *Denkschr. Math. Nat. Kl. Akad. Wiss. Wien.* 90:319–361.

4-255

Stejneger, L. 1898.

On a collection of batrachians and reptiles from Formosa and adjacent islands. *J. Coll. Sci. Imp. Univ. Tokyo* 12:215–225.

4-256

Stejneger, L. 1901.

Diagnoses of eight new batrachians and reptiles from the Riu Kiu Archipelago, Japan. *Proc. Biol. Soc. Wash.* 14:189–191.

4-257

Stejneger, L. 1907.

Herpetology of Japan and adjacent territory. *Bull. U.S. Natl Mus.* No. 58:1–557.

4-258

Stejneger, L. 1911.

The batrachians and reptiles of Formosa. *Proc. U.S. Natl Mus.* 38(1731):91–114.

4-259

Stejneger, L. 1925.

Chinese amphibians and reptiles in the United States National Museum. *Proc. U.S. Natl Mus.* 66(25):1–115.

4-260

Stejneger, L. 1936.

Types of the amphibian and reptilian genera proposed by Laurenti in 1768. *Copeia* 1936(3):133–141.

4-261
Stejneger, L.; Barbour, T. 1923.
    A check list of North American amphibians and reptiles. 2nd ed. Cambridge:
Harvard University Press.

4-262
Stejneger, L.; Barbour, T. 1933.
    A check list of North American amphibians and reptiles. 3rd ed. Cambridge:
Harvard University Press.

4-263
Stejneger, L.; Barbour, T. 1939.
    A check list of North American amphibians and reptiles. 4th ed. Cambridge:
Harvard University Press.

4-264
Stejneger, L.; Barbour, T. 1943.
    A check list of North American amphibians and reptiles. 5th ed. *Bull. Mus.
Comp. Zool.* 93(1):1–260.

4-265.5
Sternfeld, R. 1920.
    Zur tiergeographie Papuasiens und der pazifischen Inselwelt. *Abh. Senckenb.
Naturforsch. Ges.* 36(4):375–436.

4-266
Stiles, C. W. 1910.
    Summaries of six opinions (9, 11, 13, 15, 17, 18) by the International Commission
on Zoological Nomenclature. *Science (Wash. D.C.)* N.S. 31(787):150–151.

4-267
Stoliczka, F. 1870.
    Observations on some Indian and Malayan amphibia and reptilia. *Ann. Mag.
Nat. Hist.* Ser. 4, 6:105–109.

4-268
Stoliczka, F. 1870.
    Observations on some Indian and Malayan amphibia and reptilia. *J. Asiat. Soc.
Beng.* 39(3):158–228.

4-269
Stoliczka, F. 1872.
    Notes on reptiles, collected by Surgeon F. Day in Sind. *Proc. Asiat. Soc. Beng.*
1872:85–92.

4-270
Strauch, A. 1873.
   Die schlangen des Russischen reichs in systematischer und zoogeographischer beizeihung. *Mem. Acad. Imp. Sci. St. Petersbourg* Ser. 7, 21(4):1–287.

4-271
Taylor, E. H. 1922.
   The snakes of the Philippine Islands. Manila: Bureau of Printing.

4-272
Taylor, E. H. 1950.
   The snakes of Ceylon. *Univ. Kans. Sci. Bull.* 33(2):519–603.

4-273
Taylor, E. H. 1951.
   A brief review of the snakes of Costa Rica. *Univ. Kans. Sci. Bull.* 34(1):1–188.

4-274
Taylor, E. H. 1963.
   New and rare oriental serpents. *Copeia* 1963(2):429–433.

4-275
Taylor, E. H. 1965.
   The serpents of Thailand and adjacent waters. *Univ. Kans. Sci. Bull.* 45(9):609–1096.

4-276
Taylor, E. H.; Elbel, R. E. 1958.
   Contribution to the herpetology of Thailand. *Univ. Kans. Sci. Bull.* 38(2):1033–1189.

4-277
Tchang, T. L. 1932.
   Notes on some Chinese snakes. *Bull. Fan. Meml Inst. Biol. (Peiping)* 3(1):1–23.

4-278
Temminck, C. J.; Schlegel, H. 1838.
   Reptilia elaborantibus. Siebold, P. F. *Fauna Japonica.* Amstelodami: J. Mueller.

4-279
Terron, C. C. 1932.
Los coralillos mexicanos. *An. Inst. Biol. Univ. Nac. Auton. Mex.* 3(1):5–14.

4-280
Testi, F. 1934.
Nota su di un ofidio velenoso della Somalia della sottofamiglia delle "Hydrophiinae". *Arch. Sci. Med. Colon.* 14(2):81–85. (In Italian with English abstract)

4-281
Testi, F. 1935.
Un colubride proteroglifo da aggiungersi alla fauna erpetologica della Somalia Italiana. *Arch. Sci. Med. Colon.* 16:407–409.

4-282
Theobald, W. 1868.
Catalogue of reptiles in the Museum of the Asiatic Society of Bengal. *J. Asiat. Soc. Beng.* Extra Number 1868:1–88.

4-283
Tschudi, J. J. von. 1837.
Neues genus von wasserschlangen. *Arch. Nat. (Berlin)* 3:331–335.

4-284
Tweedie, M. W. F. 1953.
The snakes of Malaya. Singapore: Government Printing Office.

4-285
Tweedie, M. W. F. 1957.
The snakes of Malaya. 2nd ed. Singapore: Government Printing Office.

4-285.5
Tweedie, M. W. F. 1983.
The snakes of Malaya. 3rd ed. Singapore: Singapore National Printers.

4–286
Underwood, G. 1967.
A contribution to the classification of snakes. London: British Museum (Natural History).

4-287
Underwood, G. 1979.
Classification and distribution of venomous snakes in the world. Lee, C. Y., ed. *Snake venoms.* New York: Springer-Verlag. Pp. 15–40.

4-288
Van Denburgh, J. 1922.
The reptiles of western North America. An account of the species known to inhabit California and Oregon, Washington, Idaho, Utah, Nevada, Arizona, British Columbia, Sonora and Lower California. San Francisco: California Academy of Sciences. Volume 2.

4-289
Van Denburgh, J. 1924.
On *Laticauda schistorhynchus* and *Laticauda semifasciata. Proc. Calif. Acad. Sci.* Ser. 4, 13(16):247–248.

4-290
Van Denburgh, J.; Thompson, J. C. 1908.
Description of a new species of sea snake from the Philippine Islands, with a note on the palatine teeth in the Proteroglypha. *Proc. Calif. Acad. Sci.* Ser. 4, 3:41–48.

4-291
Volsoe, H. 1939.
The sea snakes of the Iranian Gulf and the Gulf of Oman. With a summary of the biology of the sea snakes. Jessen, K., ed. *Danish Scientific Investigations in Iran.* Copenhagen: Ejnar Munksgaard. Vol. 1, pp. 9–45.

4-292
Volsoe, H. 1958.
Additional specimens of snakes from Rennell Island. Wolff, T., ed. *The natural history of Rennell Island, British Solomon Islands.* Copenhagen: Danish Scientific Press. Vol. 1, pp. 227–228.

4-293
Volsoe, H. 1958.
Herpetology of Rennell Island. Wolff, T., ed. *The natural history of Rennell Island, British Solomon Islands.* Copenhagen: Danish Scientific Press. Vol. 1, pp. 121–134.

4-293.5
Wagler, J. 1828.
Descriptiones et icones amphibiorum. Stuttgart: Monachii. Part 1.

4-294
Wagler, J. 1830.
Natuerliches system der amphibien mit vorangehender classification der saeugthiere und voegel. Muenchen: J. G. Cotta'schen Buchhandlng.

4-295
Wall, F. 1902.
    Aids to the differentiation of snakes. *J. Bombay Nat. Hist. Soc.* 14:337–345.

4-296
Wall, F. 1904.
    Occurrence of a rare sea snake (*Distira gillespiae*) on the Malabar coast. *J. Bombay Nat. Hist. Soc.* 15:723–726.

4-297
Wall, F. 1905.
    Notes on snakes collected in Cannanore from 5th November 1903 to 5th August, 1904. *J. Bombay Nat. Hist. Soc.* 16:292–317.

4-298
Wall, F. 1906.
    A descriptive list of the sea snakes (Hydrophiidae) in the Indian Museum, Calcutta. *Mem. Asiat. Soc. Bengal* 1906:277–299.

4-299
Wall, F. 1907.
    On the Hydrophidae in the Colombo Museum. *Spolia Zeylan.* 1907:166–172.

4-300
Wall, F. 1909.
    A monograph of the sea-snakes (Hydrophiinae). *Mem. Asiat. Soc. Bengal* 2(8):169–251.

4-301
Wall, F. 1913.
    Some new snakes from the oriental region. *J. Bombay Nat. Hist. Soc.* 22:514–516.

4-302
Wall, F. 1913.
    Varieties of *Hemibungarus nigrescens* and *Hydrophis torquatus*. *J. Bombay Nat. Hist. Soc.* 22:638–639.

4-303
Wall, F. 1914.
    Remarks on the sea-snakes in our society's collection. *J. Bombay Nat. Hist. Soc.* 23:374–377.

4–304
Wall, F. 1914.
   The sea-snake, *Hydrophis caerulescens* (Shaw). *J. Bombay Nat. Hist. Soc.*
23:373.

4-305
Wall, F. 1917.
   Notes on an interesting specimen of the sea snake (*Hydrophis caerulescens*).
J. Bombay Nat. Hist. Soc. 25(2):308.

4-306
Wall, F. 1918.
   Notes on a collection of sea snakes from Madras. *J. Bombay Nat. Hist. Soc.*
25:599–607.

4-306.5
Wall, F. 1919.
   Notes on some recent additions to our society's snake collection. *J. Bombay
Nat. Hist. Soc.* 26(3):865–867.

4-307
Wall, F. 1919.
   Reply to Dr. Malcolm Smith's remarks in the last journal. *J. Bombay Nat.
Hist. Soc.* 26:864–865.

4-308
Wall, F. 1921.
   Ophidia taprobanica or the snakes of Ceylon. Colombo, Ceylon: H. R. Cottle,
Government Printer.

4-308.5
Wall, F. 1923.
   How to identify the snakes of India (including Burma and Ceylon). Karachi:
F. Wall.

4-309
Wall, F. 1924.
   A hand-list of the snakes of the Indian empire. *J. Bombay Nat. Hist. Soc.*
30(2):12–24.

4-310
Wall, F. 1924.
   Notes on two specimens of the sea snake *Lioselasma lapemidoides* (Gray).
Spolia Zeylan. 13:89–92.

4-311
Wang, C. S. 1962.
  The reptiles of Botel-Tobago. *Q. J. Taiwan Mus. (Taipei).* 15(3–4):141–191.

4-312
Wang, C. S.; Wang, Y. H. M. 1956.
  The reptiles of Taiwan. *Q. J. Taiwan Mus. (Taipei)* 9(1):1–86.

4-312.5
Welch, K. R. G. 1988.
  Snakes of the orient: a checklist. Malabar, FL: Robert E. Krieger.

4-313
Wells, R. W.; Wellington, C. R. 1983.
  A synopsis of the class reptilia in Australia. *Aust. J. Herpetol.* 1(3–4):73–129.

4-314
Werler, J. E.; Keegan, H. L. 1963.
  Venomous snakes of the Pacific area. Keegan, K. L.; MacFarlane, W. V., eds. *Venomous and poisonous animals and noxious plants of the Pacific region.* New York: Macmillan. Pp. 219–325.

4-315
Werner, F. 1895.
  Ueber eine sammlung von reptilien aus Persien, Mesopotamien und Arabien. *Verh. Zool.-Bot. Ges. Wien* 45:13–22.

4-316
Werner, F. 1900.
  Die reptilien- und batrachierfauna des Bismarck-Archipels. *Mitt. Zool. Samml. Mus. Naturk. Berl.* 1(4):1–132.

4-317
Werner, F. 1908.
  Das tierreich. Leipzig: Goeschenische. Volume 3.

4-318
Werner, F. 1924.
  Uebersicht der gattungen und arten der schlangen der familie Colubridae. II. Tiel (Dipsadomorphinae und Hydrophiinae). *Arch. Nat. (Berlin)* 90(12):108–166.

4-319

Werner, F. 1925.

Neue oder wenig bekannte schlangen; aus dem Wiener naturhistorischen Staatsmuseum. *Sitz. Deut. Akad. Wiss. Wien Math. Naturw. Kl.* Ser. 1, 134:45–66.

4-320

Westphal Castelnau, A. 1870.

Catalogue de la collection de reptiles. Montpellier: J. Martel.

4-320.5

Wiegmann, A. F. A.; Ruthe, J. F. 1832.

Handbuch der zoologie. Berlin: C. G. Leuderitz.

4-321

Wiegmann, A. F. A. 1935.

Beitrage zur zoologie, gesammelt auf einer Reise um die erde von Dr. F. J. M. Meyen. *Nova Acta Acad. Caesar. Leop. Carol.* 17:185–268.

4-322

Williams, K. L.; Wallach, V. 1989.

Snakes of the world. Malabar, FL: Krieger Pub. Corp: Volume 1.

Genus-Species Index
for Chapter 4

Chapter 5

DISTRIBUTION AND RANGES,
SIGHTINGS FROM EARLY VOYAGES

5-.05

Abe, R. 1907.

Sea snakes in Province Izumo. *Zool. Mag. (Tokyo)* 19:252–253. (In Japanese)

5-1

Acharji, M. N.; Mukherjee, A. K. 1966.

Report on a collection of snakes from lower Bengal (Reptilia: Ophidia). *J. Zool. Soc. India* 16(1–2):76–81.

5-1.5

Acosta Solis, M. 1904.

Nuevas contribuciones al conacimiento de la provincia de Esmeraldas. Quito: Editorial Ecuador.

5-2

Ahmed, S. 1975.

Sea-snakes of the Indian Ocean in the collections of the Zoological Survey of India together with remarks on the geographical distribution of all Indian Ocean species. *J. Mar. Biol. Assoc. India* 17(1):73–81.

5-3

Alvarez del Toro, M.; Smith, H. M. 1956.

Notulae herpetologicae Chiapasiae I. *Herpetologica* 12(1):3–17.

5-4

Amaral, A. do. 1925.

South American snakes in the collection of the United States National Museum. *Proc. U.S. Natl Mus.* 67(24):1–30.

5-4.3

Amaral, A. do. 1929.

Lista remissiva dos ophidios da regiao neotropica. *Mem. Inst. Butantan (Sao Paulo)* 4:129–271.

5-4.6

Amaral, A. do. 1931.

Studies of neotropical ophidia. XXVI. Ophidia of Colombia. *Bull. Antivenin Inst. Am.* 4(4): 89–94.

5-5

Anderson, J. 1871.

On some Indian reptiles. *Proc. Zool. Soc. Lond.* 1871:149–211.

5-6

Anderson, J. 1872.

On some Persian, Himalayan, and other reptiles. *Proc. Zool. Soc. Lond.* 1872:371–404.

5-7

Anderson, J. 1889.

Report on the mammals, reptiles, and batrachians, chiefly from the Mergui Archipelago, collected for the Trustees of the Indian Museum. *J. Linn. Soc. Lond. Zool.* 21:331–350.

5-7.5

Angel, F. 1927.

Liste des reptiles et des batraciens rapportes d'Indo-Chine par M. P. Chevey. Description d'une variete nouvelle de *Simotes violaceus* Cantor. *Bull. Mus. Natl Hist. Nat.* 33(6):496–498.

5-7.7

Angel M., R. 1982.

Serpientes de Colombia; guia practica para su clasificacion y tratamiento del envenenameinto causado por sus mordeduras. Medellin, Colombia: Facultad Nacional de Agronomia, Universidad Nacional de Colombia. (Also cited as *Rev. Fac. Nac. Agron. Medellin* 36(1):1–171)

5-8

Annandale, N. 1905.

Additions to the collection of oriental snakes in the Indian Museum. Part. 2. Specimens from the Andamans and Nicobars. *J. Asiat. Soc. Beng.* N.S. 1(7):173–176.

5-9
Annandale, N. 1915.
Fauna of the Chilka Lake: reptiles and batrachians. *Mem. Indian Mus.* 5(2):167–174.

5-9.5
Annandale, N. 1916.
Preliminary report on the fauna of the Tale Sap or Inland Sea of Singgora. *J. Nat. Hist. Soc. Siam (Bangkok).* Ser. 2, 2:90–102.

5-10
Anonymous. 1887.
Sea-snakes of the Gulf of Manaar. *Taprobanian* 2(2):108.

5-10.5
Anonymous. 1907.
Contributions to the museum. *J. Bombay Nat. Hist. Soc.* 17(3):851–852.

5-10.7
Anonymous. 1909.
Contributions to the museum. *J. Bombay Nat. Hist. Soc.* 19(1): 283–284.

5-11
Anonymous. 1961.
Snake's alive. *Sunday Advertiser (Honolulu).* April 2:B6.

5-12
Anonymous. 1973.
Teacher records sea snake finds in North. *Northland Holiday (New Zealand),* June 4, p. 5.

5-13
Archbold, R.; Rand, A. L. 1940.
New Guinea expedition: Fly River area, 1936–1937. New York: Robert M. McBride.

5-14
Barbour, T. 1906.
Vertebrata from the Savanna of Panama IV. Reptilia and Amphibia. *Bull. Mus. Comp. Zool.* 46(12):224–229.

5-15
Barbour, T. 1912.
A contribution to the zoogeography of the East Indian Islands. *Mem. Mus. Comp. Zool.* 44(1):1–203.

5-16
Barbour, T. 1921.
Reptiles and amphibians from the British Solomon Islands. *Proc. New Engl. Zool. Club* 7:91–112.

5-17
Bartholomew, J. G.; Clarke, W. E.; Grimshaw, P. H. 1911.
Atlas of zoogeography. Edinburgh: John Bartholomew.

5-17.5
Bartlett, E. 1896.
Notes on the snakes of Borneo and the adjacent islands. Part III. *Sar. Gaz.* 26(367):153–157.

5-18
Bavay, A. 1869
Catalogue des reptiles de la Nouvelle-Caledonie et description d'especes nouvelles. *Mem. Soc. Linn. Normandie* 15(5):1–37.

5-19
Beddome, R. H. 1862.
Notes upon the land and fresh-water snakes of the Madras Presidency. *Madras Quart. J. Med. Sci.* 5:1–31.

5-20
Bennett, G. 1860.
Gatherings of a naturalist in Australasia: being observations principally on the animal and vegetable productions of New South Wales, New Zealand and some Austral islands. London: John Van Voorst.

5-20.4
Bethencourt Ferreira, J. 1896.
Reptis da India Museu de Lisboa. *J. Sci. Math. Phys. Nat. (Lisboa)* Ser. 2, 4(16):212–234.

5-20.6
Bethencourt Ferreira, J. 1898.
Reptis de Timor no Museu de Lisboa. *J. Sci. Math. Phys. Nat. (Lisboa)* Ser. 2, 5(19):151–156.

5-21
Biswas, S.; Sanyal, D. P. 1980.
    A report on the reptilia fauna of Andaman and Nicobars Islands in the collection of the Zoological Survey of India. *Rec. Zool. Surv. India* 77:255–292.

5-22
Blanc, C. P. 1972.
    Les reptiles de Madagascar et des iles voisines. Battistini, R.; Richard-Vindard, G., eds. *Biogeography and ecology in Madagascar.* The Hague: W. Junk. Pp. 501–611.

5-23
Blanford, W. T. 1876.
    Eastern Persia: an account of the journeys of the Persian Boundary Commission 1870–71–72. London: Macmillan. Volume 2.

5-24
Blanford, W. T. 1879.
    Notes on reptilia. *J. Asiat. Soc. Beng.* 48(3):127–132.

5-25
Blanford, W. T. 1881.
    On a collection of Persian reptiles recently added to the British Museum. *Proc. Zool. Soc. Lond.* 1881:671–682.

5-26
Blanford, W. T. 1881.
    On a collection of reptiles and frogs chiefly from Singapore. *Proc. Zool. Soc. Lond.* 1881:215–227.

5-26.5
Bleeker, P. 1856.
    Reis door de Minahassa en den Molukschen Archipel. Gedaan in de maanden September en Oktober 1855 in het gevolg van den gouverneur Generaall Mr. A. J. Duymaer van Twist. Batavia: Lange. Volume 1.

5-27
Bleeker, P. 1857.
    Berigt omtrent eenige reptilien van Sumatra, Borneo, Batjan en Boero. *Natuurw. Tijdschr. Ned.-Indie* 13:470–475.

5-27.5
Bleeker, P. 1857.
   Opsomming der tot dus verre van het eiland Sumatra bekend gewordene
Reptilien. *Natuurw. Tijdschr. Ned.-Indie* 15:260–263.

5-28
Bleeker, P. 1857.
   Over eenige reptilien van Celebes. *Natuurw. Tijdschr. Ned.-Indie* 14:231–233.

5-29
Bleeker, P. 1857.
   Over eenige reptilien van het eiland Banka. *Natuurw. Tijdschr. Ned.-Indie*
14:233–235.

5-30
Bleeker, P. 1858.
   Bestuursvergadering, gehouden den 12^n November 1857, ten huize van den
heer steenstra toussaint. *Natuurw. Tijdschr. Ned.-Indie* 16:42–59.

5-31
Bleeker, P. 1858.
   Bestuursvergadering, gehouden den 13^n January 1859, ten huize van den heer
Bleeker. *Natuurw. Tijdschr. Ned.-Indie* 16:403–425.

5-32
Bleeker, P. 1859.
   Bestuursvergadering, gehouden den 11^n Augustus 1859, ten huize van den
heer De Lange. *Indones. J. Nat. Sci.* 20:215–224.

5-33
Bleeker, P. 1859.
   Bestuursvergadering, gehouden den 13^{den} Oktober 1859 ten huize van den heer
Becking. *Indones. J. Nat. Sci.* 20:321–341.

5-34
Bleeker, P. 1859.
   Bestuursvergadering, gehouden den 24^n February 1859, ten huize van den
heer Bleeker. *Natuurw. Tijdschr. Ned.-Indie* 20:85–88.

5-34.4
Bleeker, P. 1859.
   Verslag omtrent reptilien van Nieuw-Guinea, aangeboden door H. Von Rosen-
berg. *Natuurw. Tijdschr. Ned.-Indie* 16:420–423.

5-34.6
Bleeker, P. 1859.
    Verslag van een verzameling reptilien en visschen van westelijk Borneo. *Natuurw. Tijdschr. Ned.-Indie* 16:433–441.

5-35
Bleeker, P. 1860.
    Over de reptilien-fauna van Amboina. *Natuurw. Tijdschr. Ned.-Indie* 22:39–43.

5-35.5
Bleeker, P. 1860.
    Over de reptilien fauna van Sumatra. *Natuurw. Tijdschr. Ned.-Indie* 21:284–298.

5-36
Bleeker, P. 1860.
    Bestuursvergadering. Gehouden den 12^n April 1860, ten huize van den heer G. F. De Bruyn Kops. *Natuurw. Tijdschr. Ned.-Indie* 22(2):77–92.

5-36.4
Bleeker, P. 1860.
    Soorten van reptilien van het eiland Banka. *Natuurw. Tijdschr. Ned.-Indie* 21:331–334.

5-36.6
Bleeker, P. 1860.
    Thans bekende reptilien van Timor. *Natuurw. Tijdschr. Ned.-Indie* 22:86–88.

5-37
Blythe, E. 1846.
    Notes on the fauna of the Nicobar Islands. *J. Asiat. Soc. Beng.* 15:367–379.

5-37.5
Blyth, E. 1863.
    The zoology of the Andaman Islands. Mouat, F. J. *Adventures and researches among the Andaman islanders.* London: Hurst and Blackett. Pp. 345–367.

5-38
Bocourt, F. 1866.
    Notes sur les reptiles, les batraciens et les poissons recueillis pendant un voyage dans le royaume de Siam. *Nouv. Arch. Mus. Hist. Nat.* 1866:4–19.

5-38.4
Boettger, O. 1877.
   Die reptilien und amphibien von Madagascar. *Abh. Senckenb. Naturforsch. Ges.* 11:1–56.

5-38.6
Boettger, O. 1881.
   Die reptilien und amphibien von Madagascar. Dritter Nachtrag. *Abh. Senckenb. Naturforsch. Ges.* 12:435–558.

5-39
Boettger, O. 1885.
   Materialien zur herpetologischen fauna von China I. Liste der ersten moellendorff'schen sendung sud-chinesischer kriechthiere. *Ber. Offenbacher Vereins Naturkd.* 24–25:115–170.

5-40
Boettger, O. 1886.
   Aufzahlung der von den Philippinen bekannten Reptilien und Batrachier. *Ber. Senckenb. Nat. Ges. Frankfurt a. Main* 2:91–134.

5-41
Boettger, O. 1888.
   Materialien zur herpetologischen fauna von China II. *Ber. Offenbacher Vereins Naturkd.* 26–28:53–191.

5-42
Boettger, O. 1892.
   Drei neue colubriforme schlangen. *Zool. Anz.* 1892:417–420.

5-43
Boettger, O. 1892.
   Listen von kriechtieren und lurchen aus dem tropischen Asien u. aus Papuasien. *Ber. Offenbacher Vereins Naturkd.* 29–32:65–164.

5-44
Boettger, O. 1893.
   Uebersicht der von Kapt. Storm auf Borneo gesammelten reptilien und batrachier. *Mitt. Geogr. Ges. Nat. Mus. Luebeck,* Ser. 2, 5:85–89.

5-45
Boettger, O. 1894.
   Materialien zur herpetologischen fauna von China III. *Ber. Senckenb. Nat. Ges. Frankfurt a. Main* 1894:129–152.

5-46
Boettger, O. 1895.
   Liste der reptilien und batrachier der Insel Halmaheira nach den sammlungen Prof. Dr. W. Kukenthal's. *Zool. Anz.* 18(472):129–138.

5-47
Boettger, O. 1895.
   Neue froesche und schlangen von den Liukiu-Inseln. *Ber. Offenbacher Vereins Naturkd.* 33–36:101–117.

5-48
Boettger, O. 1905.
   Zoologische sammlung. *Ber. Senckenb. Nat. Ges. Frankfurt a. Main* 1905:159–223.

5-49
Boettger, O. 1913.
   Reptilien und Amphibien von Madagascar, den Inseln und dem Festland Ostafrikas. Voeltzkow, A. *Reise in Ostafrika in den Jahren 1903–1905.* Stuttgart: E. Schweizerbart'sche, 1906–1915. Vol. 3. pp. 269–375.

5-49.5
Bogert, C. M.; Oliver, J. A. 1945.
   A preliminary analysis of the herpetofauna of Sonora. *Bull. Am. Mus. Nat. Hist.* 83:297–426.

5-50
Boie, H. 1826.
   Notice sur l'erpetologie de l'ile de Java. *Bull. Sci. Nat. Geol.* 9:233–240.

5-50.5
Bolanos, R. 1982.
   Serpientes venenosas de Centro America: distribucion, caracteristicas y patrones cariologicos. *Mem. Inst. Butantan (Sao Paulo)* 46:275–291.

5-50.7
Bolanos, R. 1982.
   Las serpientes venenosas de Centro America y el problema del ofidismo. Primera parte. Aspectos zoologicos, epidemiologicos y biomedicos. *Rev. Costarric. Cienc. Med.* 3:165–184.

5-51
Bolanos-Herrera, R. 1971.
Nuevos recursos contra el ofidismo en centroamerica. 2nd. ed. Costa Rica: Instituto Clodomiro Picado.

5-51.3
Bonnemains, J.; Forsyth, E.; Smith, B., eds. 1988.
Baudin in Australian waters. The artwork of the French voyage of discovery to the Southern lands 1800–1884. Melbourne: Oxford University Press.

5-51.7
Boulenger, G. A. 1886.
On the reptiles and batrachians of the Solomon Islands. *Trans. Zool. Soc. Lond.* 12(2):35–62.

5-52
Boulenger, G. A. 1887.
A list of the reptiles and batrachians obtained near Muscat, Arabia, and presented to the British Museum by Surgeon-Major A. S. G. Jayakar. *Ann. Mag. Nat. Hist.* Ser. 5, 20:407–408.

5-53
Boulenger, G. A. 1887.
On a collection of reptiles and batrachians made by Mr. H. Pryer in the Loo Choo Islands. *Proc. Zool. Soc. Lond.* 1887:146–150.

5-53.5
Boulenger, G. A. 1888.
Third contribution to the herpetology of the Solomon Islands. *Proc. Zool. Soc. Lond.* 1888:89–90.

5-54
Boulenger, G. A. 1890.
Reptilia and batrachia. Blanford, W., ed. *The fauna of British India, including Ceylon and Burma.* London: Taylor and Francis. Part 1, Section 3.

5-55
Boulenger, G. A. 1897.
Description of a new lizard from Obok. *Ann. Mag. Nat. Hist.* 19:467–468.

5-56
Boulenger, G. A. 1898.
An account of reptiles and batrachians collected by Dr. L. Loria in British New Guinea. *Ann. Mus. Civ. Genova* Ser. 2, 18(38):694–710.

5-57
Boulenger, G. A. 1903.
Report on the batrachians and reptiles. Annandale, N.; Robinson, H. C. *Fasciculi Malayenses. Anthropological and zoological results of an expedition to Perak and the Siamese Malay states, 1901–1902, undertaken by Nelson Annandale and Herbert C. Robinson under the auspices of the University of Edinburgh and University College of Liverpool.* New York: Published for the University Press of Liverpool by Longmans, Green & Co. Vol. 1 (Zoology), pt. 1, pp. 131–176.

5-58
Boulenger, G. A. 1910.
A revised list of the South African reptiles and batrachians, with synoptic tables, special reference to the specimens in the South African Museum, and descriptions of new species. *Ann. S.A. Mus.* 5:455–549.

5-58.5
Boulenger, G. A. 1915.
A list of the snakes of East Africa, north of the Zambesi and south of the Soudan and Somaliland, and of Nyassaland. *Proc. Zool. Soc. Lond.* 1915(2):611–640.

5-59
Boulenger, G. A. 1915.
A list of the snakes of Madagascar, Comoro, Mascarenes, and Seychelles. *Proc. Zool. Soc. Lond.* 1915:369–382.

5-59.5
Boulenger, G. A. 1915.
A list of the snakes of north-east Africa, from the tropic to the Soudan and Somaliland, including Socotra. *Proc. Zool. Soc. Lond.* 1915(2):641–658.

5-60
Bourret, R. 1927.
La faune de l'Indochine: vertebres. Hanoi. Pp. 205–247. (Also cited as Societe de Geographie de Hanoi. *Inventaire general de l'Indochine* 3:205–247)

5-61
Bourret, R. 1934.
Notes herpetologiques sur l'Indochine francaise. IV. Sur une collection d'ophidiens de Cochinchine et du Cambodge. *Bull. Gen. Instr. Pub.* No. 1:1–20.

5-62
Bourret, R. 1936.
Les serpents de l'Indochine. Toulouse: Henri Basuyau. Volumes 1 and 2.

5-63
Bourret, R. 1939.
   Notes herpetologiques sur l'Indochine francaise. XVII. Reptiles et batraciens recus au Laboratoire des Sciences Naturelles de l'Universite au cours de l'annee 1938. Descriptions de trois especes nouvelles. *Annexe Bull. Gen. Instr. Pub.* 1939(Feb.):13–34.

5-64
Bourret, R. 1939.
   Notes herpetologiques sur l'Indochine francaise. XX. Liste des reptiles et batraciens actuellement connus en Indochine Francaise. *Annexe Bull. Gen. Instr. Pub.* 1939(Dec.):49–60.

5-65
Branch, W. R. 1979.
   The venomous snakes of Southern Africa, Part 2. Elapidae and Hydrophidae. *Snake* 11:199–225.

5-66
Branch, W. R. 1981.
   The venomous snakes of Southern Africa, Part 2. Elapidae and Hydrophidae. *Bull. Md Herpetol. Soc.* 17(1):1–47.

5-67
Brand, D. D. 1958.
   Coastal study of Southwest Mexico, Part II. Austin, TX: University of Texas.

5-68
Braun, M. 1859.
   Beitrage zur kenntniss zweier susswasser-Algen Pleurocladia and Pleurocarpus. *Monatsber. K. Preuss. Akad. Wiss. Berl.* 1859:275–278.

5-69
Brazenor, C. W. 1936.
   The reptiles and amphibians of Victoria. Gawler, E. O., ed. *Victorian yearbook, 1934–1935*. Melbourne: Government Printer. Pp. 25–38.

5-70
Brongersma, L. D. 1931.
   Reptilia. Resultats scientifiques du voyage aux Indes orientales neerlandaises de. ll.aa.rr. Le Prince et Princesse Leopold de Belgique. *Mem. Mus. Hist. Hat. Belg.* 5(2):32–39.

5-71
Brongersma, L. D. 1934.
Contributions to Indo-Australian herpetology. *Zool. Meded. Rijksmus. Nat. Hist. Leiden* 17(3–4):161–251.

5-72
Brongersma, L. D. 1947.
Zoological notes from Port Dickson, I. Amphibians and reptiles. *Zool. Meded. Rijksmus. Nat. Hist. Leiden* 27:300–308.

5-73
Brongersma, L. D. 1956.
Notes on New Guinean reptiles and amphibians. V. *Proc. K. Ned. Akad. Wet. Ser.C Biol. Med. Sci.* 59:599–610.

5-73.2
Brongersma, L. D. 1961.
Zoological exploration of Netherlands New Guinea. *Proc. 9th Pacif. Sci. Congr.* 19:68–71.

5-73.4
Brongersma, L. D.; Venema, G. F. 1963.
To the mountains of the stars. New York: Doubleday.

5-73.6
Brown, A. E. 1902.
A collection of reptiles and batrachians from Borneo and the Loo Choo Islands. *Proc. Acad. Nat. Sci. Phil.* 54:175–186.

5-73.8
Brown, A. E. 1902.
A list of the reptiles and batrachians in the Harrison Hiller collection from Sumatra. *Proc. Acad. Nat. Sci. Phil.* 54:693–695.

5-73.9
Brown, J. N. B. 1987.
Sea snakes in the waters of the United Arab Emirates. *Bull. Emirates Nat. Hist. Group (Abu Dhabi)* No. 32:18–20.

5-73.95
Brown, J. N. B. 1987.
Sea snakes of the Arabian Gulf. *Gazelle Dubai Nat. Hist. Group Newsl.* 4 pages.

5-74
Bryan, W. A. 1915.
   Natural history of Hawaii, being an account of the Hawaiian people, the geology and geography of the Islands, and the native and introduced plants and animals of the group. Honolulu: Hawaiian Gazette.

5-75
Burt, C. E.; Burt, M. D. 1932.
   Herpetological results of the Whitney South Sea Expedition. VI. Pacific island amphibians and reptiles in the collection of the American Museum of Natural History. *Bull. Am. Mus. Nat. Hist.* 38(2):461–597.

5-75.5
Butler, W. H. 1970.
   A summary of the vertebrate fauna of Barrow Island, W.A. *West. Aust. Nat.* 11(7):149–160.

5-76
Cahill, M.; Heatwole, H.; Goldman, B. 1973.
   Ecology of the reefs. *Occas. Pap. Dept. Geog. Univ. Papua, New Guinea* 1:209–229.

5-76.5
Campbell, J. A.; Lamar, W. W. 1989.
   The venomous reptiles of Latin America. Ithaca, NY: Comstock Publishing Asssociates.

5-77
Campo, R. M. del. 1935.
   Nota acerca de la distribucion geografica de los reptiles ponzonosos en Mexico. *An. Inst. Biol. Univ. Nac. Auton. Mex.* 6(3–4):291–300. (English abstract)

5-78
Campo, R. M. del. 1937.
   Reptiles ponzonosos de Mexico. Las viboras de Cascabel. *Folletos Divulg. Cient. Inst. Biol.* 27:1–18.

5-79
Campo, R. M. del. 1950.
   Serpientes ponzonosas de Mexico. *Rev. Mex. Cienc. Med. Biol.* 8(53):103–115.

5-80
Cantor, T. E. 1847.
Catalogue of reptiles inhabiting the Malayan Peninsula and islands. *J. Asiat. Soc. Beng.* 16:1025–1078.

5-81
Cantor, T. 1886.
Catalogue of reptiles inhabiting the Malayan Peninsula and islands. *Miscellaneous papers relating to Indo-China.* London: Truebner & Co. Vol. 2, pp. 112–257.

5-82
Carr, A. 1967.
So excellent a fishe: a natural history of sea turtles. Garden City, NY: Natural History Press.

5-82.5
Carrillo de Espinoza, N. 1970.
Contribucion al conocimiento de los reptiles del Peru. *Publ. Mus. Hist. Nat. 'Javier Prado' Ser. A. Zool.* 22:1–64.

5-83
Chang, M. L. Y. 1934.
Preliminary report on some reptiles from Chekiang. *China J. (Shanghai)* 21(5):251–253.

5-84
Chasen, F. N.; Smedley, N. 1927.
A list of reptiles from Pulau Galang and other islands of the Rhio Archipelago. *J. Malay. Branch R. Asiat. Soc.* 5:351–355.

5-85
Cheeseman, T. F. 1888.
On the flora of the Kermadec Islands; with notes on the fauna. *Trans. R. Soc. N.Z.* 20:151–181.

5-86
Cheeseman, T. F. 1908.
Notes on the occurrence of certain marine reptilia in New Zealand waters. *Trans. Proc. N.Z. Inst.* 40:267–269.

5-87
Cheesman, R. E. 1926.
In unknown Arabia. London: MacMillan.

5-87.5

Chen, R.; Lin, W. 1983.

A new record of sea snake from Fujian—*Laticauda laticaudata* (Linnaeus). *Acta Herpetol. Sin.* 2(2):75.

5-88

Cherchi, M. A. 1958.

Note sugli ofidi velenosi somali. *Boll. Mus. Ist. Biol. Univ. Genova* 28(169):79–86.

5-89

Chevey, M. P. 1927.

Liste des reptiles et des batraciens rapportes d'Indo-Chine. *Bull. Mus. Natl Hist. Nat.* 2:496–498.

5-90

Chugunov, Y. D. 1980.

A rare case of finding of a sea snake *Laticauda semifasciata* (Serpentes, Hydrophidae) in waters of the USSR. *Zool. Zh.* 59(3):470–471. (In Russian with English abstract)

5-91

Chugunov, Y. D. 1980.

A sea snake in the Peter the Great Bay. *Priroda (Mosc.)* 1980(4):49. (In Russian)

5-92

Cochran, D. M. 1930.

The herpetological collections made by Dr. Hugh M. Smith in Siam from 1923 to 1929. *Proc. U.S. Natl Mus.* 77(11):1–39.

5-93

Cochran, D. M. 1946.

Notes on the herpetology of the Pearl Islands, Panama. *Smithson. Misc. Collect.* 106(4):1–8.

5-94

Cogger, H. G. 1975.

Reptiles and amphibians of Australia. Sydney: A. H. & A. W. Reed.

5-95

Cogger, H. G. 1975.

Sea snakes of Australia and New Guinea. Dunson, W. A., ed. *Biology of sea snakes.* Baltimore: University Park Press. Pp. 59–139.

5-96
Cogger, H. G. 1979.
   Reptiles and amphibians of Australia. Rev. ed. Sydney: A. H. & A. W. Reed.

5-96.5
Cogger, H. G. 1983.
   Reptiles and amphibians of Australia. 3rd ed. Sanibel, FL: Ralph Curtis Books.

5-97
Cogger, H. G. 1983.
   Zoological catalogue of Australia. Canberra: Australian Government Publishing Service. Volume 1.

5-97.5
Cogger, H.; Heatwole, H. 1984.
   The Australian reptiles: origins, biogeography, distribution patterns and island evolution. Archer, M.; Clayton, G., eds. *Vertebrate zoogeography & evolution in Australasia (Animals in space and time).* Carlisle, Western Australia: Hesperian Press. Pp. 343–360.

5-97.7
Cogger, H. G.; Heatwole, H. 1984.
   The Australian reptiles: origins, biogeography, distribution patterns and island evolution. Keast, A., ed. *Ecological biogeography of Australia.* Boston: W. Junk. Vol. 2, pp. 1331–1373.

5-98
Constable, J. D. 1949.
   Reptiles from the Indian Peninsula in the Museum of Comparative Zoology. *Bull. Mus. Comp. Zool.* 103(2):59–160.

5-99
Cope, E. D. 1868.
   An examination of the reptilia and batrachia obtained by the Orton expedition to Equador and the Upper Amazon, with notes on other species. *Proc. Acad. Nat. Sci. Phil.* 20:96–140.

5-99.5
Cope, E. D. 1871.
   Ninth contribution to the herpetology of tropical America. *Proc. Acad. Nat. Sci. Phil.* 1871:200–224.

5-100
Cope, E. D. 1876.
On the batrachia and reptilia of Costa Rica. *J. Acad. Nat. Sci. Phil.* N.S. 8(2):93–154.

5-100.5
Cope, E. D. 1889.
Scientific results of explorations by the U.S. Fish Commission steamer Albatross. III. Report on the batrachians and reptiles collected in 1887–'88. *Proc. U.S. Natl Mus.* 12:141–147.

5-101
Corkill, N. L. 1932.
An inquiry into snake-bite in Iraq. *Indian J. Med. Res.* 20(2):599–625.

5-102
Corkill, N. L. 1932.
Snakes and snakebite in Iraq: a handbook for medical officers. London: Balliere, Tindall and Cox.

5-103
Corkill, N. L. 1932.
The snakes of Iraq. *J. Bombay Nat. Hist. Soc.* 35(3–4):552–572.

5-104
Corkill, N. L.; Cochrane, J. A. 1965.
The snakes of the Arabian Peninsula and Socotra. *J. Bombay Nat. Hist. Soc.* 62(3):475–506.

5-105
Corni, G. 1937.
Somalia Italiana. Milano: Editoriale Arte e Storia. Volume 1.

5-106
Crosswhite, F. S.; Crosswhite, C. D. 1982.
The Sonoran Desert. Bender, G. L., ed. *Reference handbook on the deserts of North America.* Westport, CT: Greenwood Press. Pp. 163–319.

5-107
Cruz, G. A.; Wilson, L. D.; Espinsoa, J. 1979.
Two additions to the reptile fauna of Honduras, *Eumeces managuae* Dunn and *Agkistrodon bilineatus* (Gunther), with comments on *Pelamis platurus* (Linnaeus). *Herpetol. Rev.* 10(1):26–27.

5-108
Daniel, H. 1949.
   Las serpientes en Colombia. *Rev. Fac. Nac. Agron. Medellin* 10(36):301–333.

5-109
Daniel, H. 1955.
   Algunos aspectos de la lucha biologica. IV: Los reptiles y la agricultura. Una serpiente bicefald. Como se han classificado las serpientes. *Rev. Fac. Nac. Agron. Medellin* 16(48):38–80.

5-110
Dao, T. V. 1981.
   List of the Vietnamese snakes at present known in Vietnam. *Tap Chi Sinh Vat Hoc* 3(1):6–10. (In Vietnamese with English abstract)

5-111
Day, J. H. 1969.
   A guide to marine life on South African shores. Cape Town: A. A. Balkema.

5-112
Deoras, P. J. 1965.
   Snakes of India. New Delhi: National Book Trust.

5-113
Deraniyagala, P. E. P. 1955.
   A colored atlas of some vertebrates from Ceylon. Colombo: Government Press. Volume 3.

5-114
Deraniyagala, P. 1960.
   Some southern temperate zone snakes, birds and whales that enter the Ceylon area. *Spolia Zeylan.* 29(1):79–85.

5-115
DeSilva, A. 1976.
   Venomous snakes of Sri Lanka. *Snake* 8:31–42.

5-116
DeSilva, P. H. D. H. 1980.
   Snake fauna of Sri Lanka with special reference to skull, dentition and venom in snakes. *Spolia Zeylan.* 34(1–2):1–472.

5-117
DeVis, C. W. 1905.
   Reptilia. *Ann. Queensl. Mus.* 6:46–52.

5-119
Ditmars, R. L. 1939.
   A field book of North American snakes. New York: Doubleday, Doran & Co.

5-120
Doederlein, L. 1881.
   Japanische seeschlangen. *Mitt. Deut. Ges. Nat. Volkerk. Ostasiens* 3(25):209–
210.

5–121
Doederlein, L. 1881.
   Die Liu-Kiu-Insel Amami Oshima. *Mitt. Deut. Ges. Nat. Voelkerk. Ostasiens*
3(24):140–156.

5-122
Donoso-Barros, R. 1962
   Los ofidios Chilenos. *Mus. Nac. Hist. Nat. Not. Mens. (Santiago)* No. 66:3–4,
8.

5-123
Donoso-Barros, R. 1966.
   Reptiles de Chile. Santiago: Universidad de Chile.

5-124
Donoso-Barros, R. 1970.
   Catalogo herpetologico chileno. *Bol. Mus. Nac. Hist. Nat. Chile* 31:49–124.

5-124.5
Doria, G. 1874.
   Enumerazione dei rettili raccolti dal Dott. O. Beccari in Amboina, alle Isole
Aru ed alle Isole Kei druante gli anni 1872–73. *Ann. Mus. Civ. Genova* 6:325–357.

5-124.7
Douglas, A. M.; Ride, W. D. L. 1962.
   Reptiles. *West. Aust. Fish. Dept. Fauna Bull.* 2:113–119.

5-125
Dowling, H. G. 1975.
   The Nearctic snake fauna. *Yearb. Herpetol.* 8(1):191–202.

5–125.5
Dowling, H. G.; Jenner, J. V. 1988.
  Snakes of Burma. Checklist of reported species & bibliography. *Smithson. Herpetol. Inf. Serv.* 76:1–19.

5-126
Duellman, W. E. 1961.
  The amphibians and reptiles of Michoacan, Mexico. *Univ. Kans. Mus. Nat. Hist. Misc. Publ.* 15(1):1–148.

5-127
Duffy, D. C. 1982.
  Seasnakes in the Galapagos. *Not. Galapagos* No. 35:23.

5-127.5
Duke, J. A. 1967.
  Herpetological dietary. Columbus: Battelle Memorial Institute.

5-128
Dumeril, A.; Bocourt, F. 1870.
  Etudes sur les reptiles et les batraciens. Part 3 of *Mission scientifique au Mexique et dans l'Amerique Central. Recherches zoologiques.* Paris: Imprimerie Imperiale.

5-129
Dunn, E. R. 1927.
  Results of the Douglas Burden expedition to the Island of Komodo. II. Snakes from the East Indies. *Am. Mus. Novitates* No. 287:1–7.

5-129.5
Dunn, E. R. 1944.
  Los generos de anfibios y reptiles de Colombia, III. Tercera Parte: reptiles: orden de las serpientes. *Caldasia* 3:155–224.

5-130
Dunn, E. R. 1957.
  Contributions to the herpetology of Colombia 1943–1946. Privately printed.

5-131
Dunson, W. A. 1975.
  Sea snakes of tropical Queensland between 18° and 20° south latitude. Dunson, W. A., ed. *The biology of sea snakes.* Baltimore: University Park Press. Pp. 151–162.

5-132
Dunson, W. A.; Minton, S. A. 1978.
Diversity, distribution, and ecology of Philippine marine snakes (Reptilia, Serpentes). *J. Herpetol.* 12(3):281–286.

5-133
Ekman, S. 1935.
Tiergeographie des meeres. Leipzig: Akademische Verlagsgesellschaft M.B.H. (English edition in 1953, *Zoogeography of the sea*, published in London by Sidgwick and Jackson.)

5-134
Elera, C. de. 1895.
Catalogo sistematico de toda la fauna de Filipinas conocida hasta el presente, y a la vez el de la coleccion zoologica del Museo de pp. Dominicos del Colegio-Universidad de Sto. Tomas de Manila. Manila: Colegio de Santo Tomas. Volume 1.

5-135
Emelianov, A. A. 1929.
Snakes of the far eastern district. Vladivostok. (In Russian with English abstract)

5-136
Ewart, J. 1878.
The poisonous snakes of India. For the use of the officials and others residing in the Indian empire. London: J. & A. Churchill.

5-137
Fayrer, J. 1871.
The thanatophidia of India. *Indian Med. Gaz.* 6(1):1–3.

5-137.5
Fayrer, J. 1871.
The thanatophidia of India. *Indian Med. Gaz.* Feb. 1871:21–24.

5-138
Fayrer, J. 1872.
The thanatophidia of India, being a description of the venomous snakes of the Indian Peninsula, with an account of the influence of their poison on life; and a series of experiments. London: Churchill.

5-139
Fayrer, J. 1874.
   The thanatophidia of India, being a description of the venomous snakes of the Indian Peninsula with an account of the influence of their poison on life; and a series of experiments. 2nd ed. rev'd and enlarged. London: Churchill.

5-140
Felix, J. 1965.
   Morsti hadi ve VDR. *Ziva* 13(5):192–193.

5-141
Ferguson, H. S. 1895.
   List of snakes taken in Travancore from 1888 to 1895. *J. Bombay Nat. Hist. Soc.* 10:68–77.

5-142
Ferguson, H. S. 1902.
   Travancore snakes. *J. Bombay Nat. Hist. Soc.* 14:386–387.

5-143
Fernando, M.; Gooneratne, W. 1983.
   Sea-snake envenoming. *Ceylon Med. J.* 28(3):131–143.

5-144
Fischer, J. G. 1856.
   Die familie der seeschlangen systematisch beschrieben. 2nd ed. Hamburg: Druck von J. A. Meissner. (Also published in *Abh. Geb. Naturwiss. Nat. Hamburg* 3:1–78)

5-145
Fischer, J. G. 1885.
   Ichthyologische und herpetologische Bemerkungen. *Jahrb. Hamb. Wiss. Anst.* 2:47–121.

5-145.5
Fischer, J. G. 1885.
   Ueber eine kollektion von amphibien und reptilien von Mindanao. *Jahrb. Hamb. Wiss. Anst.* 2:80–81.

5-146
Fischer, J. G. 1888.
   Ueber zwei von Liukiu-Insel Okinawa stammende schlangen. 1. *Platurus colubrinus* Schn. und dessen giftigkeit. *Jahrb. Hamb. Wiss. Anst.* 5:18–20.

5-147
Fish, C. J.; Cobb, M. C. 1954.
   Noxious marine animals of the central and western Pacific Ocean. *U.S. Fish Wildl. Ser. Res. Rep.* No. 36:1–45.

5-148
Fisk, G. H. R. 1885.
   Remarks upon a snake (*Pelamis bicolor*) captured at the entrance to Table Bay. *Proc. Zool. Soc. Lond.* 1885:482–483. (Letter)

5-149
Fitzinger, L. J. 1861.
   Die ausbeute der osterreichischen naturforscher an saugethieren und reptilien wahrend der weltumsgelung Sr. Majestat Fregatte Novara. *Akad. Wissenschaften, Vienna. Math.-Naturw. Situngsb.* 42(25):383–416.

5-150
FitzSimons, F. W. 1912.
   The snakes of South Africa: their venom and the treatment of snake bite. New ed. Cape Town: T. Maskew Miller.

5-151
FitzSimons, F. W. 1921.
   The snakes of South Africa: their venom and the treatment of snake bite. 3rd rev. ed. Cape Town: T. Maskew Miller.

5-152
FitzSimons, V. F. M. 1962.
   Snakes of Southern Africa. London: MacDonald.

5-153
FitzSimons, V. 1966.
   A check-list, with synoptic keys, to the snakes of Southern Africa. *Ann. Transvaal Mus.* 25(3):35–79.

5-154
FitzSimons, V. F. M. 1970.
   A field guide to the snakes of Southern Africa. London: Collins.

5-155
Flower, S. S. 1896.
   Notes on a collection of reptiles and batracians made in the Malay Peninsula in 1895–96; with a list of the species recorded from that region. *Proc. Zool. Soc. Lond.* 1896:856–914.

5-156
Flower, S. S. 1899.
   Notes on a second collection of reptiles made in the Malay Peninsula and Siam, from November 1896 to September 1898, with a list of the species recorded from those countries. *Proc. Zool. Soc. Lond.* 1899:600–696.

5-157
Forcart, L. 1953.
   Die amphibien und reptilien von Sumba, ihre zoogeographischen beziehungen und revision der unterarten von *Typhlops polygrammicus*. *Verh. Naturforsch. Ges. Basel* 64(2):356–388.

5-158
Freyvogel, T. A. 1972.
   Poisonous and venomous animals in East Africa. *Acta Trop.* 29(4):401–451.

5-159
Frith, C. B. 1974.
   Second record of the sea snake *Laticauda colubrina* in Thailand waters. *Nat. Hist. Bull. Siam Soc.* 25(3–4):209.

5-160
Frith, C. B. 1977.
   The sea snake *Hydrophis spiralis* (Shaw); a new species of the fauna of Thailand. *Nat. Hist. Bull. Siam Soc.* 26(3–4):339–341.

5-161
Frith, C. B. 1977.
   A survey of the snakes of Phuket Island and the adjacent mainland areas of Peninsular Thailand. *Nat. Hist. Bull. Siam Soc.* 26:263–316.

5-162
Garman, S. 1883.
   The reptiles and batrachians of North America. *Mem. Mus. Comp. Zool. Harv.* 8(3):1–185.

5-163
Garman, S. 1884.
   The North American reptiles and batrachians. A list of the species occurring North of the Isthmus of Tehuantepec, with references. *Bull. Essex Inst.* 16:3–46.

5-164
Garman, S. 1892.

On reptiles collected by Dr. Geo. Baur near Guayaquil, Ecuador. *Bull. Essex Inst.* 24:88–95.

5-165
Garman, S. 1901.

Some reptiles and batrachians from Australasia. *Bull. Mus. Comp. Zool.* 39(1):1–14.

5-166
Garman, S. 1908.

The reptiles of Easter Island. *Bull. Mus. Comp. Zool.* 52(1):1–14.

5-167
Gee, N. G. 1929–1930.

A contribution toward a preliminary list of reptiles recorded from China. *Bull. Dept. Biol. Yenching Univ.* 1(1):53–84.

5-168
Gibson-Hill, C. A. 1950.

A note on the reptiles occurring on the Cocos-Keeling Islands. *Bull. Raffles Mus.* No. 22:206–211.

5-169
Girard, C. 1858.

Herpetology. *United States Exploring Expedition during the years 1838, 1839, 1840, 1841, 1842. Under the Command of Charles Wilkes, U.S.N.* Philadelphia: J. B. Lippincott. Volume 20.

5-170
Glauert, L. 1928.

The vertebrate fauna of Western Australia. *J. R. Soc. West. Aust.* 14:61–77.

5-171
Glauert, L. 1950.

A handbook of the snakes of Western Australia. Perth: Western Australian Naturalists' Club.

5-172
Gogorza y Gonzales, D. J. 1888.

Datos para la fauna Filipina vertebrados. *Anal. Soc. Esp. Hist. Nat.* 17:247–303.

5-173
Goris, R. C. 1965.
   A herpetological survey of Tsushima Island. *Acta Herpetol. Jpn.* 2(2):9–15.

5-174
Gow, G. F. 1977.
   Snakes of the Darwin area. Darwin: Museums and Art Galleries Board of the
Northern Territory.

5-175
Gray, J. E. 1827.
   Reptilia. King, P. P. *Narrative of a survey of the intertropical and western
coasts of Australia. Performed between the years 1818 and 1822.* London: John
Murray. Vol. 2, pp. 424–434.

5-176
Gray, J. E. 1841.
   A catalogue of the species of reptiles and amphibia hitherto described as
inhabiting Australia, with a description of some new species from Western Australia, and some remarks on their geographical distribution. Grey, G. *Journals of
two expeditions of discovery in north-west and western Australia, during the years
1837, 38, and 39, under the authority of Her Majesty's government. Describing many
newly discovered, important, and fertile districts, with observations on the moral
and physical condition of the aboriginal inhabitants, &c, &c.* London: T. and W.
Boone. Vol. 2, pp. 422–449.

5-177
Gray, J. E. 1842.
   Australian reptiles and amphibia: catalogue of the species of reptiles and
amphibia hitherto described as inhabiting Australia. *Tasman. J. Nat. Sci. Agric.
Stat.* 1:385–387.

5-178
Great Britain. Challenger Office. 1895.
   Summary of results. *Report on the scientific results of the voyage of the H. M.
S. Challenger during the years 1872–1876. Prepared under superintendence of Sir
C. W. Thomson.* Edinburgh: Printed for H. M. Stationery Office by Neill and
Company. Volumes 49 and 50. New York: Johnson Reprint; 1965.

5-179
Griffin, L. E. 1909.
   A list of snakes found in Palawan. *Philipp. J. Sci. Sect. A,* 4:595–601.

5-180
Griffin, L. E. 1911.
   A check-list and key of Philippine snakes. *Philipp. J. Sci. Sect.D. Gen. Biol., Ethnol. Anthropol.* 6(5):253–269.

5-181
Guenther, A. 1858.
   On the geographical distribution of reptiles. *Proc. Zool. Soc. Lond.* 1858:373–397.

5-182
Guenther, A. C. L. G. 1864.
   The reptiles of British India. London: Printed for the Ray Society by Robert Hardwicke.

5-183
Guenther, A. 1869.
   Report on two collections of Indian reptiles. *Proc. Zool. Soc. Lond.* 1869:500–502.

5-184
Guenther, A. 1872.
   On the reptiles and amphibians of Borneo. *Proc. Zool. Soc. Lond.* 1872:586–600.

5-185
Guenther, A. 1874.
   A contribution to the fauna of Savage Island. *Proc. Zool. Soc. Lond.* 1874(20):295–297.

5-185.5
Guenther, A. 1877.
   On a collection of reptiles and fishes from Duke-of-York Island, New Ireland, and New Britain. *Proc. Zool. Soc. Lond.* 1877:127–132

5–185.7
Guenther, A. C. L. G. 1895.
   Biololgia centrali-americana. Reptilia and batrachia. London: Taylor & Frances. Athens, OH: Society for the Study of Amphibians and Reptiles; 1985.

5-186
Guibe, J. 1958.
   Les serpents de Madagascar. *Mem. Inst. Scient. Madagascar,* Ser.A, 12:189–260.

5-187
Guibe, J. 1966.
   Reptiles et amphibiens recoltes par la Mission Franco-Iranienne. *Bull. Mus. Natl Hist. Nat.* 38(2):97–98.

5-188
Guinea, M. L. 1981.
   The sea snakes of Fiji. *Proc. 4th Int. Coral Reef Symp.*, Manila, *1981*. Vol. 2, pp. 581–585.

5-188.5
Guppy, H. B. 1887.
   The Solomon Islands and their natives. London: Swan, Sonnenschein, Lowrey & Co.

5-189
Gyldenstolpe, N. 1916.
   Zoological results of the Swedish zoological expeditions to Siam 1911–1912 & 1914–1915. 1. Snakes. *K. Svenska Vetensk.-Akad. Handl.* Ser. 4, 55(3):1–28.

5-190
Haas, C. P. J. de. 1950.
   Checklist of the snakes of the Indo-Australian Archipelago (Reptiles, Ophidia). *Treubia* 20(3):511–625. Amsterdam: Linnaeus Press; 1972.

5-191
Haas, G. 1957.
   Some amphibians and reptiles from Arabia. *Proc. Calif. Acad. Sci.* Ser. 4, 29(3):47–86.

5-191.5
Haas, G. 1961.
   On a collection of Arabian reptiles. *Ann. Carnegie Mus.* 36(3):19–28.

5-192
Haile, N. S. 1958.
   The snakes of Borneo, with a key to the species. *Sar. Mus. J.* Ser. 12, 8:743–771.

5-193
Hallowell, E. 1860.
   Report upon the reptilia of the North Pacific Exploring Expedition, under command of Capt. John Rogers, U.S.N. *Proc. Acad. Nat. Sci. Phil.* 1860:480–510.

5-194
Halstead, B. W. 1970.
Poisonous and venomous marine animals of the world. Washington DC: Government Printing Office. Volume 3.

5-194.5
Halstead, B. W. 1988.
Poisonous and venomous marine animals of the world. 2nd rev. ed. Princeton: Darwin Press.

5-195
Haly, A. 1887.
On a remarkable sea snake from Colombo. *Taprobanian* 2:107–108.

5-196
Harding, K. A.; Welch, K. R. G. 1980.
Venomous snakes of the world: a checklist. New York: Pergamon.

5-197
Hardy, L. M.; McDiarmid, R. W. 1969.
The amphibians and reptiles of Sinola, Mexico. *Univ. Kans. Mus. Nat. Hist. Misc. Publ.* 18(3):39–252.

5-198
Hattori, Z.; Inoue, T.; Okonogi, T. 1974.
In front of a pile of sea snake carcasses. *Snake* 6(2):94–98. (In Japanese with English abstract)

5-199
Heatwole, H. 1975.
Biogeography of reptiles on some of the islands and cays of eastern Papua-New Guinea. *Atoll Res. Bull.* 180:1–32.

5-200
Heatwole, H. 1975.
Sea snakes found on reefs in the Southern Coral Sea (Saumarez, Swains, Cato Island). Dunson, W. A., ed. *The biology of sea snakes*. Baltimore: University Park Press. Pp. 163–171.

5-201
Heatwole, H. 1975.
Sea snakes of the Gulf of Carpentaria. Dunson, W. A., ed. *The biology of sea snakes*. Baltimore: University Park Press. Pp. 145–149.

5-201.5
Heatwole, H.; Limpus, C. 1984.
    Class reptilia: sea snakes, turtles and lizards. Mather, P.; Bennett, I., eds. *A coral reef handbook: a guide to the fauna, flora, and geology of Heron Island and adjacent reefs and cays.* 2nd ed. Brisbane: Australian Coral Reef Society.

5-202
Hecht, M. K.; Kropach, C.; Hecht, B. M. 1974.
    Distribution of the yellow-bellied sea snake, *Pelamis platurus*, and its significance in relation to the fossil record. *Herpetologica* 30(4):387–396.

5-202.3
Hediger, H. 1933.
    Ueber die von herrn Dr. A. Buehler auf der Admiralitats-Gruppe und einigen benachbarten inseln gesammelten reptilien und amphibien. *Verh. Naturforsch. Ges. Basel* 44(2):28–52.

5-202.6
Hediger, H. 1934.
    Beitrage zur herpetologie und zoogeographie Neu Brittanniens und einiger umliegender gebiete. *Zool. Jahrb. Abt. Syst. Oekol. Geogr. Tiere* 65(5–6):441–582.

5-203
Herklots, G. A. C. 1951.
    The Hong Kong countryside. Hong Kong: South China Morning Post, Ltd.

5-204
Herre, A. W. C. T. 1942.
    Notes on Philippine sea-snakes. *Copeia* 1942(1):7–9.

5-205
Hewitt, J. 1937.
    A guide to the vertebrate fauna of the eastern Cape Province, South Africa Part II: Reptiles, amphibians, and freshwater fishes. Grahamstown: Albany Museum.

5-206
Hidalgo, H. 1980.
    Occurrence of *Pelamis platurus* (Linnaeus) in El Salvador. *Herpetol. Rev.* 11(4):117.

5-207
Hilgendorf, F. 1876.
    Die Japanischen schlangen. *Mitt. Deut. Ges. Nat. Voelkerk. Ostasiens* 1(10):29–34.

5-208

Hill, F. L. 1955.

Notes on the natural history of the Monte Bello Islands. *Proc. Linn. Soc. Lond.* 165(2):113–124.

5-208.5

Hirosaki, Y. 1964.

Ecological study on fishes with the drifting sea weeds. III. Accompanying animals excluded fishes. *Misc. Rep. Res. Inst. Nat. Resour. (Tokyo)* 62:63–70. (In Japanese)

5-209

Hoge, A. R.; Romano, S. A. R. W. D. L. 1971.

Neotropical pit vipers, sea snakes, and coral snakes. Buecherl, W.; Buckley, E. E., eds. *Venomous animals and their venoms.* Academic Press: New York. Vol. 2, pp. 211–293.

5-210

Holmes, R. W.; Blackburn, M. 1960.

Physical, chemical, and biological observations in the eastern tropical Pacific Ocean: SCOT Expedition, April-June 1958. Washington. (Also cited as *U.S. Fish Wildl. Ser. Sp. Sci. Rep.-Fish.* No. 345:1–106)

5-211

Honma, Y.; Kitami, T. 1970.

Records of occurrence of the sea snakes in the coasts of Niigata Perfecture, facing the Japan Sea. *Collect. Culture* 32(5):177–179. (In Japanese)

5-212

Honma, Y.; Kitami, T. 1976.

Records of the marine reptiles in adjacent waters of Niigata and Sado Island, based partially on the old documents. *Bull. Sado Mus.* No. 25:18–24. (In Japanese)

5-213

Honma, Y.; Kitami, T. 1979.

The second record of a sea snake, *Pelamis platurus*, caught off the coast of Sado Island in summer, with the additional records of marine reptiles in the coasts of Niigata and Sado. *Nippon Herpetol. J.* No. 15:7–10. (In Japanese)

5-214

Honma, Y.; Kitami, T. 1983.

Additional records of the marine reptiles in the waters adjacent to Niigata and Sado Island. *Nippon Herpetol. J.* No. 27:27–31. (In Japanese)

5-215
Honma, Y. 1984.
   A record of the sea snake, *Laticauda semifasciata* (Reinwardt, 1837), from the Peter the Great Bay of the Soviet Far East, Sea of Japan. *Nippon Herpetol. J.* No. 28:15–18. (In Japanese)

5-216
Horikawa, Y. 1930.
   Reptilia and amphibians in the island of Pescadore (Hoko-to). *Trans. Nat. Hist. Soc. Taiwan* 20(106):19–23. (In Japanese)

5-217
Horn, M. J. 1964.
   Water snakes: Indian waters. *Mar. Observer* 34(203):15–16.

5-218
Hosmer, W. 1958.
   List of snakes recorded from Cairns and Hinterland. *North Queensl. Nat.* 27(120):5.

5-219
Hubbs, C. L. 1960.
   The marine vertebrates of the outer coast. *Syst. Zool.* 9(3–4):134–147.

5-220
Hubbs, C. L.; Roden, G. I. 1964.
   Oceanography and marine life along the Pacific Coast of Middle America. Wauchope, R. ed. *Handbook of Middle American Indians*. Austin, TX: University of Texas Press. Vol. 1, pp. 143–186.

5-221
Hubrecht, A. A. W. 1882.
   List of reptiles and amphibians brought from British India by Mr. Francis Day. *Notes Leyden Mus.* 4:138–144.

5-221.5
Hutton, F. W., ed. 1904.
   Index faunae Novae Zealandiae. London: Dulau.

5-222
In Den Bosch, H. A. J. 1985.
   Snakes of Sulawesi: checklist, key and additional biogeographical remarks. *Zool. Verh.* (*Leiden*) No. 217:3–50.

5-224
Isemonger, R. M. 1962.
   Snakes of Africa: southern, central and east. Johannesburg: Nelson.

5-225
Jagor, F. 1873.
   Reisen in den Philippinen. Berlin: Weidmannsche Buchhandlung.

5-226
Jagor, F. 1875.
   Travels in the Philippines. London: Chapman and Hall.

5-227
Jennings, M. R. 1983
   An annotated check list of the amphibians and reptiles of California. *Calif. Fish Game* 69(3):151–171.

5-228
Jerdon, T. C. 1854.
   Catalogue of reptiles inhabiting the Peninsula of India. *J. Asiat. Soc. Beng.* 22(6):462–534.

5-229
Joger, U. 1984.
   The venomous snakes of the near and middle east. *Tubinger Atlas des Vorderen Orients.* Weisbaden: Reichert.

5-230
Jong, J. K. de. 1925.
   Fauna Buruana. Reptiles. *Boeroe-Expeditie 1921–1922. Resultats zoologiques de l'Expedition Scientifique Neerlandaise a l'ile de Buru en 1921 et 1922.* Buitenzorg: Archipel Drukkerj. Vol. 1, pt. 1, pp. 1–12.

5-231
Jong, J. K. de. 1925.
   Fauna Buruana. Reptiles. *Treubia* 7(2):85–96.

5-232
Jong, J. K. de. 1930.
   Notes on some reptiles from the Dutch-East-Indies. *Treubia* 12(1):115–119.

5-232.5
Jouan, H. 1863.
Notes sur quelques animaux observes a la Nouvelle-Caledonie pendant les annees 1861 et 1862. *Mem. Soc. Nat. Sci. Nat. Math. Cherbourg* 9:89–127.

5-232.7
Jouan, H. 1864.
Additions a la faune de la Nouvelle-Caledonie. *Mem. Soc. Nat. Sci. Nat. Math. Cherbourg* 10:301–311.

5-232.8
Kamita, T. 1962.
Visits by unusual aquatic animals to the sea coasts of San-in District, Japan Sea. *Mem. Nat. Cult. Res. San-in Reg.* 2:1–35. (In Japanese)

5-233
Kengragsat, L. S. 1934.
The Hydrophiidae (sea-snakes) of the Gulf of Siam. *Proc. Pacif. Sci. Congr.,* 5th, Canada 4:3079–3081.

5-234
Kennedy, W. P. 1937.
Some additions to the fauna of Iraq. *J. Bombay Nat. Hist. Soc.* 39(3):745–749.

5-235
Kershaw, J. A. 1918.
Two snakes new to Victoria, with a list of the Victorian species. *Victorian Nat.* 35:30–32.

5-236
Kershaw, J. A. 1927.
Victorian reptiles. *Victorian Nat.* 43:335–344.

5-237
Khalaf, K. T. 1959.
Reptiles of Iraq with some notes on the amphibians. Baghdad: Ar-Rabbita Press.

5-238
Khan, M. S. 1982.
An annotated checklist and key to the reptiles of Pakistan. Part III: Serpentes (Ophidia). *Biologia* 28(2):215–254.

5-239

Kharin, V. E. 1984.

The first record of three species of sea snakes from Vietnam with a note on rare variety of *Praescutata viperina*. *Biol. Morya (Vladivost.)* 10(2):26–30. (In Russian)

5-240

Kharin, V. E. 1984.

A review of sea snakes of the group *Hydrophis* sensu lato (Serpentes, Hydrophiidae). 3. The genus *Leioselasma*. *Zool. Zh.* 63(10):1535–1546. (In Russian with English abstract)

5-241

Kharin, V. E. 1984.

Sea snakes of the genus *Hydrophis* sensu lato (Serpentes, Hydrophiidae). On taxonomic status of the New Guinea *H. obscurus*. *Zool. Zh.* 63(4):630–632. (In Russian with English abstract)

5-242

Kharin, V. E. 1984.

Three species of sea snakes first discovered in Vietnamese waters, with a comment on a rare form *Praescutata viperina*. *Sov. J. Mar. Biol.* 10(2):80–84. (Translated from *Biol. Morya (Vladivost.)* 10(2):26–30, 1984.)

5-243

Kharin, V. E. 1985.

A new species of sea snakes of the genus *Enhydrina* (Serpentes, Hydrophiidae) from waters of New Guinea. *Zool. Zh.* 64(5):785–787. (In Russian with English abstract)

5-244

Kinghorn, J. R. 1928.

Herpetology of the Solomon Islands. *Rec. Aust. Mus.* 16(3):123–178.

5-245

Kinghorn, J. R. 1929.

Snakes of Australia. Sydney: Angus & Robertson.

5-246

Kinghorn, J. R. 1942.

Herpetological notes No. 4. *Rec. Aust. Mus.* 21(2):118–121.

5-247
Kinghorn, J. R. 1957.
   The snakes of Australia. 2nd ed. Cape Town: Angus and Robertson.

5-248
Kitami, T.; Honma, Y.; Yoshie, S. 1975.
   Records of the marine reptiles in adjacent waters of Niigata and Sado Island, based partially on the old documents. *Jpn. J. Herpetol.* 6(2):50–51. (In Japanese)

5-249
Kitami, T.; Honma, Y. 1979.
   A record of a sea snake, *Pelamis platurus*, caught off the coast of Sotokaifu village, Sado Island, in summer. *Nippon Herpetol. J.* No. 13:13–16. (In Japanese)

5-250
Klemmer, K. 1963.
   Liste der rezenten giftschlangen: Elapidae, Hydropheidae, Viperidae und Crotalidae. *Die giftschlangen der erde*. Marburg: N. G. Elwert Universitats und Verlags-Buchhandlung. Pp. 255–464.

5-251
Koba, K. 1959.
   Herpetofauna of the Amami Group of the Loo Choo Islands (III). *Mem. Fac. Educ. Kumamoto Univ.* 7:187–202. (In Japanese)

5-252
Koba, K. 1973.
   The venomous snakes of southeast Asia. *Snake* 5(1–2):77–115. (In Japanese with English abstract)

5-253
Kopstein, F. 1926.
   Reptilien von den Molukken und den Benach-Barten Inseln. *Zool. Meded. Rijksmus. Nat. Hist. Leiden* 9(2–3):71–112.

5-254
Kopstein, F. 1930.
   De javaansche gifslangen en haar beteekenis voor den mensch. Weltevreden: Nederlandsch-Indische Natuurhistorische Vereeniging.

5-255
Kopstein, F. 1930.
   Herpetologische notizen II: Oologische beobachtungen an West-Javanischen reptilien. *Treubia* 11(3):301–307.

5-255.5
Kopstein, F. 1937.
  Schlangen von Enggano. *Treubia* 16(2):239–244.

5-256
Kopstein, F. 1938.
  Ein beitrag zur eierkunde und zur fortpflanzung der malaiischen reptilien. *Bull. Raffles Mus.* 14:81–167.

5-256.5
Kopstein, F. 1938.
  Herpetologische notizen. XIX. Ueber pigmentierungsanomalien bei malaiischen reptilien. *Treubia* 16(3):361–364.

5-257
Krefft, G. 1866.
  On snakes observed in the neighbourhood of Sydney. *Trans. Phil. Soc. N.S. W.* 1866: 34–60.

5-258
Krefft, G. 1869.
  The snakes of Australia: an illustrated and descriptive catalogue of all the known species. Sydney: Thomas Richards, Government Printer.

5-259
Kropach, C. 1972.
  A field study of the sea snake Pelamis platurus (Linnaeus) in the Gulf of Panama. New York, NY: City University of New York. 200 pp. (Dissertation)

5-260
Kuntz, R. E. 1963.
  Snakes of Taiwan. *Q.J. Taiwan Mus.* (Taipei) 16(1–2):1–79. (Reprinted as *U.S. Naval Medical Research Unit Publication No. 2*)

5-261
Kuntz, R. E.; Ming, D. Z. 1970.
  Vertebrates of Taiwan taken for parasitological and biomedical studies by U.S. Naval Medical Research Unit No. 2, Taipei, Taiwan Republic of China. *Q.J. Taiwan Mus.* (Taipei) 23:1–37.

5-262
Laidlaw, F. F. 1901.

List of a collection of snakes, crocodiles, and chelonians from the Malay Peninsula, made by members of the "Skeat Expedition", 1899–1900. *Proc. Zool. Soc. Lond.* 2:575–586.

5-263
Laidlaw, F. F. 1903.

Amphibia and reptilia. Gardiner, J. S. *Fauna and geography of the Maldive and Laccadive Archipelagoes.* Cambridge, England: University Press, 1903–1906. Vol. 1, pp. 119–122.

5-264
Lanza, B. 1954.

Su un esemplare di *Pelamis platurus* (L.) catturato in un fiume della piccola Andaman (Serpentes; Hydrophiidae). *Monit. Zool. Ital.* 62:67–70. (In Italian with English abstract)

5-265
Latifi, M.; Hoge, A. R.; Eliazan, M. 1966.

The poisonous snakes of Iran. *Mem. Inst. Butantan (Sao Paulo)* 33(3):735–744.

5-266
Laurent, R. 1948.

Notes sur quelques reptiles appartenant a la collection du Musee Royal d'Historie Naturelle de Belgique. II. Formes asiatiques et neo-guineennes. *Bull. Mus. R. Hist. Nat. Belg.* 24(17):1–12.

5-266.5
Lesson, R. P. 1828.

Observations generales sur les reptiles observes dans le voyage autour du monde de la corvette La Coquille. *Ann. Sci. Nat.* 13:369–394.

5-267
Leviton, A. E.; Anderson, S. C. 1967.

Survey of the reptiles of the Sheikhdom of Abu Dhabi, Arabian Peninsula, Part II. Systematic account of the collection of reptiles made in the Sheikhdom of Abu Dhabi by John Gasperetti. *Proc. Calif. Acad. Sci.* Ser. 4, 35(9):157–192.

5-268
Liat, L. B.; Balasingam, E. 1969.

A collection of sea snakes from Port Swettenham, Selangor and Tumpat, Kelantan. *J. Fed. Malay States Mus.* 14:123–126.

5-269
Liat, L. B.; Sawai, Y. 1975.
   Ecology and distribution of some seasnakes in Malay Peninsula. *Snake* 7:28–32. (In Japanese with English abstract)

5-269.5
Lidth de Jeude, T. W. van. 1890.
   On a collection of reptiles from Nias, and on *Calamaria virgulata*, Boie. *Notes Leyden Mus.* 12:253–256.

5-270
Lidth de Jeude, T. W. van. 1890.
   On a collection of snakes from Dehli. *Notes Leyden Mus.* 12:17–27.

5-270.3
Lidth de Jeude, T. W. van. 1890.
   Reptilia from the Malay Archipelago. II. Ophidia. Weber, M. *Zoologische ergebnisse einer reise in Nederlaendisch Ost-Indien*. Leiden: E. J. Brill. Vol. 1, pp. 178–192.

5-270.4
Lidth de Jeude, T. W. van. 1897.
   Reptiles and batrachians from New Guinea. *Notes Leyden Mus.* 18:249–257.

5-270.5
Lidth de Jeude, T. W. van. 1911.
   Reptilien (schlangen). *Nova Guinea. Resultats de l'expedition scientifique Neerlandaise a la Nouvelle-Guinee en 1903*. Leide: E. J. Brill. Vol. 5, pt. 4, pp. 519–530.

5-271
Lim, B. L.; Sawai, Y. 1975.
   Ecology and distribution of sea snakes in Peninsular Malaysia. *Toxicon* 13:107. (Abstract)

5-272
Lim, B. L.; Sawai, Y. 1976.
   Ecology and distribution of some sea snakes in Peninsular Malaysia. Ohsaka, A.; Hayashi, K.; Sawai Y., eds. *Animal, plant, microbial toxins*. New York: Plenum. Vol. 2, pp. 515–520.

5-273
Limpus, C. J. 1975.
   Coastal sea snakes of subtropical Queensland waters (23° to 28° south latitude).
Dunson, W. A., ed. *The biology of sea snakes*. Baltimore: University Park Press.
Pp. 173–182.

5-274
Limpus, C. J.; Lyon, B. J. 1979.
   Two additional sea snake species from South Queensland. *Herpetofauna*
11(1):10–11.

5-275
Lindsay, G. E. 1962.
   The Belvedere expedition to the Gulf of California. *Trans. San Diego Soc. Nat.
Hist.* 13(1):1–44.

5-276
Loennberg, E.; Andersson, L. G. 1913.
   Results of Dr. E. Mjoebergs Swedish Scientific Expeditions to Australia 1910–
13. III. Reptiles. *K. Svenska Vetensk.-Akad. Handl.* N.F. 52(3):1–17.

5-277
Longman, H. A. 1918.
   Notes on some Queensland and Papuan reptiles. *Mem. Queensl. Mus.* 6:37–
44.

5-278
Loomis, R. B.; Bennett, S. G.; Sanborn, S. R.; Barbour, C. H.; Wiener, H. 1974.
   A handlist of the herpetofauna of Baja California, Mexico & adjacent islands.
Privately printed in Long Beach, CA: California State University, Long Beach.

5-279
Lord, C. E. 1919.
   Notes on the snakes of Tasmania. *Pap. Proc. R. Soc. Tasm.* 1918:76–81.

5-280
Lord, C. E. 1920.
   On the occurrence in Tasmania of *Hydrus platurus* Linn. *Pap. Proc. R. Soc.
Tasm.* 1919:22.

5-281
Lord, C. E.; Scott, H. H. 1924.
   A synopsis of the vertebrate animals of Tasmania. Hobart: Oldham, Beddome
& Meredith.

5-282
Loveridge, A. 1928.
Notes on snakes and snake-bites in East Africa. *Bull. Antivenin Inst. Am.* 1(4):106–119.

5-283
Loveridge, A. 1934.
Australian reptiles in the Museum of Comparative Zoology, Cambridge, Massachusetts. *Bull. Mus. Comp. Zool.* 77(6):241–385.

5-284
Loveridge, A. 1948.
New Guinean reptiles and amphibians in the Museum of Comparative Zoology and United States National Museum. *Bull. Mus. Comp. Zool.* 101(2):305–430.

5-285
Loveridge, A. 1957.
Check list of the reptiles and amphibians of East Africa (Uganda, Kenya, Tanganyika, Zanzibar). *Bull. Mus. Comp. Zool.* 117(2):153–362.

5-286
Loveridge, A. 1959.
On a fourth collection of reptiles, mostly taken in Tanganyika Territory by Mr. C. J. P. Ionides. *Proc. Zool. Soc. Lond.* 133:29–44.

5-287
Lucas, A. H. S.; LeSouef, W. H. D. 1909.
The animals of Australia. Mammals, reptiles & amphibians. Melbourne: Whitcombe and Tombs.

5-288
Luetken, C. F. 1862.
Nogle nye krybdyr og padder. *Vidensk Medd. Dan. Naturhist. Foren.* 1862:292–311.

5-289
Macleay, W. 1878.
The Ophidians of the Chevert Expedition. *Proc. Linn. Soc. N.S.W.* 1877(2):33–41.

5-290
Macleay, W. 1884.
Census of Australian snakes with descriptions of two new species. *Proc. Linn. Soc. N.S.W.* 9:548–568.

5-291
Maki, M. 1931.
   A monograph of the snakes of Japan. Tokyo: Dai-Ichi Shobo.

5-292
Mao, S. H.; Chen, B. Y. 1975.
   Disteira stokesi (Gray): an addition to the sea snake fauna of Taiwan. *Herpetologica* 31(4):456–458.

5-293
Mao, S. H.; Chen, B. Y. 1980.
   Sea snakes of Taiwan: a natural history of sea snakes. Taipei, Taiwan: National Science Council.

5-294
Marinkelle, C. J. 1966.
   Accidents by venomous animals in Colombia. *Indust. Med. Surg.* 1966:988–992.

5-295
Martens, E. von. 1867.
   Die Preussische expedition nach ost-asien. Berlin: R.v. Decker. Volume 1.

5-296
Martin, J. P. 1877.
   Curator's report. *J.N. China Br. R. Asiat. Soc.* 11:viii–ix.

5-297
Matsui, T. 1969.
   The snakes of Japan. *Snake* 1:50–55. (In Japanese)

5-298
McCann, C. 1966.
   The marine turtles and snakes occurring in New Zealand. *Rec. Dom. Mus. (Wellington)* 5(21):201–215.

5-299
McCoy, C. J.; Hahn, D. E. 1979.
   The yellow-bellied sea snake, *Pelamis platurus* (Reptilia: Hydrophiidae), in the Philippines. *Ann. Carnegie Mus.* 48(14):231–234.

5-300

McCoy, M. 1980.

Reptiles of the Solomon Islands. Hong Kong: Sheck Wah Tong Printing Press Limited. (*WAU Ecology Institute Handbook* No. 7)

5-301

McLain, R. B. 1899.

Contributions to Neotropical herpetology. Wheeling: West Virginia.

5-302

McPhee, D. R. 1959.

Some common snakes and lizards of Australia. Brisbane: Jacaranda Press.

5-303

Medem, F. 1968.

El desarrollo de la herpetologia en Colombia. *Rev. Acad. Colomb. Cienc. Exactas Fis. Nat.* 13(50):149–199.

5-303.5

Medem, F. 1979.

Los anfibios y reptiles de las islas Gorgona y Gorgonilla. Prahl, H. von; Guhl, F.; Grogl, M., eds. *Gorgona.* Bogota: Universidad de Los Andes.

5-304

Mell, R. 1922.

Beitrage zur fauna sinica. I. Die vertebraten Sudchinas; Feldlisten und Feldnoten der sauger, vogel, reptilien, batrachier. *Arch. Nat. (Berlin)* 88A:1–134.

5-305

Mell, R. 1928.

List of Chinese snakes. *Lingnan Sci. J.* 8:199–219.

5-306

Mertens, R. 1930.

Die amphibien und reptilien der Inseln Bali, Lombok, Sumbawa und Flores (Beitrage zur fauna der Kleinen Sunda-Inseln. I.) *Abh. Senckenb. Naturforsch. Ges.* 42(3):115–344.

5-307

Mertens, R. 1934.

Ueber die verbreitung und das verbreitungszentrum der seeschlangen (Hydrophidae). *Zoogeographica* 2:305–319.

5-308
Mertens, R. 1957.
    Zur herpetofauna von Ostjava und Bali. *Senckenb. Biol.* 38(1–2):23–31.

5-309
Mertens, R. 1969.
    Die amphibien und reptilien West-Pakistans. *Stuttg. Beitr. Naturkd.* 197:1–96.

5-310
Mertens, R. 1974.
    Die amphibien und reptilien West-Pakistans. *Senckenb. Biol.* 55(1–3):35–38.

5-310.5
Meyer, A. B. 1874.
    Mittheilung ueber die von ihm auf Neu-Guinea und den Inseln Jobi, Mysore
und Mafoor im Jahre 1873 gesammelten amphibien. *Monatsber. K. Preuss. Akad.
Wiss. Berl.* 1874:128–140.

5-311
Mialaret, T. 1897.
    L'Ile des Pins, son passe, son present, son avenir. Paris: Librairie Africaine
& Coloniale Paris.

5-312
Minton, S. A. 1962.
    An annotated key to the amphibians and reptiles of Sind and Las Bela, West
Pakistan. *Am. Mus. Novitates* No. 2081:1–60.

5-313
Minton, S. A. 1966.
    A contribution to the herpetology of West Pakistan. *Bull. Am. Mus. Nat.
Hist.* 134(2):27–184.

5-314
Minton, S. A. 1974.
    Observations on sea snakes at Ashmore Reef, Timor Sea. *Proc. Indiana Acad.
Sci.* 83:467–468.

5-315
Minton, S. A. 1975.
    Geographic distribution of sea snakes. Dunson, W. A., ed. *The biology of sea
snakes.* Baltimore: University Park Press. Pp. 21–31.

5-316

Minton, S. A.; Heatwole, H. 1975.

Sea snakes from three reefs of the Sahul Shelf. Dunson, W. A., ed. *The biology of sea snakes*. Baltimore: University Park Press. Pp. 141–144.

5-317

Minton, S. A.; Dunson, W. W. 1985.

Sea snakes collected at Chesterfield Reefs, Coral Sea. *Atoll Res. Bull.* No. 292:101–108.

5-318

Mitchell, F. J. 1955.

Preliminary account of the reptilia and amphibia collected by the National Geographic Society-Commonwealth Government-Smithsonian Institution Expedition to Arnhem Land (April to November, 1948). *Rec. S. Aust. Mus.* 11:373–407.

5-319

Mitchell, F. J. 1964.

Reptiles and amphibians of Arnhem Land. New York: Cambridge University Press. (also cited as *Records of the American Australian Scientific Expedition to Arnhem Land* 4:309–343)

5-320

Mocquard, F. 1890.

Recherches sur la faune herpetologique des iles de Borneo et de Palawan. *Nouv. Arch. Mus. Hist. Nat.* Ser. 3, 2:115–168.

5-321

Mocquard, F. 1899.

Contribution a la faune herpetologique de la Basse-Californie. *Nouv. Arch. Mus. Hist. Nat.* Ser. 4, 1:297–343.

5-322

Mocquard, F. 1904.

Serpents recueillis par M. A. Pavie en Indo-Chine. *Mission Pavie Indo-Chine 1879–1895: Etudes diverses. III. Recherches sur l'histoire naturelle de l'Indo-Chine orientale.* Paris: Ernest Leroux. Pp. 481–486.

5-323

Mocquard, F. 1909.

Synopsis des familles, genres et especes des reptiles ecailleux et des batraciens de Madagascar. *Nouv. Arch. Mus. Hist. Nat.* Ser. 5, 1:1–110.

5-324
Montaquim, M. A.; Sarker, A. H.; Khan, M. A. R.; Husain, K. Z. 1980.
    List of the snakes of Bangladesh. *Bangladesh J. Zool.* 8(2):127–129.

5-325
Morrison, C. 1939.
    Port Phillip sea-serpent. *Victorian Nat.* 56:16.

5-326
Mueller, F. 1881.
    Nachtrag I zum katalog der herpetologischen sammlung des Basler museums.
Basel: J. G. Bauer. Pp. 122–165.

5-326.5
Mueller, F. 1889.
    Sechster nachtrag zum katalog der herpetologischen sammlung des Basler
museums. *Verh. Naturforsch. Ges. Basel* 8:685–705.

5-326.7
Mueller, S. 1857.
    *Reizen en onderzoekingen in den indischen archipel, gedaan op last der ned-
erlandsche indische regering, tusschen de jahren 1828 en 1836.* Amsterdam: Fred-
erik Muller.

5-327
Murphy, R. C. 1936.
    *Oceanic birds of South America: a study of species of the related coasts and
seas, including the American quadrant of Antarctica. Based upon the Brewster-
Sanford collection in the American Museum of Natural History.* New York: Macmillan. Volume 1.

5-328
Murphy, R. C. 1956.
    The vertebrates of SCOPE, November 7-December 16, 1956. *U.S. Fish Wildl.
Ser. Sp. Sci. Rep.-Fish.* No. 279(2):101–111.

5-329
Murray, J. A. 1884.
    *The vertebrate zoology of Sind. A systematic account, with descriptions of all
the known species of mammals, birds, and reptiles inhabiting the province; observations on their habits, &c; tables of their geographical distribution in Persia, Be-
loochistan, and Afghanistan; Punjab, North-West provinces, and the peninsula of
India generally.* London: Richardson & Co.

5-330
Murray, J. A. 1886.
   The reptiles of Sind. London: Richardson & Co.

5-331
Murthy, T. S. N. 1977.
   The herpetofauna of Madras and its vicinity. *Newsl. Zool. Surv. India* 3(6):372–378.

5-332
Murthy, T. S. N. 1977.
   On sea snakes occurring in Madras waters. *J. Mar. Biol. Assoc. India* 19(1):68–72.

5-333
Nakamura, K.; Ueno, S. I. 1971.
   Japanese reptiles and amphibians in colour. Osaka: Hoikusha. (Reprint of 1963 edition.) (In Japanese)

5-334
Niceforo, M. H. 1942.
   Los ofidios de Colombia. *Rev. Acad. Colomb. Cienc. Exactas Fis. Nat.* 5(17):84–101.

5-335
Nikolskii, A. M. 1905.
   Herpetologia Rossica. St. Petersbourg. (Also in *Mem. Acad. Imp. Sci. St. Petersbourg. Cl. Phys.-Math.* Ser. 8, 17(1):1–517) (In Russian)

5-336
Nikolskii, A. M. 1916.
   Faune de la Russie et des pays limitrophes. Petrograd. Volume 2. (In Russian) (Translated 1964 by the Israel Program for Scientific Translations: *Fauna of Russia and adjacent countries.*)

5-337
Nishimura, S. 1965.
   The zoogeographical aspects of the Japan Sea, Part 1. *Publ. Seto Mar. Biol. Lab.* 13(1):35–79.

5-338
Nishimura, S. 1969.
   The zoogeographical aspects of the Japan Sea, Part V. *Publ. Seto Mar. Biol. Lab.* 17(2):67–142.

5-339
Novikov, N. P.; Khomenko. L. P. 1974.
About distribution of sea snakes (Hydrophiidae) off the Indian west coast. *Zool. Zh.* 53(6):951–953. (In Russian with English abstract)

5-339.5
Ochoterena, I. 1934.
Distribucion geografica de los animales de Mexico. *Resen. Cient. R. Soc. Esp. Hist. Nat.* 9:23–41

5-340
Ofuchi, M. 1934.
Pelamydrus platurus (Linnaeus) caught off Kinkazan Pt. *Dobutsugaku Zasshi* 46(554):537–539. (In Japanese)

5-341
Ogilby, J. D. 1898.
Ophidia and Pices, Pt. 2 of Thomas Steel's Contributions to a knowledge of the fauna of British New Guinea. *Proc. Linn. Soc. N.S. W.* 23:357–369.

5-342
Okada, Y. 1932.
Notes on the reptilia and amphibia of Kotosho (Botel Tobago). *Bull. Biogeogr. Soc. Jpn* 3(1):13–23.

5-343
Okada, Y. 1938.
A catalogue of vertebrates of Japan. Tokyo: Maruzen.

5-343.3
Okada, Y. 1938.
Reptiles of the Tohoku-district, the northern part of Honsyu, Japan. *Saito Ho-on Kai Mus. Nat. Hist. Res. Bull.* 15:67–84.

5-343.5
Okada, Y. 1953.
Reptiles. Uchida, K. *Illustrated encyclopedia of the fauna of Japan (exclusive of insects).* Tokyo: Hokuryukan. Pp. 223–271. (In Japanese)

5-344
Oliver, J. A.; Shaw, C. E. 1953.
The amphibians and reptiles of the Hawaiian Islands. *Zoologica (N.Y.)* 38(2):65–95.

5-345
Oliver, W. R. B. 1911.
   Notes on reptiles and mammals in the Kermadec Islands. *Trans. R. Soc. N.Z.* 43:535–539.

5-346
Orces, G. 1942.
   Los ofidios venenosos del Ecuador. *Flora (Quito)* 2(5–6):147–155.

5-346.5
Orces, G. 1948.
   Notas sobre los ofidios venenosos del Ecuador. *Filos. Let. Quito Univer. Cent. Ecuador* 1948(3):231–250.

5-347
Oshima, M. 1910.
   An annotated list of Formosan snakes, with descriptions of four new species and one new subspecies. *Annot. Zool. Jpn.* 7:185–207.

5-348
Oshima, M. 1914.
   Sea snakes of Formosa. *Zool. Mag. (Tokyo)* 26:423–434. (In Japanese)

5-348.5
Oshima, M. 1915.
   Additional notes on Formosan sea snakes. *Dobutsugaku Zasshi* 26:423–434.

5-348.7
Oshima, M. 1916.
   Key to the snakes of Formosa and the Riu-Kius. *Trans. Nat. Hist. Soc. Taiwan* 6(23):1–12. (In Japanese)

5-349
Oshima, M. 1920.
   Notes on the venomous snakes from the islands of Formosa and Riu-Kiu. *Taiwan Sotokufu Kenkyujo Hokoku (Annual Report of the Institute of Science, Government of Formosa)* 8(2):1–11.

5-350
Oshima, S. 1944.
   Explanations of poisonous snakes in the greater East Asia coprosperity sphere. Tokyo: Hokuryu-Kan. (In Japanese)

5-351
Ota, H.; Takahashi, H.; Kamezaki, N. 1985.
   On the specimens of yellow lipped sea krait *Laticauda colubrina* from the Yaeyama group, Ryukyu Archipelago. *Snake* 17(2):156–159.

5-352
Ouwens, P. A. 1916.
   De voornaamste giftslangen van Nederlandsch Oost-Indie. Leiden: E. J. Brill.

5-353
Palacky, J. 1898.
   La distribution des ophidiens sur le globe. *Mem. Soc. Zool. Fr.* 9:88–125.

5-354
Parenti, P.; Picaglia, L. 1886.
   Rettili ed anfibi raccolti da P. Parenti nel viaggio di circumnavigazione della R. Corvetta "Vettor Pisani", Comandante G. Palumbo, negli anni 1882–85 e da V. Ragazzi sulle coste del Mar Rosso e dell'America meridionale negli anni 1879–84. *Atti Soc. Nat. Mat. Modena* Ser. 3, 5:26–96.

5-355
Parker, H. W. 1926.
   The reptiles and batrachians of Gorgona Island, Colombia. *Ann. Mag. Nat. Hist.* Ser. 9, 17:549–554.

5-356
Patil, A. M. 1953.
   Study of the marine fauna of the Karwar Coast and the neighbouring islands, Part. IV: Echinodermata and other groups. *J. Bombay Nat. Hist. Soc.* 51(2):429–434.

5-356.3
Peron, F. 1815.
   Voyage de decouvertes au terres Australes, execute sur les corvettes le Geographe, le Naturaliste, et la goelette le Casuarina, pendant les annees 1800, 1801, 1802, 1803, et 1804; sous le commandement du capitaine de vaisseau N. Baudin. Paris: Imprimerie Royale. 2 volumes.

5-356.5
Peters, J. A. 1960.
   The snakes of Ecuador. A check list and key. *Bull. Mus. Comp. Zool.* 122(9):491–541.

5-357
Peters, J. A.; Orejas Miranda, B. 1970.
   Catalogue of the neotropical squamata, Part I. Snakes. *Bull. U.S. Natl Mus.*
No. 297:1–347. (Reprinted in 1986)

5-358
Peters, W. 1859.
   Ueber die von Hrn. Dr. Hoffmann in Costa Rica gesammelten und an das
konigl. zoologische museum gesandten schlangen. *Monatsber. K. Preuss. Akad.
Wiss. Berl.* 1859:275–278.

5-359
Peters, W. 1861.
   Eine zweite uebersicht (vergl. Monatsberichte 1859, p.269) der von Hrn. F.
Jagor auf Malacca, Java, Borneo und den Philippinen gesammelten und dem kgl.
zoologischen museum uebersandten schlangen. *Monatsber. Deut. Akad. Wiss. Berl.*
1861:683–691.

5-360
Peterson, H. W.; Smith, H. M. 1973.
   Observations on sea snakes in the vicinity of Acapulco, Guerrero, Mexico.
Bull. Chic. Herpetol. Soc. 8(3–4):29.

5-361
Phillipps, W. J. 1941.
   Recent occurrence of *Pelamis platurus* Linn. in New Zealand. *Trans. R. Soc.
N.Z.* 71(1):23–24.

5-362
Phillips, W. W. A. 1958.
   Some observations on the fauna of the Maldive Islands. Part IV. Amphibians
and reptiles. *J. Bombay Nat. Hist. Soc.* 55(2):217–220.

5-363
Phipson, H. M. 1887.
   The poisonous snakes of the Bombay Presidency. *J. Bombay Nat. Hist. Soc.*
2:244–250.

5 363.5
Phisalix, M.; Houdemer, E. 1934.
   Contribution a la faune venimeuse du Tonkin. *Bull. Mus. Natl Hist. Nat.*
6(2):171–176.

5-364

Picado Twight, C. 1931.

Serpientes venenosas de Costa Rica, sus venenos seroterapia anti-ofidica. San Jose, Costa Rica: Sauter, Arias & Co. (Reissued in 1976 by Editorial Universidad de Costa Rica)

5-365

Pickwell, G. V.; Bezy, R. L.; Fitch, J. E. 1983.

Northern occurrences of the sea snake, *Pelamis platurus*, in the eastern Pacific, with a record of predation on the species. *Calif. Fish Game* 69(3):172–177.

5-366

Ping, C. 1926.

A sea snake from Yenting *Disteira* sp. *Trans. Sci. Soc. China* 4:45–47.

5-367

Pope, C. H. 1929.

Notes on reptiles from Fukien and other Chinese provinces. *Bull. Am. Mus. Nat. Hist.* 58:335–487.

5-368

Pope, C. H. 1934.

List of Chinese turtles, crocodilians, and snakes, with keys. *Am. Mus. Novitates* No. 733:1–29.

5-369

Pope, C. H. 1935.

The reptiles of China. New York: American Museum of Natural History. (Also published as *Central asiatic expeditions of the American Museum of Natural History.* New York. Vol. 10:1–604)

5-370

Prater, S. H. 1924.

The snakes of Bombay Island and Salsette. *J. Bombay Nat. Hist. Soc.* 30:151–176.

5-371

Rai, M. 1969.

Sur deux serpents recoltes en Iran *Hydrophis spiralis* Gray et *Hydrophis ornatus* Gunther. *Actes Soc. Linn. Bordeaux* Ser.A. 106(4):3 pages. (Reprinted in *Bull. Stn Biol. Arcachon* (N. S.) No. 22:unpaged)

5-372
Raj, B. S. 1927.
   The vertebrate fauna of Krusadai Island. *Bull. Madras Gov. Mus. (Nat. Hist. Sect.)* N.S. 1(1):181–183.

5-372.3
Rasmussen, A. R. 1987.
   Persian Gulf sea snake *Hydrophis lapemoides* (Gray): new records from Phuket Island, Andaman Sea, and the southern part of the Straits of Malacca. *Nat. Hist. Bull. Siam Soc.* 35:57–58.

5-372.5
Rasmussen, A. R. 1989.
   An analysis of *Hydrophis ornatus* (Gray), *H. lamberti* Smith, and *H. inornatus* (Gray) (Hydrophiidae, Serpentes) based on samples from various localities, with remarks on feeding and breeding biology of *H. ornatus*. *Amphib.-Reptilia* 10:397–417.

5-372.7
Rasmussen, A. R. 1989.
   Sea snakes *Thalassophina viperina* (Schmidt) and *Laticauda laticaudata* (Linne): new records from Phuket Island, Andaman Sea, with remarks on subspecies of *L. laticaudata*. *Nat. Hist. Bull. Siam Soc.* 37(1):99–103.

5-373
Redfield, J. A.; Holmes, J. C.; Holmes, R. D. 1978.
   Sea snakes of the Eastern Gulf of Carpentaria. *Aust. J. Mar. Freshwater Res.* 29:325–334.

5-373.5
Restrepo, J. H. 1986.
   Las serpientes. Prahl, H. von; Alberico, M., et al. *Isla d Gorgona*. Bogota: Textos Universitorios.

5-374
Reynolds, R. P.; Pickwell, G. V. 1984.
   Records of the yellow-bellied sea snake, *Pelamis platurus* from the Galapagos Islands. *Copeia* 1984(3):786–789.

5-375
Rodas C., J. T. 1938.
   Contribucion al estudio de las serpientes venenosas de Guatemala. Guatemala: Tipografia Nacional.

5-376
Romer, J. D. 1954.
   Notes on sea snakes (Hydrophiidae) occurring in or near Hong Kong territorial waters. *Hong Kong Univ. Fish. J.* No. 1:35–37.

5-377
Romer, J. D. 1958.
   Occurrence of the sea snake *Hydrophis ornatus ornatus* (Gray) near Hong Kong. *Hong Kong Univ. Fish. J.* 2:129.

5-378
Romer, J. D. 1961.
   Annotated checklist with keys to the snakes of Hong Kong. *Mem. Hong Kong Nat. Hist. Soc.* 5:1–14.

5-379
Romer, J. D. 1970.
   Revised annotated checklist with keys to the snakes of Hong Kong. *Mem. Hong Kong Nat. Hist. Soc.* No. 8:1–22.

5-380
Romer, J. D. 1979.
   Second revised annotated checklist with keys to the snakes of Hong Kong. *Mem. Hong Kong Nat. Hist.* Soc. No. 14:1–23.

5-381
Rooij, N. de. 1913.
   Praeda itineris a L. F. de Beaufort in Archipelago indico facti annis 1909–1910. III. Reptilien. *Bijdr. Dierkd.* 19:15–20.

5-381.5
Rooij, N. de. 1915.
   Reptiles. Kleiweg de Zwaan, J. P. *Die insel Nias bei Sumatra.* Haag: Martinus Nijhoff. Vol. 3, pp. 282–307.

5-382
Rooij, N. de. 1917.
   The reptiles of the Indo-Australian Archipelago. Leiden: E. J. Brill.

5-383
Rooij, N. de. 1922.
   Fauna Simalurensis. Reptilia. *Zool. Meded. (Leiden)* 6(4):217–238.

5-383.5
Rosenfeld, G. 1963.
   Unfaelle durch giftschlangen. Die Giftschlangen der erde wirkungen und antigenitat der gifte therapie von giftschlangenbissen. Der farbwerke hoechst AG zum 100 jahrigen bestehen gewidmet. Marburg-Lahn: Elwert. Pp. 161–202.

5-384
Roux, J. 1910.
   Reptilien und amphibien den Aru- und Kei-Inseln. *Abh. Senckenb. Naturforsch. Ges.* 33(3):211–247.

5-385
Roux, J. 1913.
   Les reptiles de la Nouvelle-Caledonie et des iles Loyalty. Sarasin, F.; Roux, J. *Nova Caledonia.* Wiesbaden: C. W. Kreidels Verlag, 1913–1926. Vol. 1A, pt. 1, pp. 79–160.

5-386
Russell, P. 1796.
   An account of Indian serpents, collected on the coast of Coromandel; containing descriptions and drawings of each species; together with experiments and remarks on their several poisons. London: George Nicol.

5-387
Russell, P. 1801.
   A continuation of An account of Indian serpents, containing descriptions and figures, from specimens and drawings, transmitted from various parts of India. London: G. and W. Nicol.

5-388
Sachet, M. H. 1962.
   Monographie physique et biologique de l'ile Clipperton. *Ann. Inst. Oceanogr.* 40(1):1–108.

5-389
Saint Girons, H. 1964.
   Notes sur l'ecologie et la structure des populations des Laticaudinae (Serpents, Hydrophidae) en Nouvelle Caledonie. *Terre et la Vie* 2:185–214.

5-390
Saint Girons, H. 1972.
   Les serpents du Cambodge. *Mem. Mus. Natl Hist. Nat. Ser.A Zool.* 74:1–170.

5-390.3
Sauvage, M. H. E. 1878,
    Essai sur la faune herpetologique de la Nouvelle-Guinee, suivi de la description de quelques especes nouvelles ou peu connues. *Bull. Soc. Philomath. Paris* 2(7):25–44.

5-390.5
Sauvage, M. H. E. 1879.
    Note sur les geckotiens de la Nouvelle-Caledonie. *Bull. Soc. Philomath. Paris* 3(7):63–73.

5-390.7
Sauvage, M. H. E. 1879.
    Notice sur quelques reptiles nouveaux ou peu connus de la Nouvelle-Guinee. *Bull. Soc. Philomath. Paris* 3(7):47–48.

5-391
Savage, J. M. 1970.
    On the trail of the golden frog: with Warszewicz and Gabb in Central America. *Proc. Calif. Acad. Sci.* Ser. 4, 38(14):273–288.

5-392
Sawai, Y.; Koba, K. 1971.
    The venomous snakes of Malaysia. *Snake* 3:129–152. (In Japanese with English abstract)

5-393
Sayed, S. M. 1972.
    Check list of marine fauna of Pakistan. *Rec. Zool. Surv. Pakistan. Marine Fauna Suppl.* 4(1–2):1–68.

5-394
Schleich, H. H. 1977.
    Distributional maps of reptiles of Iran. *Herpetol. Rev.* 8(4):126–129.

5-395
Schmidt, K. P. 1927.
    The reptiles of Hainan. *Bull. Am. Mus. Nat. Hist.* 54:395–465.

5-396
Schmidt, K. P. 1932.
    Reptiles and amphibians from the Solomon Islands. *Publ. Field. Mus. Nat. Hist. Zool. Ser.* 18(9):175–190.

5-397
Schmidt, K. P. 1933.
Amphibians and reptiles collected by the Smithsonian Biological Survey of the Panama Canal Zone. *Smithson. Misc. Collect.* 89(1):1–20.

5-398
Schmidt, K. P. 1939.
Reptiles and amphibians from Southwestern Asia. *Field Mus. Nat. Hist. Publ. Zool. Ser.* 24(7):49–92.

5-398.3
Schnee, S. 1902.
Die kriechtiere der marshallinseln. *Zool. Garten* 43(11):354–362.

5-398.5
Schnee, S. 1906.
Die schlangeninsel bei Herbertshoehe. *Aus der Natur (Berlin)* 1(21):669–670.

5-399
Scortecci, G. 1939.
Gli ofidi velenosi dell'Africa Italiana. Milano: Istituto Sieroterapico Milanese.

5-399.5
Scott, E. O. G. 1932.
On the occurrence in Tasmania of *Hydrophis ornatus*, variety *ocellatus*; with a note on *Pelamis platurus* (=*Hydrus platurus*). *Pap. Proc. R. Soc. Tasm.* 1931:111.

5-400
Scott, F.; Parker, F.; Menzies, J. I. 1977.
A checklist of the amphibians and reptiles of Papua New Guinea. *Papua New Guinea Wildl. Br. Dept. Nat. Res. Wildl. Publ.* 77(3):1–18.

5-401
Scott, N. J. 1969.
A zoogeographic analysis of the snakes of Costa Rica. Los Angeles, CA: University of Southern California. 390 pp. (Dissertation)

5-401.5
Sefton, A. R. 1964.
Natural history of Illawarra. 5. Snakes of the Illawarra District. *Illawarra Nat. Hist. Soc.* 1964:3 pages.

5-402
Shannon, F. A. 1956.
    The reptiles and amphibians of Korea. *Herpetologica* 12:22–49.

5-403
Sharma, R. C. 1976.
    Records of the reptiles of Goa. *Rec. Zool. Surv. India* 71:149–167.

5-404
Sharma, R. C. 1982.
    Taxonomic and ecological studies on the reptiles of Gujarat. *Rec. Zool. Surv. India* 80:85–106.

5-405
Shelford, R. 1901.
    A list of the reptiles of Borneo. *J. Straits Br. R. Asiat. Soc.* 35:43–68.

5-406
Shibata, Y. 1960.
    Amphibia and reptilia collected from Tokara Islands. *Bull. Osaka Mus. Nat. Hist.* No. 12:57–62.

5-407
Shockley, C. H. 1949.
    Herpetological notes for Ras Jiunri, Baluchistan. *Herpetologica* 5(6):121–123.

5-408
Shuntov, V. P. 1965.
    Morskie zmei (Hydrophiidae) Tonkinskogo (Severo-V'etnamskogo) Zaliva. Sea snakes (Hydrophiidae) of the Gulf of Tonkin (Northern Viet-Nam). (Trans-240 by the U.S. Naval Oceanographic Office of *Zool. Zh.* 41(8):1203–1209, 1962)

5-409
Shuntov, V. P. 1966.
    On the distribution of sea snakes in the South-Chinese Sea and East Indian Ocean. *Zool. Zh.* 45:1882–1886. (In Russian with English abstract)

5-410
Shuntov, V. P. 1972.
    Sea snakes of the North Australian shelf. *Sov. J. Ecol.* 4:338–344. (Translation of *Ekologiya* 4:65–72, 1971)

5-410.3
Siebold, P. F. 1824.
   De historiae naturalis in Japonia statu, nec non de augmento emolumentisque in decursu perscrutationum exspectandis dissertatio, cui accedunt spicilegia faunae Japonicae. Bataviae.

5-410.4
Siebold, P. F. 1826.
   De historiae naturalis in Japonia statu, nec non de augmento emolumentisque in decursu perscrutationum exspectandis dissertatio, cui accedunt spicilegia faunae Japonicae. Wirceburgi: Ex officina literaria C. P. Bonitas.

5-410.5
Sinclair, S. 1890.
   Guide to the contents of the Australian Museum. Sydney.

5-411
Slevin, J. R. 1926.
   Expedition to the Revillagigedo Islands, Mexico, in 1925. III. Notes on a collection of reptiles and amphibians from the Tres Marias and Revillagigedo Islands, and west coast of Mexico, with description of a new species of *Tantilla*. *Proc. Calif. Acad. Sci.* Ser. 4, 15(3):195–207.

5-412
Slevin, J. R. 1934.
   The Templeton Crocker Expedition to western Polynesian and Melanesian islands, 1933. No. 15. Notes on the reptiles and amphibians, with the description of a new species of sea-snake. *Proc. Calif. Acad. Sci.* Ser. 4, 21(15):183–188.

5-413
Slevin, J. R. 1935.
   An account of the reptiles inhabiting the Galapagos Islands. *Bull. N.Y. Zool. Soc.* 38(1):2–24.

5-414
Slevin, J. R. 1937.
   Contributions to oriental herpetology. V. Honshu or Hondo, the neighboring islands of Sado and Awaji, and the seven islands of Idzu. *Proc. Calif. Acad. Sci.* Ser. 4, 23(11):175–190.

5-415
Smedley, N. 1931.
   Notes on some Malaysian snakes. *Bull. Raffles Mus.* 5:49–54.

5-416
Smith, H. M. 1943.
Summary of the collections of snakes and crocodilians made in Mexico under the Walter Rathbone Bacon Traveling Scholarship. *Proc. U.S. Natl Mus.* 93(3169):393–504.

5-416.5
Smith, H. M. 1958.
Handlist of the snakes of Panama. *Herpetologica* 14:222–224.

5-417
Smith, H. M.; Mittleman, M. B. 1943.
Notes on the Mansfield Museum's Mexican reptiles collected by Wilkinson. *Trans. Kans. Akad. Sci.* 46:243–249.

5-418
Smith, L. A. 1974.
The sea snakes of Western Australia (Serpentes: Elapidae, Hydrophiinae) with a description of a new subspecies. *Rec. West. Aust. Mus.* 3(2):93–110.

5-419
Smith, M. 1915.
List of the snakes at present known to inhabit Siam. *J. Nat. Hist. Soc. Siam (Bangkok)* 1:211–215.

5-420
Smith, M. A. 1915.
Notes on some snakes from Siam. *J. Bombay Nat. Hist. Soc.* 23:784–789.

5-421
Smith, M. 1915.
The snakes of Bangkok. *J. Nat. Hist. Soc. Siam Suppl.* 1(3):173–185.

5-422
Smith, M. A. 1916.
Note on a rare sea snake (*Thalassophis anomalus*) from the coast of Siam. *J. Nat. Hist. Soc. Siam Suppl.* 2(2):176–177.

5-423
Smith, M. A. 1920.
On sea snakes from the coasts of the Malay Peninsula, Siam and Cochin-China. *J. Fed. Malay. Mus.* 10(1):1–63.

5-424

Smith, M. 1926.

Monograph of the sea snakes (Hydrophiidae). London: British Museum (Natural History).

5-425

Smith, M. A. 1930.

The reptilia and amphibia of the Malay Peninsula from the Isthmus of Kra to Singapore including the adjacent islands. *Bull. Raffles Mus.* No. 3:1–149 (A supplement to Boulenger, G. A. *Reptilia and Batrachia*, 1912)

5-426

Smith, M. 1931.

Description of a new genus of sea-snake from the Coast of Australia, with a note on the structures providing for complete closure of the mouth in aquatic snakes. *Proc. Zool. Soc. Lond.* 1931:397–398.

5-427

Smith, M. A. 1935.

The sea snakes (Hydrophiidae). *Dana-Rep. Carlsberg Found.* No. 8:1–6.

5-428

Smith, M. A. 1941.

The herpetology of the Andaman and Nicobar Islands. *Proc. Linn. Soc. Lond.* 1941:150–158.

5-429

Smith, M. A. 1943.

The fauna of British India, Ceylon and Burma, including the whole of the Indo-Chinese sub-region. London: Taylor and Francis. Volume 3.

5-430

Smith, M.; Kloss, C. B. 1915.

On reptiles and batrachians from the coast and islands of South-East Siam. *J. Nat. Hist. Soc. Siam (Bangkok)* 1(4):237–249.

5-430.5

Soderberg, P. 1967.

Notes on a collection of herpetological specimens recently donated to the Centre for Thai National Reference Collections. *Nat. Hist. Bull. Siam. Soc.* 22(1–2):151–166.

5-430.7
Sowerby, A. de C. 1943.
   Amphibians and reptiles recorded from or known to occur in the Shanghai area.
Shanghai: Universite l'Aurore. (also cited as Musee Heude, Chang-Hai, *University
L'Aurore Notes d'Herpetologie* 1:1–16)

5-431
Stanley, A. 1914.
   The collection of Chinese reptiles in the Shanghai Museum. *J. N. China Br.
R. Asiat. Soc.* 45:21–31.

5-431.5
Stebbins, R. C. 1985.
   A field guide to western reptiles and amphibians. Boston: Houghton Mifflin.

5-432
Stejneger, L. 1898.
   On a collection of batrachians and reptiles from Formosa and adjacent islands.
J. Coll. Sci. Imp. Univ. Tokyo 12:215–225.

5-433
Stejneger, L. 1899.
   The land reptiles of the Hawaiian Islands. *Proc. U.S. Natl Mus.* 21(1174):783–
813.

5-434
Stejneger, L. 1901.
   Diagnoses of eight new batrachians and reptiles from the Riu Kiu Archipelago,
Japan. *Proc. Biol. Soc. Wash.* 14:189–191.

5-435
Stejneger, L. 1907.
   Herpetology of Japan and adjacent territory. *Bull. U.S. Natl Mus.* No. 58:1–
557.

5-436
Stejneger, L. 1911.
   The batrachians and reptiles of Formosa. *Proc. U.S. Natl Mus.* 38(1731):91–
114.

5-436.5
Sternfeld, R. 1913.
   Beitrage zur schlangenfauna Neuguineas und der benachbarten inselgruppen.
Sitz. Ges. Nat. Freunde Berlin 1913:384–389.

5-437.5
Sternfeld, R. 1920.
Zur tiergeographie Papuasiens und der pazifischen Inselwelt. *Abh. Senckenb. Naturforsch. Ges.* 36(4):375–436.

5-438
Stoliczka, F. 1870.
Observations on some Indian and Malayan amphibia and reptilia. *J. Asiat. Soc. Beng.* 39(3):159–228.

5-439
Stoliczka, F. 1872.
Notes on reptiles, collected by Surgeon F. Day in Sind. *Proc. Asiat. Soc. Beng.* 1872:85–92.

5-440
Strauch, A. 1873.
Die schlangen des Russischen Reichs in systematischer und zoogeographischer beiziehung. *Mem. Acad. Imp. Sci. St. Petersbourg* Ser. 7, 21(4):1–287.

5-441
Stuart, L. C. 1963.
A checklist of the herpetofauna of Guatemala. *Misc. Publ. Mus. Zool. Univ. Mich.* No. 122:1–150.

5-442
Sutherland, S. K. 1983.
Australian animal toxins: the creatures, their toxins and care of the poisoned patient. New York: Oxford University Press.

5-443
Suvatti, C. 1950.
Fauna of Thailand. Bangkok, Thailand: Department of Fisheries.

5-444
Swinhoe, R. 1863.
A list of the Formosan reptiles; with notes on a few of the species, and some remarks on a fish (*Orthagoriscus*, sp.). *Ann. Mag. Nat. Hist.* Ser. 3, 12:219–226.

5-445
Swinhoe, R. 1870.
Note on reptiles and batrachians collected in various parts of China. *Proc. Zool. Soc. Lond.* 1870:409–412.

5-446
Sworder, G. H. 1923.
    A list of the snakes of Singapore Island. *Singapore Nat.* 2:55–73.

5-447
Takahashi, H. 1981.
    The frequency of sea snakes in the Ryukyu Islands. *Jpn. J. Herpetol.* 9(2):63.
(In Japanese) (Abstract)

5-448
Takahashi, H. 1984.
    The number and distribution of the sea snakes observed in the Ryukyu Islands,
southern Japan. *Snake* 16:71–74. (In Japanese with English abstract)

5-449
Takahashi, S. 1935.
    Notes on Hydridae of Kotosho (Botel Tobago Island). *Trans. Nat. Hist. Soc.
Taiwan* 25:142–144. (In Japanese)

5-449.5
Takashima, H. 1958.
    A synopsis of the reptiles of Japan. *Yamashina Chorui Kenkyujo, Tokyo*
12:486–493.

5-450
Talukdar, S. K.; Dattagupta, B. 1980.
    Notes on the occurrence of the sea snakes, *Hydrophis mamillaris* (Daudin)
and *Microcephalophis gracilis* (Shaw) from West Bengal. *J. Mar. Biol. Assoc. India*
18(2):389–391.

5-451
Tanner, V. M. 1951.
    Pacific Islands herpetology, No. V. Guadalcanal, Solomon Islands: a check list
of species. *Great Basin Nat.* 11(3–4):53–86.

5-452
Tayless, J. 1968.
    Serpents in paradise. *Pac. Discovery* 21(3):24–26.

5-452.5
Taylor, E. H. 1917.
    Snakes and lizards known from Negros, with descriptions of new species and
new subspecies. *Philipp. J. Sci. Sect. D. Gen. Biol., Ethnol., Anthropol.* 12(6):353–
381.

5-453

Taylor, E. H. 1918.

Reptiles of Sulu Archipelago. *Philipp. J. Sci. Sec. D. Gen. Biol., Ethnol., Anthropol.* 13:233–276.

5-454

Taylor, E. H. 1922.

The snakes of the Philippine Islands. Manila: Bureau of Printing.

5-455

Taylor, E. H. 1923.

Additions to the herpetological fauna of the Philippine Islands, III. *Philipp. J. Sci.* 22:515–555.

5-455.5

Taylor, E. H. 1928.

Amphibians, lizards, and snakes of the Philippines. Dickerson, R. E. *Distribution of life in the Philippines.* Manila: Bureau of Printing. Pp. 214–241.

5-456

Taylor, E. II. 1950.

The snakes of Ceylon. *Univ. Kans. Sci. Bull.* 33(2):519–603.

5-457

Taylor, E. H. 1951.

A brief review of the snakes of Costa Rica. *Univ. Kans. Sci. Bull.* 34(1):1–188.

5-458

Taylor, E. H. 1953.

Early records of the seasnake *Pelamis platurus* in Latin America. *Copeia* 1953(2):124.

5-459

Taylor, E. H. 1965.

The serpents of Thailand and adjacent waters. *Univ. Kans. Sci. Bull.* 45(9):609–1096.

5-460

Taylor, E. H.; Elbel, R. E. 1958.

Contribution to the herpetology of Thailand. *Univ. Kans. Sci. Bull.* 38(2):1033–1189.

5-461
Taylor, R. T.; Flores, G.; Flores, A.; Bolanos, R. 1974.
Geographical distribution of Viperidae, Elapidae and Hydrophidae in Costa Rica. *Rev. Biol. Trop.* 21(2):383–397.

5-462
Tchang, T. L. 1932.
Notes on some Chinese snakes. *Bull. Fan. Meml Inst. Biol. (Peiping)* 3(1):1–23.

5-463
Temminck, C. J.; Schlegel, H. 1838.
Reptilia elaborantibus. Siebold, P. F. *Fauna Japonica.* Amstelodami: J. Mueller.

5-464
Tennent, J. E. 1860.
Ceylon: an account of the island: physical, historical, and topographical with notices of its natural history, antiquities and productions. London: Longman, Green, Longman, and Roberts. Volume 1.

5-465
Terron, C. C. 1932.
Los coralillos mexicanos. *An. Inst. Biol. Univ. Nac. Auton. Mex.* 3(1):5–14.

5-466
Testi, F. 1934.
Nota su di un ofidio velenoso della Somalia della sottofamiglia delle "Hydrophiinae". *Arch. Sci. Med. Colon.* 14(2):81–85. (In Italian with English abstract)

5-467
Theobald, W. 1868.
Catalogue of the reptiles of British Birma, embracing the provinces of Pegu, Martaban, and Tenasserim; with descriptions of new or little-known species. *J. Linn. Soc. Lond. Zool.* 10:4–67.

5-468
Tinker, S. W. 1980.
A list of the amphibians, reptiles and mammals of the Hawaiian Islands (exclusive of the whales). Honolulu, HI: Published by the author.

5-469

Tirant, G. 1884.

Notes sur les reptiles de la Cochinchine et du Cambodge. *Excur. Reconn.* No. 20:387–428.

5-470

Tiwari, S. K. 1985.

Readings in Indian zoogeography. New Delhi: Today & Tomorrow's Printers and Publishers. Volume 1.

5-471

Toriba, M.; Sawai, Y. 1981.

New record of the Persian Gulf seasnake, *Hydrophis lapemoides* from Malaysia. *Jpn. J. Herpetol.* 9(2):63–64. (In Japanese) (Abstract)

5-472

Toriba, M.; Sawai, Y. 1981.

New record of Persian Gulf seasnake, *Hydrophis lapemoides* (Gray), from Penang, Malaysia. *Snake* 13:134–136.

5-473

Troschel, H. 1865.

Amphibia. Mueller, J. W. von. *Reisen in den Vereinigten staaten, Canada, und Mexico: III. Beitrage zur geschichte, statistik und zoologie von Mexico.* Leipzig: Brodhaus. Pp. 595–620.

5-473.5

Tu, A. T. 1970.

Sea snake (Hydrophiidae) collection and venoms. *Herpetol. Rev.* 2:7. (Abstract)

5-474

Tu, A. T. 1974.

Sea snake investigation in the Gulf of Thailand. *J. Herpetol.* 8(3):201–210.

5-475

Tu, A. T. 1976.

Investigation of the sea snake, *Pelamis platurus* (Reptilia, Serpentes, Hydrophiidae), on the Pacific coast of Costa Rica, Central America. *J. Herpetol.* 10(1):13–18.

5-476

Tu, A. T.; Stringer, J. M. 1973.

Three species of sea snakes not previously reported in the Strait of Formosa. *J. Herpetol.* 7(4):384–386.

5-477
Tuck, R. G. 1971.
   Amphibians and reptiles from Iran in the United States National Museum collection. *Bull. Md Herpetol. Soc.* 7(3):48–86.

5-477.5
Tweedie, M. W. F. 1940.
   Notes on Malayan reptiles. *Bull. Raffles Mus.* 16:83–87.

5-478
Tweedie, M. W. F. 1953.
   The snakes of Malaya. Singapore: Government Printing Office.

5-479
Tweedie, M. W. F. 1957.
   The snakes of Malaya. 2nd ed. Singapore: Government Printing Office.

5-480
Tweedie, M. W. F. 1983.
   The snakes of Malaya. 3rd ed. Singapore: Singapore National Printers.

5-481
Underwood, G. 1979.
   Classification and distribution of venomous snakes in the world. Lee, C. Y., ed. *Snake venoms.* New York: Springer-Verlag. Pp. 15–40.

5-482
Van Denburgh, J. 1912.
   Expedition of the California Academy of Sciences to the Galapagos Islands, 1905–1906. IV. The snakes of the Galapagos Islands. *Proc. Calif. Acad. Sci.* Ser. 4, 1:323–374.

5-483
Van Denburgh, J. 1922.
   The reptiles of western North America. An account of the species known to inhabit California and Oregon, Washington, Idaho, Utah, Nevada, Arizona, British Columbia, Sonora and Lower California. San Francisco: California Academy of Sciences. Volume 2.

5-484
Van Denburgh, J.; Slevin, J. R. 1914.
   Reptiles and amphibians of the Islands of the West Coast of North America. *Proc. Calif. Acad. Sci.* Ser. 4, 4:129–152.

5-485
Villa, J. 1962.
Las serpientes venenosas de Nicaragua. Managua.

5-486
Villa, J. 1983.
Peces, anfibios y reptiles nicaraguenses: Lista y bibliografia; Nicaraguan fishes, amphibians & reptiles: a checklist and bibliography. Managua, Nicaragua: Universidad Centroamericana.

5-487
Villa, J. 1984.
The venomous snakes of Nicaragua: a synopsis. *Milw. Public Mus. Contrib. Biol. Geol.* 59:1–41.

5-488
Viquez S., C. 1935.
Animales venenosos de Costa Rica. San Jose, Costa Rica: Imprenta Nacional.

5-489
Viquez Segreda, C. 1940.
Nuestros animales venenosos. San Jose, Costa Rica: Imprenta Nacional.

5-490
Viquez Segreda, C. 1942.
Distribucion geografica de nuestras serpientes venenosas. *Rev. Inst. Def. Cafe Costa Rica* 11(87):608–611.

5-491
Visser, J. 1966
Poisonous snakes of southern Africa and the treatment of snakebite. Cape Town: Howard Timmons.

5-491.5
Vogt, T. 1913.
Ueber die reptilien- und amphibienfauna der Insel Hainan. *Sitz. Ges. Nat. Freunde Berlin* 1913:222–229.

5-492
Volsoe, H. 1939.
The sea snakes of the Iranian Gulf and the Gulf of Oman. With a summary of the biology of the sea snakes. Jessen, K., ed. *Danish Scientific Investigations in Iran.* Copenhagen: Ejnar Munksgaard. Vol. 1, pp. 9–45.

5-493
Volsoe, H. 1958.
Additional specimens of snakes from Rennell Island. Wolff, T., ed. *The natural history of Rennell Island, British Solomon Islands*. Copenhagen: Danish Scientific Press. Vol. 1, pp. 227–228.

5-494
Volsoe, H. 1958.
Herpetology of Rennell Island. Wolff, T., ed. *The natural history of Rennell Island, British Solomon Islands*. Copenhagen: Danish Scientific Press. Vol. 1, pp. 121–134.

5-494.5
Volz, W. 1905.
Schlangen von Palembang (Sumatra). *Zool. Jahrb. Abt. Syst. Geogr. Biol. Thiere* 20(5):491–508.

5-495
Voris, H. K. 1964.
Notes on the sea snakes of Sabah. *Sabah Soc. J.* 2(3):140–141.

5-496
Voris, H. 1970.
Survey of potential study areas. San Diego, CA: *Alpha Helix Research Program: 1969–1970*. Page 36. (Abstract)

5-497
Voris, H. 1985.
Population size estimates for a marine snake (*Enhydrina schistosa*) in Malaysia. *Copeia* 1985(4):955–961.

5-498
Voris, H.; Kropach, C.; Henderson, B. 1970.
Population biology of *Pelamis*. San Diego, CA: *Alpha Helix Research Program: 1969–1970*. Page 36. (Abstract)

5-499
Wagner, P., 1984.
Poisonous sea snake comes ashore at Lanikai, then dies. *Honolulu Star Bull.* Sept. 20:A-12.

5-500
Waite, E. R. 1929.
The reptiles and amphibians of South Australia. Adelaide: Harrison Weir.

5-501

Walczak, P. 1978.

First record of a black and yellow sea snake in the Yemen Arab Republic. *Br. Herpetol. Soc. Newsl.* No. 18:7.

5-502

Wall, F. 1903.

A prodromus of the snakes hitherto recorded from China, Japan, and the Loo Choo Islands; with some notes. *Proc. Zool. Soc. Lond.* 1903:84–102.

5-503

Wall, F. 1904.

Occurrence of a rare sea snake (*Distira gillespiae*) on the Malabar coast. *J. Bombay Nat. Hist. Soc.* 15:723–726.

5-504

Wall, F. 1905.

Notes on a collection of snakes from Japan and the Loo Choo Islands. *Proc. Zool. Soc. Lond.* 1905(2):511–517.

5-505

Wall, F. 1905.

Notes on snakes collected in Cannanore from 5th November 1903 to 5th August, 1904. *J. Bombay Nat. Hist. Soc.* 16:292–317.

5-506

Wall, F. 1906.

A descriptive list of the sea snakes (Hydrophiidae) in the Indian Museum, Calcutta. *Mem. Asiat. Soc. Bengal* 1906:277–299.

5-507

Wall, F. 1909

A monograph of the sea-snakes (Hydrophiinae). *Mem. Asiat. Soc. Bengal* 2(8):169–215.

5-508

Wall, F. 1913.

Some new snakes from the oriental region. *J. Bombay Nat. Hist. Soc.* 22:514–516.

5-509

Wall, F. 1918.

Notes on a collection of sea snakes from Madras. *J. Bombay Nat. Hist. Soc.* 25:599–607.

5-510
Wall, F. 1921.
   Ophidia taprobanica or the snakes of Ceylon. Colombo, Ceylon: H. R. Cottle,
Government Printer.

5-511
Wall, F. 1924.
   A hand-list of the snakes of the Indian empire. *J. Bombay Nat. Hist. Soc.*
30(2):12–24.

5-512
Wall, F. 1926.
   Snakes collected in Burma in 1925. *J. Bombay Nat. Hist. Soc.* 31:558–566.

5-513
Wall, F.; Evans, G. H. 1900.
   Notes on ophidia collected in Burma from May to December, 1899. *J. Bombay
Nat. Hist. Soc.* 13:343–354.

5-514
Wall, F.; Evans, G. H. 1901.
   Burmese snakes. Notes on specimens including 45 species of ophidian fauna
collected in Burma from 1st January to 30th June, 1900. *J. Bombay Nat. Hist. Soc.*
13:611–620.

5-516
Wallace, A. R. 1876.
   *The geographical distribution of animals with a study of the relations of living
and extinct faunas as elucidating the past changes of the earth's surface.* New York:
Harper & Brothers. Volumes 1 and 2.

5-517
Wang, C. S. 1962.
   The reptiles of Botel-Tobago. *Q.J. Taiwan Mus.* (*Taipei*) 15(3–4):141–191.

5-518
Wang, C. S.; Wang, Y. H. M. 1956.
   The reptiles of Taiwan. *Q.J. Taiwan Mus.* (*Taipei*) 9(1):1–86.

5-518.2
Welch, K. R. G. 1982.
   *Herpetology of Africa: a checklist and bibliography of the orders Amphisbaenia,
Sauria and Serpentes.* Melbourne, FL: Krieger Publishing Co.

5-518.3
Welch, K. R. G. 1983.
   Herpetology of Europe and southwest Asia: a checklist and bibliography of the orders of Amphisbaenia, Sauria and Serpentes. Malabar, FL: Krieger Publishing Co.

5-518.5
Welch, K. R. G. 1989.
   Snakes of the Orient: a checklist. Malabar, FL: Robert E. Krieger.

5-519
Wells, R. W.; Wellington, C. R. 1983.
   A synopsis of the class Reptilia in Australia. *Aust. J. Herpetol.* 1(3–4):73–129.

5-520
Werler, J. E.; Keegan, H. L. 1963.
   Venomous snakes of the Pacific area. Keegan, H. L.; MacFarlane, W. V., eds. *Venomous and poisonous animals and noxious plants of the Pacific region.* New York: Macmillan. Pp. 219–325.

5-521
Werner, F. 1895.
   Ueber eine sammlung von reptilien aus Persien, Mesopotamien und Arabien. *Verh. Zool.-Bot. Ges. Wien* 45:13–22.

5-522
Werner, F. 1899.
   Beitrage zur herpetologie der Pacifischen Inselwelt und von Kleinasien. *Zool. Anz.* 22:371–378.

5-523
Werner, F. 1900.
   Die reptilien- und batrachierfauna des Bismarck-Archipels. *Mitt. Zool. Samml. Mus. Naturk. Berl.* 1(4):1–132.

5-524
Werner, F. 1909.
   Reptilia exkl. Geckonidae und Scincidae. Michaelson, W.; Hartmeyer, R. *Die fauna sudwest-Australiens. Ergebnisse der hamburger sudwest-Australischen forschungsreise 1905.* Jena: Gustav Fischer. Vol. 2, pt. 16, pp. 251–278

5-525
Werner, F. 1913.
Neue oder seltene reptilien und frosche des naturhistorischen museums in Hamburg. *Mitt. Naturh. Mus. Hamburg.* 30(2):1–51.

5-525.5
West, J. A. 1979.
The occurrence of some exotic reptiles and amphibians in New Zealand. *Herpetofauna* 10(2):4–9.

5-526
Westerman, J. H. 1942.
Snakes from Bangka and Billiton. *Treubia* 18(3):611–619.

5-527
Whitaker, R. 1978.
Common Indian snakes: a field guide. New Delhi: Macmillan.

5-528
Wilson, L. D.; McCranie, J. R.; Porras, L. 1978.
Two snakes, *Leptophis modestus* and *Pelamis platurus,* new to the herpetofauna of Honduras. *Herpetol. Rev.* 9(2):63–64.

5-529
Wilson, L. D.; Meyer, J. R. 1982.
The snakes of Honduras. Milwaukee: Milwaukee Pub. Mus.

5-530
Witte, G. F. de. 1933.
Liste des batraciens et des reptiles d'extreme-orient et des Indes orientales recueillis, en 1932, par S. A. R. Le Prince Leopold de Belgique. *Bull. Mus. R. Hist. Nat. Belg.* 9(24):1–8.

5-530.5
Wolff, T. 1958.
Et havslange-problem fra sydhavet. *Nord. Med.* 60L:1149–1150.

5-531
Wolff, T. 1969.
The fauna of Rennell and Bellona, Solomon Islands. *Philos. Trans. R. Soc.* B 255:321–343.

5-532
Wolff, T. 1970.
     Lake Tegano on Rennell Island, the former lagoon of a raised atoll. *Nat. Hist. Rennell Isl. Br. Solomon Isl.* 6:7–29.

5-533
Wright, A. H.; Wright, A. A. 1952.
     List of the snakes of the United States and Canada by states and provinces. *Am. Midl. Nat.* 48(3):574–603.

5-534
Zhejiang Medical College; Chengdu Institute of Biology, Academia Sinica; Shanghai Natural History Museum; Zhejiang Sheng Institute of Chinese Medicine, editors. 1980.
     Illustrated book of the snakes of China. Shanghai: Shanghai Kexuejishu Chubanshe. (In Chinese)

5-535
Zweifel, R. G. 1960.
     Results of the Puritan-American Museum of Natural History Expedition to Western Mexico. 9. Herpetology of the Tres Marias Islands. *Bull. Am. Mus. Nat. Hist.* 119(2):77–128.

Genus-Species Index
for Chapter 5

Hydrophis jerdonii 5-5, 5-10, 5-36, 5-137.5, 5-138, 5-139, 5-220, 5-252, 5-419, 5-446

Hydrophis kingi 5-170, 5-171, 5-194, 5-245, 5-247, 5-283, 5-293, 5-410, 5-424, 5-481

Hydrophis kingii 5-250, 5-276, 5-287, 5-418

Hydrophis klossi 5-2, 5-62, 5-194, 5-194.5, 5-196, 5-239, 5-250, 5-252, 5-293, 5-315, 5-392, 5-419, 5-423, 5-424, 5-425, 5-427, 5-429, 5-443, 5-446, 5-459, 5-460, 5-478, 5-479, 5-480, 5-481, 5-518.5

Hydrophis lamberti 5-372.5, 5-423, 5-443

Hydrophis laevis 5-288

Hydrophis lapemoides 5-2, 5-10, 5-24, 5-73.9, 5-73.95, 5-101, 5-103, 5-104, 5-112, 5-113, 5-115, 5-116, 5-137.5, 5-138, 5-139, 5-143, 5-182, 5-194, 5-196, 5-234, 5-238, 5-250, 5-252, 5-293, 5-309, 5-310, 5-312, 5-313, 5-315, 5-372.3, 5-372.5, 5-393, 5-424, 5-429, 5-456, 5-467, 5-470, 5-471, 5-472, 5-481, 5-492, 5-518.3, 5-518.5, 5-527

Hydrophis latifasciata 5-7, 5-24, 5-137.5, 5-138, 5-139

Hydrophis latifasciatus 5-54, 5-467

Hydrophis (Lieoselasma) cyanocinctus 5-239, 5-242

Hydrophis lindsayi 5-5, 5-6, 5-39, 5-41, 5-60, 5-138, 5-139, 5-182, 5-329, 5-330, 5-363, 5-467, 5-469

Hydrophis lindsayii 5-137.5

Hydrophis lindsays 5-322

Hydrophis longiceps 5-137.5, 5-138, 5-139, 5-182

Hydrophis loreata 5-40, 5-134, 5-145, 5-145.5, 5-172, 5-182, 5-184, 5-320

Hydrophis macfarlani 5-382, 5-519

Hydrophis major 5-137.5, 5-138, 5-139, 5-170, 5-171, 5-182, 5-194, 5-194.5, 5-198, 5-245, 5-247, 5-250, 5-252, 5-283, 5-293, 5-410, 5-418, 5-424, 5-481

Hydrophis mamillaris 5-2, 5-54, 5-98, 5-104, 5-112, 5-194, 5-196, 5-238, 5-250, 5-252, 5-293, 5-309, 5-310, 5-312, 5-315, 5-339, 5-386, 5-392, 5-424, 5-429, 5-450, 5-459, 5-460, 5-470, 5-481, 5-506, 5-518.5

Hydrophis mammilaris 5-527

Hydrophis melanocephala 5-517, 5-518, 5-343.5

Hydrophis melanocephalus 5-94, 5-95, 5-96, 5-96.5, 5-97, 5-188, 5-194, 5-196, 5-239, 5-242, 5-251, 5-293, 5-297, 5-305, 5-315, 5-316, 5-317, 5-333, 5-368, 5-369, 5-402, 5-424, 5-442, 5-448, 5-476, 5-481, 5-518.5, 5-519, 5-520

Hydrophis melanocinctus 5-506

Hydrophis melanosoma 5-94, 5-95, 5-96, 5-96.5, 5-97, 5-147, 5-182, 5-190, 5-192, 5-194, 5-194.5, 5-196, 5-222, 5-250, 5-293, 5-315, 5-392, 5-410, 5-423, 5-424, 5-425, 5-442, 5-446, 5-478, 5-479, 5-480, 5-481, 5-497, 5-502, 5-504, 5-518.5, 5-519, 5-520, 4-534

Hydrophis mertoni 5-190, 5-194, 5-250, 5-283, 5-284, 5-382, 5-424

Hydrophis microcephala 5-144

Hydrophis (Microcephalophis) gracilis 5-41

Hydrophis mjobergi 5-519

Hydrophis neglectus 5-506

Hydrophis nigra 5-6, 5-136, 5-138, 5-139

Hydrophis nigrocincta 5-20.4, 5-137.5, 5-138, 5-139, 5-144, 5-182, 5-481

Hydrophis nigrocinctus 5-2, 5-54, 5-57, 5-60, 5-112, 5-125.5, 5-134, 5-136, 5-155, 5-156, 5-190, 5-194, 5-194.5, 5-196, 5-250, 5-252, 5-262, 5-293, 5-382, 5-387, 5-419, 5-424, 5-429, 5-443, 5-446, 5-467, 5-470, 5-506, 5-518.5, 5-527

Hydrophis obscurus 5-1, 5-2, 5-9, 5-17.5, 5-43, 5-54, 5-57, 5-94, 5-95, 5-96, 5-96.5, 5-97, 5-98, 5-112, 5-125.5, 5-156, 5-196, 5-241, 5-250, 5-252, 5-293, 5-304, 5-315, 5-387, 5-400, 5-405, 5-424, 5-429, 5-442, 5-470, 5-481, 5-502, 5-506, 5-518.5, 5-519, 5-527

Hydrophis ocella 5-290

Hydrophis ocellata 5-182, 5-258

Hydrophis ocellatus 5-418

Hydrophis ornata 5-39, 5-62, 5-87, 5-137.5, 5-138, 5-139, 5-182

Hydrophis ornata godeffroi 5-517, 5-343.5, 5-518

Hydrophis ornatus 5-41, 5-64, 5-73.9, 5-94, 5-95, 5-96, 5-96.5, 5-97, 5-101, 5-103, 5-110, 5-116, 5-125.5, 5-132, 5-143, 5-147, 5-170, 5-194, 5-201, 5-218, 5-238, 5-250, 5-252, 5-274, 5-293, 5-300, 5-305, 5-312, 5-313, 5-315, 5-317, 5-371, 5-372.5, 5-373, 5-380, 5-382, 5-390, 5-400, 5-408, 5-410, 5-412, 5-419, 5-424, 5-425, 5-442, 5-443, 5-446, 5-459, 5-470, 5-474, 5-476, 5-478, 5-479, 5-480, 5-481, 5-492, 5-495, 5-509, 5-519, 5-520, 5-527, 5-530, 5-534

Hydrophis ornatus godeffroyi 5-250

Hydrophis ornatus inornatus 5-250

Hydrophis ornatus maresianus 5-250

Hydrophis ornatus ocellatus 5-124.7, 5-171, 5-190, 5-196, 5-208, 5-245, 5-246, 5-247, 5-250, 5-283, 5-284, 5-317, 5-399.5, 5-424, 5-518.5

Hydrophis ornatus ornatus 5-2, 5-98, 5-104, 5-112, 5-113, 5-115, 5-190, 5-196, 5-238, 5-250, 5-284, 5-297, 5-309, 5-310, 5-333, 5-368, 5-369, 5-376, 5-377, 5-378, 5-379, 5-392, 5-429, 5-459, 5-460, 5-518.3, 5-518.5

Hydrophis pachycercos 5-144

Hydrophis pachycercus 5-137.5, 5-138, 5-139, 5-182, 6-467

Hydrophis pacificus 5-94, 5-95, 5-96, 5-96.5, 5-97, 5-196, 5-240, 5-315, 5-400, 5-442, 5-518.5, 5-523

Hydrophis parviceps 5-2, 5-62, 5-64, 5-147, 5-196, 5-250, 5-293, 5-427, 5-429, 5-481, 5-518.5

Hydrophis pelamidoides 5-28, 5-31, 5-34.4, 5-35, 5-36, 5-38, 5-295, 5-390.3, 5-463

Hydrophis pelamidoides var. unimaculatus 5-295

Hydrophis pelamis 5-35, 5-326.7, 5-463

Hydrophis (Pelamis) bicolor 5-144

Hydrophis (Pelamis) bicolor var. *alternans.* 5-144

Hydrophis (Pelamis) bicolor var. *sinuata* 5-144

Hydrophis (Pelamis) pelamidoides 5-144

Hydrophis (Pelamis) pelamidoides var. *annulata* 5-144

Hydrophis pelamoides 5-207, 5-326.7

Hydrophis pelamidoides 5-34.4

Hydrophis pemoides 5-238

Hydrophis phipsoni 5-363

Hydrophis platycercus 5-382

Hydrophis polyodontus 5-382

Hydrophis protervus 5-18

Hydrophis rhombifer 5-57

Hydrophis robusta 5-7, 5-52, 5-136, 5-137.5, 5-138, 5-139, 5-182, 5-329, 5-330, 5-363

Hydrophis robustus 5-467

Hydrophis saravacensis 5-382

Hydrophis schistosa 5-27.5, 5-35.5, 5-144

Hydrophis schistosus 5-17.5, 5-20.4, 5-27, 5-34.6, 5-54, 5-172, 5-184, 5-320

Hydrophis semperi 5-40, 5-133, 5-134, 5-147, 5-162, 5-194, 5-196, 5-204, 5-240, 5-250, 5-251, 5-252, 5-293, 5-307, 5-315, 5-424, 5-518.5, 5-520

Hydrophis siamensis 5-443

Hydrophis schlegelii 5-144

Hydrophis spiralis 5-2, 5-10, 5-40, 5-43, 5-54, 5-62, 5-98, 5-101, 5-103, 5-112, 5-113, 5-115, 5-116, 5-125.5, 5-134, 5-137.5, 5-138, 5-139, 5-142, 5-143, 5-147, 5-157, 5-160, 5-161, 5-182, 5-190, 5-191.5, 5-192, 5-194, 5-194.5, 5-196, 5-204, 5-222, 5-238, 5-250, 5-252, 5-254, 5-261, 5-269, 5-272, 5-293, 5-309, 5-310, 5-312, 5-313, 5-315, 5-329, 5-330, 5-331, 5-332, 5-371, 5-372, 5-382, 5-387, 5-392, 5-393, 5-394, 5-424, 5-425, 5-429, 5-470, 5-476, 5-478, 5-479, 5-480, 5-481, 5-492, 5-509, 5-513, 5-518.3, 5-518.5, 5-520, 5-527

Hydrophis spiralis melanocephalus 5-250

Hydrophis spiralis spiralis 5-104, 5-250

Hydrophis stewartii 5-6, 5-136, 5-138, 5-139, 5-329, 5-330

Hydrophis stokesi 5-26, 5-39, 5-41, 5-382, 5-481

Hydrophis stokesii 5-10, 5-137.5, 5-138, 5-139, 5-182, 5-258, 5-290, 5-446

Kerilia jerdoni siamensis 5-190, 5-196, 5-424, 5-425, 5-430.5, 5-443, 5-518-5

Kerilia jerdonii 5-250, 5-252, 5-390, 5-392, 5-443, 5-459, 5-460, 5-476

Kerilia jerdonii siamensis 5-239, 5-242, 5-459, 5-474

Kerril pattee 5-387

Kolpophis 5-383.5

Kolpophis annandalei 5-2, 5-62, 5-64, 5-110, 5-147, 5-190, 5-194, 5-194.5, 5-196, 5-239, 5-242, 5-250, 5-252, 5-293, 5-315, 5-390, 5-392, 5-424, 5-425, 5-429, 5-443, 5-459, 5-460, 5-478, 5-479, 5-480, 5-481, 5-518.5, 5-520

Kolpophis anomalus 5-443

Lapemis 5-383.5, 5-455.5

Lapemis curtus 5-2, 5-101, 5-103, 5-104, 5-112, 5-113, 5-115, 5-116, 5-125.5, 5-143, 5-194, 5-194.5, 5-196, 5-198, 5-234, 5-238, 5-250, 5-252, 5-293, 5-309, 5-312, 5-313, 5-315, 5-324, 5-332, 5-350, 5-370, 5-393, 5-403, 5-404, 5-424, 5-429, 5-446, 5-456, 5-470, 5-481, 5-492, 5-510, 5-511, 5-518.3, 5-518.5, 5-527

Lapemis hardwickei 5-198, 5-269, 5-272, 5-510, 5-511

Lapemis hardwicki 5-110, 5-140, 5-194, 5-194.5, 5-196, 5-268, 5-410, 5-452.5, 5-518.5

Lapemis hardwickii 5-2, 5-17.5, 5-62, 5-63, 5-64, 5-92, 5-95, 5-96, 5-96.5, 5-97, 5-132, 5-143, 5-147, 5-157, 5-167, 5-174, 5-180, 5-190, 5-192, 5-194.5, 5-201, 5-201.5, 5-204, 5-218, 5-222, 5-239, 5-242, 5-245, 5-250, 5-252, 5-255.5, 5-273, 5-277, 5-283, 5-284, 5-291, 5-293, 5-297, 5-305, 5-314, 5-315, 5-316, 5-317, 5-333, 5-343, 5-343.5, 5-350, 5-368, 5-369, 5-373, 5-376, 5-380, 5-390, 5-392, 5-400, 5-408, 5-424, 5-425, 5-427, 5-429, 5-435, 5-442, 5-443, 5-446, 5-454, 5-459, 5-460, 5-473.5, 5-474, 5-476, 5-478, 5-479, 5-480, 5-495, 5-497, 5-518, 5-519, 5-520, 5-534

Lapemis viperina 5-125.5

Lapemis viperinus 5-481

Laticauda 5-76, 5-383.5, 5-455.5

Laticauda colubrina 5-2, 5-16, 5-21, 5-62, 5-64, 5-70, 5-71, 5-72, 5-75, 5-94, 5-95, 5-96, 5-96.5, 5-97.5, 5-98, 5-112, 5-125.5, 5-129, 5-132, 5-159, 5-160, 5-161, 5-167, 5-174, 5-179, 5-180, 5-188, 5-190, 5-192, 5-194, 5-194.5, 5-196, 5-199, 5-202.3, 5-204, 5-218, 5-222, 5-232, 5-239, 5-242, 5-245, 5-247, 5-250, 5-252, 5-260, 5-261, 5-269, 5-271, 5-272, 5-277, 5-283, 5-284, 5-291, 5-293, 5-294, 5-297, 5-298, 5-299, 5-300, 5-302, 5-305, 5-306, 5-308, 5-315, 5-317, 5-333, 5-343, 5-343.5, 5-348.7, 5-349, 5-350, 5-351, 5-357, 5-389, 5-392, 5-396, 5-400, 5-412, 5-424, 5-425, 5-427, 5-428, 5-429, 5-435, 5-436, 5-437.5, 5-442, 5-443, 5-446, 5-448, 5-449, 5-451, 5-453, 5-454, 5-459, 5-460, 5-470, 5-473.5, 5-476, 5-478, 5-479, 5-480, 5-493, 5-494, 5-510, 5-511, 5-517, 5-518, 5-518.5, 5-519, 5-520, 5-525.5, 5-526, 5-527, 5-530.5, 5-531, 5-532, 5-534

Laticauda columbrina 5-147

Laticauda crockeri 5-194, 5-196, 5-250, 5-293, 5-300, 5-315, 5-412, 5-481, 5-493, 5-494, 5-520, 5-532

Laticauda ijimae 5-342

Laticauda laticauda 5-389, 5-527

Laticauda laticauda affinis 5-518, 5-343

Laticauda laticauda wolffi 5-493, 5-494

Laticauda laticaudata 5-2, 5-15, 5-21, 5-62, 5-64, 5-70, 5-71, 5-75, 5-94, 5-95, 5-96, 5-96.5, 5-132, 5-147, 5-167, 5-180, 5-188, 5-190, 5-194, 5-194.5, 5-196, 5-202.6, 5-204, 5-222, 5-239, 5-242, 5-245, 5-247, 5-250, 5-251, 5-252, 5-283, 5-284, 5-293, 5-297, 5-300, 5-305, 5-306, 5-315, 5-317, 5-333, 5-348.7, 5-349, 5-350, 5-372.7, 5-392, 5-400, 5-424, 5-429, 5-435, 5-436, 5-442, 5-446, 5-448, 5-454, 5-459, 5-460, 5-470, 5-476, 5-481, 5-493, 5-495, 5-510, 5-511, 5-512, 5-518.5, 5-519, 5-520, 5-530.5, 5-531, 5-532, 5-534

Laticauda laticaudata affinis 5-250, 5-252, 5-291, 5-517

Laticauda laticaudata crockeri 5-531

Laticauda laticaudata laticaudata 5-250, 5-252

Laticauda laticaudata wolffi 5-250, 5-252

Laticauda laticaudatus 5-192, 5-443

Laticauda schistorhyncha 5-518.5

Platurus schistorhynchus 5-50, 5-185, 5-230, 5-231, 5-341, 5-382, 5-504, 5-507, 5-525
Platurus scutatus 5-26, 5-137, 5-138, 5-139, 5-182, 5-257, 5-258, 5-270.3, 5-289, 5-290, 5-322, 5-410.5, 5-467, 5-469
Platurus semifasciatus 5-50
Polyodontognathus caerulescens 5-370, 5-510, 5-511
Polypholophis neglectus 5-510, 5-511
Porrecticollis obscurus 5-510, 5-511
Praescutata 5-383.5
Praescutata viperina 5-2, 5-104, 5-110, 5-112, 5-113, 5-115, 5-116, 5-143, 5-190, 5-192, 5-194.5, 5-239, 5-242, 5-250, 5-252, 5-309, 5-310, 5-312, 5-313, 5-315, 5-332, 5-376, 5-378, 5-379, 5-380, 5-390, 5-392, 5-394, 5-429, 5-459, 5-460, 5-470, 5-474, 5-476, 5-478, 5-479, 5-480, 5-510, 5-520, 5-527, 5-534
Praescutata viperine 5-393
Pseudolaticauda semifasciata 5-239, 5-242
Sea snake 5-11, 5-447
Serpens de mer 5-356.3
Serpens marins 5-356.3
Shiddil 5-387
Shootur sun 5-387
Tatta pam 5-386
Thalassophina 5-383.5
Thalassophina viperina 5-62, 5-64, 5-147, 5-194, 5-194.5, 5-233, 5-252, 5-293, 5-305, 5-343.5, 5-372.7, 5-395, 5-408, 5-425, 5-427, 5-460, 5-492
Thalassophis annandalei 5-350, 5-382, 5-423, 5-446, 5-460, 5-507
Thalassophis annandalii 5-56, 5-419
Thalassophis anomalus 5-2, 5-17.5, 5-62, 5-110, 5-147, 5-190, 5-192, 5-194, 5-196, 5-233, 5-239, 5-242, 5-250, 5-254, 5-255, 5-256, 5-269, 5-272, 5-293, 5-315, 5-350, 5-352, 5-353, 5-382, 5-389, 5-390, 5-392, 5-422, 5-423, 5-424, 5-429, 5-446, 5-459, 5-460, 5-474, 5-478, 5-479, 5-480, 5-481, 5-507, 5-518.5, 5-520
Thalassophis viperinus 5-196, 5-511, 5-518.3, 5-518.5
Thallasophina viperina 5-101, 5-103, 5-368, 5-369, 5-424
Valakadyen 5-387
Water snakes 5-217
White and blue snake 5-178
Yellow bellied sea snake 5-12, 5-13, 5-401.5, 5-499

Chapter 6

EVOLUTION AND GENETICS

6-1
Anthony, J. 1955.
 Essai sur l'evolution anatomique de l'appareil venimeux des Ophidiens. *Ann. Sci. Nat. Zool. Biol. Anim.* Ser. 11, 17:7–53.

6-1.5
Becak, W. 1965.
 Constituicao cromossomica e mechanismo de determinacacao do sexo em ofidios sul-americanos. I. Aspectos cariotipicos. *Mem. Inst. Butantan (Sao Paulo)* 32:37–78.

6-2
Becak, W. 1968.
 Karyotypes, sex chromosomes, and chromosomal evolution in snakes. Bucherl, W.; Buckley, E. E.; Deulofeu, V., eds. *Venomous animals and their venoms.* New York: Academic. Vol. 1, pp. 53–95.

6-3
Bellairs, A. d'A.; Underwood, G. 1951.
 The origin of snakes. *Biol. Rev. Camb. Philos. Soc.* 26(2):193–237.

6-4
Blackburn, D. G. 1985.
 Evolutionary origins of viviparity in the reptilia. II. Serpentes, Amphisbaenia, and Ichthyosauria. *Amphib.-Reptilia* 6(3):259–291.

6-4.5
Bolanos, R. 1982.
 Serpeintes venenosas de Centro America: distribucion, caracteristicas y patrones cariologicos. *Mem. Inst. Butantan (Sao Paulo)* 46:275–291.

6-4.7
Bolanos, R. 1984.
   Serpeintes venenos y ofidismo en Centroamerica. San Jose, Costa Rica: Editorial Universidad de Costa Rica.

6-4.8
Brown, A. E. 1905.
   The utility principle in relation to specific characters. *Proc. Acad. Nat. Sci. Phil.* 57:206–209.

6-4.9
Buth, D. G.; Murphy, R. W.; Miyamoto, M. M.; Lieb, C. S. 1985.
   Creatine kinases of amphibians and reptiles: evolutionary and systematic aspects of gene expresssion. *Copeia* 1985(2):279–284.

6-5
Cadle, J. E.; Gorman, G. C. 1981.
   Albumin immunological evidence and the relationships of sea snakes. *J. Herpetol.* 15(3):329–334.

6-6
Cadl, J. E.; Sarich, V. M. 1981.
   An immunological assessment of the phylogenetic position of New World coral snakes. *J. Zool. (Lond.)* 195:157–167.

6-6.5
Cogger, H.; Heatwole, H. 1984.
   The Australian reptiles: origins, biogeography, distribution patterns and island evolution. Archer, M.; Clayton, G., eds. *Vertebrate zoogeography & evolution in Australasia (Animals in space and time).* Carlisle, Western Australia: Hesperian Press. Pp. 343–360.

6-7
Cohen, E. 1955.
   Immunological studies of the serum proteins of some reptiles. *Biol. Bull. (Woods Hole)* 109:394–403.

6-7.3
Cuenot, L. 1951.
   L'evolution biologique; les faits, les incertitudes. Paris: Maisson et Cie.

6-7.5
Dunson, W. A.; Mazzotti, F. J. 1989.
Salinity as a limiting factor in the distribution of reptiles in Florida Bay: a theory for the estuarine origin of marine snakes and turtles. *Bull. Mar. Sci.* 44(1):229–244.

6-8
Feder, J. H.; Gorman, G. C. 1979.
Genetic divergence in two subfamilies of seasnakes (family Hydrophiidae). *Am. Zool.* 19(3):872. (Abstract)

6-9
Gilboa, I. 1974.
Karyotypes of amphibians & reptiles: a bibliographic review. *Yearb. Herpetol.* 8(1):91–156.

6-10
Gorman, G. C. 1981.
The chromosomes of *Laticauda* and a review of karyotypic evolution in the Elapidae. *J. Herpetol.* 15(2):225–233.

6-11
Gutierrez, J. M.; Bolanos, R. 1980.
Karyotype of the yellow-bellied sea snake, *Pelamis platurus*. *J. Herpetol.* 14(2):161–165.

6-12
Hatai, K.; Masuda, K.; Noda, H. 1974.
Marine fossils from the Moniwa formation along the Natori River, Sendai, Northeast Honshu, Japan, Part 2. Problematica from the Moniwa Formation. *Trans. Proc. Palaeontol. Soc. Jpn. New Ser.* 25:364–370.

6-13
Hoffstetter, R. 1967.
Revue des recentes acquisitions concernant l'histoire et la systematique des squamates. *Colloq. Int. Cent. Natl Rech. Sci.* 104:243–279.

6-14
Hseu, T. H.; Jou, E. D.; Wang, C.; Yang, C. C. 1977.
Molecular evolution of snake venom toxins. *J. Mol. Evol.* 10(2):167–182.

6-15

Ivanov, O. C. 1978.

Application of the 'profiles of relationship' method to distantly related proteins. *J. Mol. Evol.* 12(1):1–10.

6-16

Janensch, W. 1906.

Ueber *Archaeophis proavus* Mass., eine schlange aus dem eocan des Monte Bolca. *Beitr. Palaeont. Geol. Oest.-Ung.* 19:1–33.

6-17

Johnson, R. G. 1955.

The adaptive and phylogenetic significance of vertebral form in snakes. *Evolution* 9:367–388.

6-18

Kiran, U. 1979.

Evolutionary significance of the relative disposition of the bones in the jaw complex of snakes and lizards. *Ann. Zool. (Agra)* 15(2):59–78.

6-19

Lombard, R. E.; Marx, H.; Rabb, G. B. 1986.

Morphometrics of ectopterygoid in advanced snakes (Colubroidea): a concordance of shape and phylogeny. *Biol. J. Linn. Soc.* 27(2):133–164.

6-20

Makino, S. 1951.

An atlas of the chromosome numbers in animals. 2nd ed. Ames, IA: Iowa State College Press.

6-21

Mao, S. H.; Chen, B. Y.; Chang, H. M. 1977.

The evolutionary relationships of sea snakes suggested by immunological cross-reactivity of transferrins. *Comp. Biochem. Physiol. A Comp. Physiol.* 57:403–406.

6-22

Mao, S. H.; Dessauer, H. C.; Chen, B. Y. 1978.

Fingerprint correspondence of hemoglobins and the relationships of sea snakes. *Comp. Biochem. Physiol. B Comp. Biochem.* 59(4):353–361.

6-23

Mao, S. H.; Guo, Y. W.; Yin, F. Y. 1985.

Physical properties of some snake plasma albumins. *Comp. Biochem. Physiol. B Comp. Biochem.* 82(4):655–658.

6-24

Marx, H.; Rabb, G. B. 1970.

Character analysis: an empirical approach applied to advanced snakes. *J. Zool. (Lond.)* 161:525–548.

6-25

Marx, H.; Rabb, G. B. 1972.

Phyletic analysis of fifty characters of advanced snakes. *Fieldiana Zool.* 63:1–321.

6-26

Matthey, R. 1949.

Les chromosomes des vertebres. Lausanne: F. Rouge.

6-27

Matthey, R. 1954.

Les chromosomes des vertebres. Grasse, P. P., ed. *Traite de zoologie.* Paris: Masson et Cie. Vol. 12, pp. 1044–1063.

6-28

Matthey, R. 1970.

Les chromosomes des reptiles. Grasse, P. P., ed. *Traite de zoologie.* Paris: Masson et Cie. Vol. 14, no. 3, pp. 829–858.

6-28.4

McCarthy, C. J. 1985.

Monophyly of elapid snakes (Serpentes: Elapidae); an assessment of the evidence. *Zool. J. Linn. Soc.* 83(1):79–93.

6-28.6

McCarthy, C. J. 1986.

Relationships of the laticaudine sea snakes (Serpentes: Elapidae: Laticaudinae). *Bull. Br. Mus. (Nat. Hist.) Zool.* 50(2):127–161.

6-29

McDowell, S. B. 1969.

Notes on the Australian sea-snake *Ephalophis greyi* M. Smith (Serpentes: Elapidae, Hydrophiinae) and the origin and classification of sea-snakes. *Zool. J. Linn. Soc.* 48:333–349.

6-30

McDowell, S. B. 1970.

On the status and relationships of the Solomon Island elapid snakes. *J. Zool. (Lond.)* 161:145–190.

6-31
McDowell, S. B. 1974.

Additional notes on the rare and primitive sea-snake, *Ephalophis greyi*. *J. Herpetol.* 8(2):123–128.

6-31.5
McDowell, S. B. 1985.

The terrestrial Australian elapids: general summary. Grigg, G.; Shine, R.; Ehmann, H., eds. *Biology of Australasian frogs and reptiles*. Chipping Norton, N.S.W.: Surrey Beatty & Sons. Pp. 261–264.

6-32
Minton, S. A. 1978.

Serological relationships of some Philippine sea snakes. *Copeia* 1978(1):151–154.

6-33
Minton, S. A. 1981.

Evoluion and distribution of venomous snakes. Banks, C. B.; Martin, A. A., eds. *Proceedings of the Melbourne Herpetological Symposium*. Melbourne: Zoological Board. Pp. 55–59.

6-33.5
Minton, S. A. 1986.

Origins of poisonous snakes: evidence from plasma and venom proteins. Harris, J. B. *Natural toxins: animal, plant, and microbial*. Oxford: Clarendon Press. Pp. 1–21.

6-34
Minton, S. A.; Da Costa, M. S. 1975.

Serological relationships of sea snakes and their evolutionary implication. Dunson, W. A., ed. *The biology of sea snakes*. Baltimore, MD: University Park Press. Pp. 33–55.

6-35
Nakamura, K. 1935.

Studies on reptilian chromosomes. VI: Chromosomes of some snakes. *Mem. Col. Sci. Kyoto Imp. Univ. Series B* 10(5, Art.17):361–402.

6-35.5
Oguma, K.; Makino, S. 1937.

A new list of the chromosome numbers in vertebrata (March 1937). *J. Fac. Sci. Hokkaido Univ. Zool. Ser. 6*, 5(4):297–356.

6-36

Pickwell, G. V. 1979.

Possible evolutionary affinities seen in sea snake venom-antivenom reactions. *Toxicon* 17(Suppl.1):141. (Abstract)

6-37

Radovanovic, M. 1967.

Phylogenie und evolution der giftschlangen. *Zool. Anz.* 179:199–229.

6-38

Ray-Chaudhuri, S. P.; Singh, L.; Sharma, T. 1971.

Evolution of sex-chromosomes and formation of W-chromatin in snakes. *Chromosoma (Berl.)* 33:239–251.

6-39

Ray-Chaudhuri, S. P.; Singh, L. 1972.

DNA replication pattern in sex-chromosomes of snakes. *Nucleus (Calcutta)* 15(3):200–210.

6-39.5

Romer, A. S. 1966.

Vertebrate paleontology. 3rd ed. Chicago: University of Chicago Press.

6-40

Schmidt, K. P. 1950.

Modes of evolution discernible in the taxonomy of snakes. *Evolution* 4:79–86.

6-40.5

Schwaner, T. D.; Baverstock, P. R.; Dessauer, H. C.; Mengden, G. A. 1985.

Immunological evidence for the phylogenetic relationships of Australian elapid snakes. Grigg, G.; Shine, R.; Ehmann, H., eds. *Biology of Australian frogs and reptiles.* Chipping Norton, N.S.W.: Surrey Beatty & Sons. Pp. 177–184.

6-41

Singh, L. 1972.

Evolution of karyotypes in snakes. *Chromosoma (Berl.)* 38:185–236.

6-42

Singh, L. 1972.

Multiple W chromosome in a sea snake, *Enhydrina schistosa* Daudin. *Experientia (Basel)* 28(1):95–97.

6-43
Singh, L. 1974.
   Chromosomes of six species of Indian snakes. *Herpetologica* 30(4):419–429.

6-44
Storr, G. M. 1964.
   Some aspects of the geography of Australian reptiles. *Senckenb. Biol.* 45(3–5):577–589.

6-45
Stromer, E. 1911.
   Neue forschungen ueber fossile lungenatmende meeresbewohner. *Fortschr. Naturw. Forsch.* 2:83–114.

6-46
Strydom, D. J. 1972.
   Phylogenetic relationships of proteroglyphae toxins. *Toxicon* 10:39–45.

6-47
Strydom, D. J. 1973.
   Snake venom toxins: structure-function relationships and phylogenetics. *Comp. Biochem. Physiol. B Comp. Biochem.* 44:269–281.

6-48
Strydom, D. J. 1973.
   Snake venom toxins: the evolution of some of the toxins found in snake venoms. *Syst. Zool.* 22:596–608.

6-48.5
Tamiya, N. 1985.
   A comparison of amino acid sequences of neurotoxins and of phospholipases of some Australian elepid snakes with those of other proteroglyphous snakes. Grigg, G.; Shine, R.; Ehmann, H., eds. *Biology of Australasian frogs and reptiles.* Chipping Norton, N.S.W.: Surrey Beatty & Sons. Pp. 209–219.

6-48.7
Tamiya, N. 1988.
   Non-divergence theory of evolution: amino acid sequence comparison of proteroglyphous snake venom components. *Toxicon* 26(1):42. (Abstract)

6-49
Tamiya, N.; Maeda, N. 1978.
Chemical taxonomy of snake neurotoxins. Matsubara, H.; Yamanaka, T., eds. *Evolution of protein molecules.* Tokyo: Japan Scientific Societies Press. Pp. 297–310.

6-49.5
Tamiya, N.; Yagi, T. 1985.
Non-divergence theory of evolution: sequence comparison of some proteins from snakes and bacteria. *J. Biochem. (Tokyo)* 98(2):298–303.

6-50
Taylor, R.; Bolanos, R. 1975.
Descripcion de un metodo simple y economico para el estudio de cariotipos en serpientes. *Rev. Biol. Trop.* 23(2):177–183. (In Spanish with English abstract)

6-51
Toriba, M.; Yosida, T. H. 1986.
Karyotypes of the Laticaudinae sea snakes from Ryukyu Archipelago. *Snake* 18:70–76.

6-51.5
Toriba, M. 1989.
Karyotypes of two species of elapid snakes. *Snake* 21:87–89.

6-51.7
Underwood, G. 1957.
On lizards of the family Pygopodidae. *J. Morphol.* 100(2):207–268.

6-52
Voris, H. K. 1969.
The evolution of the Hydrophiidae with a critique on methods of approach. Chicago, IL: University of Chicago. 429 pp. (Dissertation)

6-53
Voris, H. K. 1971.
New approaches to character analysis applied to the sea snakes (Hydrophiidae). *Syst. Zool.* 20(4):442–458.

6-54
Voris, H. K. 1977.
A phylogeny of the sea snakes (Hydrophiidae). *Fieldiana Zool.* 70(4):79–169.

6-55
Zerova, G. A.; Chkhikvadze, V. M. 1984.
   Review of cenozoic lizards and snakes of the USSR. *Izv. Akad. Nauk Gruz. S.S.R. Ser. Biol.* 10(5):319–326. (In Russian with English abstract)

Genus-Species Index
for Chapter 6

Chapter 7

ECOLOGY AND FEATURES OF ADAPTATION TO THE ENVIRONMENT, SPECIFIC HABITATS

7-1

Burns, G. W. 1984.

Aspects of population movements and reproductive biology of Aipysurus laevis, the olive sea snake. Armidale, N.S.W.: University of New England. 171 pp. (Dissertation)

7-2

Cahill, M.; Heatwole, H.; Goldman, B. 1973.

Ecology of the reefs. *Occas. Pap. Dept. Geog. Univ. Papua, New Guinea* 1:209–229.

7-3

Cott, H. B. 1940.

Adaptive coloration in animals. New York: Oxford University Press.

7-4

Cozzi, C. A. 1980.

The absence of sea snakes in the Atlantic Ocean. *Bull. Md Herpetol. Soc.* 16(3):113–118.

7-5

Dunson, W. A. 1975.

Adaptations of sea snakes. Dunson, W. A., ed. *The biology of sea snakes.* Baltimore, MD: University Park Press. Pp. 3–19.

7-6
Dunson, W. A.; Ehlert, G. 1970.
    Oceanographic limiting factors in aggregation, dispersal and migration of sea snakes. San Diego, CA: *Alpha Helix Research Program 1969–1970*. Page 35. (Abstract)

7-7
Dunson, W. A.; Ehlert, G. W. 1971.
    Effects of temperature, salinity, and surface water flow on distribution of the sea snake *Pelamis*. *Limnol. Oceanogr.* 16(6):845–853.

7-8
Dunson, W. A.; Minton, S. A. 1978.
    Diversity, distribution, and ecology of Philippine marine snakes (Reptilia, Serpentes). *J. Herpetol.* 12(3):281–286.

7-8.5
Dunson, W. A.; Mazzotti, F. J. 1989.
    Salinity as a limiting factor in the distribution of reptiles in Florida Bay: a theory for the estuarine origin of marine snakes and turtles. *Bull. Mar. Sci.* 44(1):229–244.

7-9
Guibe, J. 1970.
    Donnees ecologiques. Grasse, P. P., ed. *Traite de zoologie*. Paris: Masson et Cie. Vol. 14, no. 3, pp. 987–1063.

7-9.5
Heatwole, H. 1976.
    Reptile ecology. St. Lucia, Queensland: Queensland University Press.

7-10
Hecht, M. K.; Kropach, C.; Hecht, B. M. 1974.
    Distribution of the yellow-bellied sea snake, *Pelamis platurus*, and its significance in relation to the fossil record. *Herpetologica* 30(4):387–396.

7-11
Hirosaki, Y. 1964.
    Ecological study on fishes with the drifting sea weeds. III. Accompanying animals excluded fishes. *Misc. Rep. Res. Inst. Nat. Resour.* (*Tokyo*) 62:63–70. (In Japanese)

7-12
Kropach, C. 1971.
   Sea snake (*Pelamis platurus*) aggregations on slicks in Panama. *Herpetologica* 27(2):131–135.

7-13
Kropach, C. 1972.
   A field study of the sea snake Pelamis platurus (Linnaeus) in the Gulf of Panama. New York, NY: City University of New York. 200 pp. (Dissertation)

7-14
Liat, L. B.; Sawai, Y. 1975.
   Ecology and distribution of some seasnakes in Malay Peninsula. *Snake* 7:28–32. (In Japanese with English abstract)

7-15
Lim, B. L.; Sawai, Y. 1975.
   Ecology and distribution of sea snakes in Peninsular Malaysia. *Toxicon* 13(2):107.

7-16
Lim, B. L.; Sawai, Y. 1976.
   Ecology and distribution of some sea snakes in Peninsular Malaysia. Ohsaka, A.; Hayashi, K.; Sawai, Y., eds. *Animal, plant, microbial toxins.* New York: Plenum. Vol. 2, pp. 515–520.

7-17
McCoy, M. 1980.
   Reptiles of the Solomon Islands. Hong Kong: Sheck Wah Tong Printing Press Limited. (*WAU Ecology Institute Handbook No. 7*)

7-17.5
McCoy, M. 1985.
   I bought 200 poisonous snakes from children. *Geo* 7(1):90–101.

7-18
Minton, S. A. 1974.
   Observations on sea snakes at Ashmore Reef, Timor Sea. *Proc. Indiana Acad. Sci.* 83:467–468.

7-19
Saint Girons, H. 1964.
   Notes sur l'ecologie et la structure des populations des Laticaudinae (Serpents, Hydrophidae) en Nouvelle Caledonie. *Terre et la Vie* 2:185–214.

7-19.3
Wolff, T. 1956.
    Renell—an out of the way coral island. Bruun, A. F.; Greve, S.; Mielche, H.; Sparck, R., eds. *The Galathea Deep Sea Expedition 1950–1952*. London: George Allen and Unwin. Pp. 211–223.

7-19.5
Wolff, T. 1958.
    Et havslange—problem fra sydhavet. *Nord. Med.* 60:1149–1150.

7-20
Zug, D. A.; Dunson, W. A. 1979.
    Salinity preference in fresh water and estaurine snakes (*Nerodia sipedon* and *N. fasciata*). *Fla. Sci.* 42(1):1–8.

Genus-Species Index
for Chapter 7

Chapter 8

FEEDING, FOOD ITEMS

8-.05
Breder, C. M. 1946.
   An analysis of the deceptive resemblances of fishes to plant parts, with critical remarks on protective coloration, mimicry and adaptation. *Bull. Bingham Ocean. Coll.* 10(2):1–49.

8-1
Glodek, G. S.; Voris, H. K. 1982.
   Marine snake diets: prey composition, diversity and overlap. *Copeia* 1982(3):661–666.

8-2
Hunter, J. R.; Mitchell, C. T. 1967.
   Association of fishes with flotsam in the offshore waters of Central America. *U.S. Fish. Wildl. Ser. Fish. Bull.* 66(1):13–29.

8-3
Hwang, M. H.; Ho, S. S.; Chou, S. A.; Hsieh, T. T.; Hu, B.C. 1965.
   Feeding habits of snakes from Chekiang. *Acta Zool. Sin.* 17(2):137–146. (In Chinese with English abstract)

8-4
Kailola, P. J. 1975.
   Notes on some fishes of the families Uranoscopidae, Scorpaenidae, Ophichthidae and Muraenidae from Torres Strait. *Proc. Linn. Soc. N.S.W.* 100(2):110–117.

8-5
Kelso, J. E. H. 1888.
   Notes on an Indian water-snake (*Enhydrina valakadyen*). *Proc. Phys. Soc. Edinb.* 9:385–386.

8-6
Klauber, L. M. 1935.
The feeding habits of a sea snake. *Copeia* 1935(4):182.

8-7
Klawe, W. L. 1963.
Observations on the spawning of four species of tuna *Neothunnus macropterus, Katsuwonus pelamis, Auxis thazard* and *Euthynnus lineatus*) in the eastern Pacific Ocean, based on the distribution of their larvae and juveniles. *Inter-Am. Trop. Tuna Comm. Bull.* 6(9):449–540. (In English and Spanish)

8-8
Klawe, W. L. 1964.
Food of the black-and-yellow sea snake, *Pelamis platurus*, from Ecuadorian coastal waters. *Copeia* 1964(4):712–713.

8-9
Kropach, C. 1971.
Sea snake (*Pelamis platurus*) aggregations on slicks in Panama. *Herpetologica* 27(2):131–135.

8-10
Kropach, C. 1972.
A field study of the sea snake <u>Pelamis platurus</u> (Linnaeus) in the Gulf of Panama. New York, NY: City University of New York. 200 pp. (Dissertation)

8-11
McCosker, J. E. 1975.
Feeding behavior of Indo-Australian Hydrophiidae. Dunson, W. A., ed. *The biology of sea snakes.* Baltimore, MD: University Park Press. Pp. 217–232.

8-12
McCoy, M. 1980.
Reptiles of the Solomon Islands. Hong Kong: Sheck Wah Tong Printing Press Limited. (*WAU Ecology Institute Handbook No. 7*)

8-12.3
McCoy, M. 1985.
I bought 200 poisonous snakes from children. *Geo* 7(1):90–101.

8-12.5
Moriguchi, H. 1988.
A case of food item of a seasnake, *Laticauda laticaudata. Snake* 20:163.

8-13

Pernetta, J. C. 1977.

Observations on the habits and morphology of the sea snake *Laticauda colubrina* (Schneider) in Fiji. *Can. J. Zool.* 55(10):1612–1619.

8-14

Pickwell, G. V. 1972.

The venomous sea snakes. *Fauna* (4):17–32.

8-15

Pitman, C. R. S. 1962.

More snake and lizard predators of birds. *Bull. Br. Ornithol. Club* 82(2):33–40.

8-16

Radcliffe, C. W.; Chiszar, D. A. 1980.

A descriptive analysis of predatory behavior in the yellow lipped sea krait (*Laticauda colubrina*). *J. Herpetol.* 14(4):422–424.

8-16.5

Rasmussen, A. R. 1989,

An analysis of *Hydrophis ornatus* (Gray), *H. lamberti* Smith, and *H. inornatus* (Gray) (Hydrophiidae, Serpentes) based on samples from various localities, with remarks on feeding and breeding biology of *H. ornatus. Amphib.-Reptilia* 10:397–417.

8-17

Shine, R.; Schwaner, T. 1985.

Prey constriction by venomous snakes: a review, and new data on Australian species. *Copeia* 1985(4):1067–1071.

8-17.5

Smythies, B. E. 1960.

The birds of Borneo. London: Oliver and Boyd.

8-18

Sternfeld, R. 1913.

Die erscheinungen der mimicry bei den schlangen. *Sitz. Ges. Nat. Freunde Berlin* 1913:98–117.

8-19

Takahashi, H. 1981.

The feeding behaviour of a seasnake, *Hydrophis melanocephalus* Gray. *Snake* 13:158–159.

8-20

Voris, H. K. 1966.

Fish eggs as the apparent sole food item for a genus of sea snake, *Emydocephalus* (Krefft). *Ecology* 47(1):152–154.

8-21

Voris, H. K. 1972.

The role of sea snakes (Hydrophiidae) in the trophic structure of coastal ocean communities. *Abstracts of the Symposium on the Indian Ocean and Adjacent Seas, Cochin, India, 12–18 January 1971.* Cochin, India: Marine Biological Association of India. Pp. 120–121.

8-22

Voris, H. K. 1972.

The role of sea snakes (Hydrophiidae) in the trophic structure of coastal ocean communities. *J. Mar. Biol. Assoc. India* 14(2):429–442.

8-23

Voris, H. K.; Voris, H. H.; Liat, L. B. 1978.

The food and feeding behavior of a marine snake, *Enhydrina schistosa* (Hydrophiidae). *Copeia* 1978(1):134–146.

8-24

Voris, H. K.; Moffett, M. W. 1981.

Size and proportion relationship between the beaked sea snake and its prey. *Biotropica* 13(1):15–19.

8-25

Voris, H. K.; Voris, H. H. 1983.

Feeding strategies in marine snakes: an analysis of evolutionary, morphological, behavioral and ecological relationships. *Am. Zool.* 23(2):411–425.

Genus-Species Index
for Chapter 8

Chapter 9

PREDATORS

9-.05
Ali, S.; Ripley, S. D. 1987.
   Compact handbook of the birds of India and Pakistan together with those of Bangladesh, Nepal, Bhutan and Sri Lanka. 2nd ed. New York: Oxford University Press.

9-1
Allen, G. R. 1974.
   The marine crocodile, *Crocodylus porosus*, from Ponape, Eastern Caroline Islands, with notes on food habits of crocodiles from the Palau Archipelago. *Copeia* 1974(2):553.

9-2
Banfield, E. J. 1908.
   The confessions of a beachcomber. London: T. Fisher, Unwin.

9-3
Barrett, C. 1954.
   Wild life of Australia and New Guinea. Toronto: William Heinemann.

9-4
Bruggen, A. C. van. 1961.
   Pelamis platurus, an usual item of food of *Octopus* spec. *Basteria* 25(4–5):73–74.

9-4.4
Bucknill, J. A. S.; Chasen, F. N. 1927.
   The birds of Singapore island. Singapore: Government Printing Office.

9-5
Caldwell, G. S.; Rubinoff, R. W. 1983.
  Avoidance of venomous sea snakes by naive herons and egrets. *Auk* 100:195–198.

9-6
Campbell, A. J.; White, S. A. 1910.
  Birds identified on the Capricorn group during expedition of R.A.O.U., 8th to 17th October, 1910. *Emu* 10:195–204.

9-7
Campbell, H. 1905.
  The diet of the Precibiculturists. *Br. Med. J.* 1905:979–981.

9-8
Cayley, N. W. 1939.
  What bird is that? a guide to the birds of Australia. 7th ed. London: Angus & Robertson.

9-9
Chacko, P. I. 1949.
  Food and feeding habits of the fishes of the Gulf of Manaar. *Proc. Indian Acad. Sci. Sect. B* 29:83–97.

9-10
Dharmakumarsinhji, R. S.; Lavkumar, K. S. 1956.
  The whitebellied sea eagles of Karwar (*Haliaetus leucogaster* (Gmelin)). *J. Bombay Nat. Hist. Soc.* 53(4):569–580.

9-11
Gibson-Hill, C. A. 1949.
  The coastal birds. Kesteven, G. L., ed. *Malayan fisheries.* Singapore: Malaya Publishing House. Pp. 42–44.

9-12
Greene, H. W. 1973.
  Defensive tail display by snakes and amphisbaenians. *J. Herpetol.* 7(3):143–161.

9-13
Greene, H. W.; Pyburn, W. F. 1973.
  Comments on aposematism and mimicry among coral snakes. *Biologist* 55(4):144–148.

9-13.5
Greene, H. W. 1988.
    Antipredator mechanisms in reptiles. Gans, C.; Huey, R. B., eds. *Biology of the reptilia: defense and life history*. New York: Alan R. Liss. Vol. 168, pp. 1–152.

9-14
Heatwole, H. 1975.
    Predation on sea snakes. Dunson, W. A., ed. *The biology of sea snakes*. Baltimore, MD: University Park Press. Pp. 233–249.

9-15
Heatwole, H.; Heatwole, E.; Johnson, C. R. 1974.
    Shark predation on sea snakes. *Copeia* 1974(3):780–781.

9-16
Heatwole, H.; Finnie, E. P. 1980.
    Seal predation on a sea snake. *Herpetofauna* 11:24.

9-16.5
Henry, G. M. 1955.
    A guide to the birds of Ceylon. London: Oxford University Press. (second edition published in 1971)

9-17
Jerdon, T. C. 1877.
    The birds of India; being a natural history of all the birds known to inhabit continental India: with descriptions of the species, genera, families, tribes, and orders, and a brief notice of such families as are not found in India, making it a manual of ornithology specially adapted for India. Calcutta: P. S. D'Rozario. Vol. 1.

9-17.5
Lyle, J. M.; Timms, G. J. 1982.
    Predation on aquatic snakes by sharks from Northern Australia. *Copeia* 1987(3):802–803.

9-18
Pickwell, G. V.; Bezy, R. L.; Fitch, J. E. 1983.
    Northern occurrences of the sea snake, *Pelamis platurus*, in the eastern Pacific, with a record of predation on the species. *Calif. Fish Game* 69(3):172–177.

9-18.5
Pough, R. H. 1988.
   Mimicry and related phenomena. Gans, C.; Huey, R. B., eds. *Biology of the reptilia.* New York: Alan R. Liss. Vol. 16B, pp. 153–234.

9-19
Rancurel, P.; Intes, A. 1982.
   Le requin tigre, *Galeocerdo cuvieri* Lacepede, des eaux neocaledoniennes examen des contenus stomacaux. *Tethys* 10(3):195–199.

9-19.5
Robinson, H. C. 1927.
   The birds of the Malay Peninsula: a general account of the birds inhabiting the region from the Isthmus of Kra to Singapore with the adjacent islands. London: H. F. & G. Witherby. Volume 1.

9-20
Rubinoff, I.; Kropach, C. 1970.
   Differential reactions of Atlantic and Pacific predators to sea snakes. *Nature (Lond.)* 228:1288–1290.

9-20.5
Suckling, H. J. 1876.
   Ceylon, a general description of the island, historical, physical, statistical. Containing the most recent information. London: Chapman & Hall.

9-21
Van der Meer Mohr, J. C. 1927.
   Notiz ueber seeschlangen. *Misc. Zool. Sumatr.* 23:1–2.

9-22
Van der Meer Mohr, J. C. 1930.
   Notes on the fauna of Pulau Berhala. *Treubia* 12(3–4):277–298.

9-23
Wall, F. 1906.
   The snake and its natural foes. *J. Bombay Nat. Hist. Soc.* 17(2):375–395.

9-23.5
Weldon, P. J.; Vallarino, O. 1988.
   Wounds on the yellow-bellied sea snake (*Pelamis platurus*) from Panama; evidence of would-be predators? *Biotropica* 20(2):174–176.

9-24
Wetmore, A. 1965.

The birds of the Republic of Panama. Part 1. Tinamidae (Tinamous) to Rynchopidae (Skimmers). *Smithson. Misc. Collect.* 150:1–483.

Genus-Species Index
for Chapter 9

Chapter 10

PARASITES AND COMMENSALS

10-.05
Annandale, N. 1906.
Note on a rare Indo-Pacific barnacle. *J. Asiat. Soc. Beng.* 2(6):207–208.

10-.07
Annandale, N. 1909.
An account of the Indian Cirripedia Pedunculata. *Mem. Indian Mus.* 2:61–137.

10-1
Annandale, N. 1906.
Stalked barnacles (*Cirripedia pedunculata*) in the Colombo Museum. *Spolia Zeylan.* 3(11):193–195.

10-1.3
Anonymous. 1958.
Enterobacteries. *Inst. Pasteur Viet-Nam. Rapp. Annu. Fonct. Tech.* 1958:16–23.

10-1.5
Anonymous. 1960.
Enterobacteries. *Inst. Pasteur Viet-Nam. Rapp. Annu. Fonct. Tech.* 1960:3–16.

10-1.7
Anonymous. 1960.
Microbiologie animale. *Inst. Pasteur Viet-Nam Rapp. Annu. Fonct. Tech.* 196:135.

10-2
Ash, L. R. 1968.
The occurrence of *Angiostrongylus cantonensis* in frogs of New Caledonia with observations on paratenic hosts of metastrongyles. *J. Parasitol.* 54(3):432–436.

10-3
Audy, J. R. 1960.
Parasites as 'ecological labels' in vertebrate ecology. Purchon, R. D., ed. *Proceedings of the Centenary and Bicentenary Congress of Biology, Singapore, 1958.* Singapore: University of Malaya Press. Pp. 123–127.

10-4
Audy, J. R.; Nadchatram, M.; Lim, B. L. 1960.
Malaysian parasites. XLIX. Host distribution of Malayan ticks (Ixodoidea). *Stud. Inst. Med. Res. Kuala Lumpur* No. 29:225–246.

10-5
Audy, J. R.; Nadchatram, M.; Vercammen Grandjean, P. H. 1963.
La "Neosomie", un phenomene inedit de neoformation en acarologie, alliee a un cas remarquable de tachygenese. *Bull. Acad. R. Belgique. Class Sciences*, Ser. 5, 49:1015–1027.

10-6
Audy, J. R.; Radovsky, F. J.; Vercammen Grandjean, P. H.. 1972.
Neosomy: radical intrastadial metamorphosis associated with arthropod symbioses. *J. Med. Entomol.* 9(6):487–494.

10-7
Beebe, W. 1926.
The Arcturus adventure: an account of the New York Zoological Society's First Oceanic Expedition. New York: G. P. Putnam's Sons.

10-8
Boulenger, G. A. 1898.
Exhibition of a large female specimen of the sea-snake *Distira stokesii. Proc. Zool. Soc. Lond.* 1898:851–852.

10-9
Brooks, D. R.; Overstreet, R. M. 1978.
The family Liolopidae (Digenea) including a new genus and two new species from crocodilians. *Int. J. Parasitol.* 8:267–273.

10-10

Brooks, D. R.; Caira, J. N. 1982.

Atrophecaecum lobacetabulare n. sp. (Digenea: Cryptogonimidae: Acanthostominae) with discussion of the generic status of *Paracanthostomum* Fischthal and Kuntz, 1965, and *Ateuchocephala* Coil and Kuntz, 1960. *Proc. Biol. Soc. Wash.* 95(2):223–231.

10-11

Bush, A. O.; Holmes, J. C. 1979.

Sterrhurus carpentariae nov. sp. (Digenea: Hemiuridae) from *Lapemis hardwickii* (Serpentes: Hydrophiidae) from the eastern Gulf of Carpentaria, Australia. *Int. J. Parasitol.* 9:189–192.

10-12

Bush, A. O.; Holmes, J. C. 1984.

Plicatrium visayanensis new species (Digenea: Hemiuridae) from *Hydrophis ornatus* (Serpentes: Hydrophiidae) from the Visayan Sea, Philippines. *Int. J. Parasitol.* 14(1):35–38.

10-13

Chattopadhyaya, D. R. 1970.

Studies on the trematode parasite of reptiles found in India (Digenetic flukes from the marine snakes of the Coast of Bombay). *Helminthologia (Bratisl.)* 11(1–4):35–62.

10-14

Coil, W. H. 1965.

Observations on egg shell formation in *Hydrophitrema gigantica* Sandars, 1960 (Hemiuridae: Digenea). *Z. Parasitenkd.* 25:510–517.

10-15

Coil, W. H.; Kuntz, R. E. 1960.

Three new genera of trematodes from Pacific sea serpents, *Laticauda colubrina* and *L. semifasciata*. *Proc. Helminthol. Soc. Wash.* 27(2):145–150.

10-16

Darwin, C. 1851.

A monograph on the sub-class Cirripedia. London: Ray Society. Weinheim: J. Cramer reprint; 1964.

10-17

Dean, B. 1938.

Note on the sea-snake, *Pelamas platurus* (Linnaeus). *Science (Wash. D.C.)* New Ser. 88(2276):145.

10-17.5
Diesing, C. M. 1851.
   Systema helminthum. Vindobonae: Brumueller. New York: Hafner Publishing Co.; 1960.

10-17.7
Dobell, C. C. 1911.
   On some parasitic protozoa from Ceylon. *Spolia Zeylan.* 7(26):65–87.

10-17.8
Dollfus, R. P. 1942.
   Etudes critiques sur les tetrarhynques du Museum de Paris. *Arch. Mus. Nat. Hist. Nat.* Ser. 6, 19:1–466.

10-18
Fischthal, J. H.; Kuntz, R. E. 1965.
   Digenetic trematodes of amphibians and reptiles from North Borneo (Malaysia). *Proc. Helminthol. Soc. Wash.* 32(2):124–136.

10-19
Fischthal, J. H.; Kuntz, R. E. 1975.
   Some trematodes of amphibians and reptiles from Taiwan. *Proc. Helminthol. Soc. Wash.* 42(1):1–13.

10-20
Harmer, S. F. 1931.
   Presidential address. *Proc. Linn. Soc. Lond.* 1930–1931:113–168.

10-21
Hirst, S.; Hirst, L. F. 1910.
   Descriptions of five new species of ticks (Ixodidae). *Ann. Mag. Nat. Hist.* Ser. 6, 8:299–308.

10-21.5
Hoek, P. P. C. 1887.
   On *Dichelapsis pellucida*, Darwin, from the scales of an hydrophid obtained at Mergui. *J. Linn. Soc. Lond. Zool.* 2(129):154–155.

10-21.6
Hughes, R. C.; Baker, J. R.; Dawson, C. B. 1941.
   The tapeworms of reptiles. Part I. *Am. Midl. Nat.* 25(2):454–468.

10-21.7
Hughes, R. C.; Baker, J. R.; Dawson, C. B. 1941.
   The tapeworms of reptiles. Part II. Host catalog. *Wasmann Collector* 4(3):97–104.

10-21.8
Hughes, R. C.; Higginbotham, J. W.; Clary, J. W. 1942.
   The trematodes of reptiles, Part 1, systematic section. *Am. Midl. Nat.* 27(1):109–134.

10-22
Institut Pasteur du Viet-Nam. 1960.
   Enterobacteries. *Rapport annuel sur le fonctionnement technique.* Hanoi. Pp. 3–15.

10-23
Jeffries, W. B.; Voris, H. K. 1977.
   The relationship between *Octolasmis grayii* (Darwin, 1851) (Cirripedia) and certain marine snakes. *Am. Zool.* 17(4):968. (Abstract)

10-24
Jeffries, W. B.; Voris, H. K. 1979.
   Observations on the relationship between *Octolasmis grayii* (Darwin, 1851) (Cirripedia, Thoracica) and certain marine snakes (Hydrophiidae). *Crustaceana* 37(2):123–132.

10-25
Jeffries, W. B.; Voris, H. K.; Man, Y. C. 1982.
   The distribution of the barnacle *Octolasmis* on crustacean and ophidian hosts in the seas near Singapore. *Am. Zool.* 22(4):922. (Abstract)

10-26
Johnston, T. H.; Mawson, P. M. 1948.
   Some new records of nematodes from Australian snakes. *Rec. S. Aust. Mus.* 9(1):101–106.

10-27
Kagei, N. 1973.
   The endoparasites of snakes in Japan. *Snake* 5(1–2):141–150. (In Japanese with English abstract)

10-28
Ko, R. C.; Lance, V.; Duggan, R. T. 1975.

Note on the biology of *Hydrophitrema gigantica* Sandars, 1960 (Trematoda: Hemiuridae) from the lung of sea snakes (*Hydrophis cyanocinctus*). *Can. J. Zool.* 53(8):1181–1184.

10-29
Kropach, C.; Soule, J. D. 1973.

An unusual association between an ectoproct and a sea snake. *Herpetologica* 29(1):17–19.

10-29.4
Kruger, P. 1911.

Beitraege zur Cirripedien fauna Ostasiens. *Abh. K. Bayer. Akad. Wiss. Math.-Phys. Kl.* Supp. 2(6):1–72.

10-29.6
Kruger, P. 1912.

Ueber ostasiatische Rhizocephalen. *Abh. K. Bayer. Akad. Wiss. Math.-Phys. Kl.* Supp. 2(8):1–16.

10-30
Lanchester, W. F. 1902.

On the crustacea collected during the "Skeat Expedition" to the Malay Peninsula. Part II. Anomura, Cirripedia, and Isopoda. *Proc. Zool. Soc. Lond.* 2:363–380.

10-31
Loomis, R. B.; Wrenn, W. J. 1984.

Systematics of the pest chigger genus *Eutrombicula* (Acari: Trombiculidae). Griffiths, D. A.; Bowman, C. E., eds. *Acarology VI*. Chichester: Ellis Horwood. Vol. 1, pp. 152–159.

10-32
Madhavi, R.; Rao, K. H. 1973.

Record of *Hydrophitrema gigantica* Sandars, 1960 (Trematoda: Hemiuridae) from sea snakes of Waltair Coast Bay of Bengal. *Curr. Sci.* 42(12):425–427.

10-33
Mao, S. H.; Chen, B. Y. 1979.

The stalked barnacle *Concoderma virgatum* var. *hunteri* (Owen) found on the sea snake *Pelamis platurus* (Linnaeus). *Snake* 11(2):242–244.

10-34
Nadchatram, M. 1980.
   The genus *Iguanacarus*, new status (Acari: Prostigmata: Trombiculidae), with description of a new species from the tracheae of amphibious sea snakes. *J. Med. Entomol.* 17(6):529–532.

10-35
Nadchatram, M.; Audy, J. R. 1965.
   The unusual life-history of *Vatacarus ipoides* Southcott (Acarina, Trombiculidae). *Med. J. Malaya* 20:80–81.

10-36
Nadchatram, M.; Radovsky, F. J. 1971.
   A second species of *Vatacarus* (Prostigmata, Trombiculidae) infesting the trachea of amphibious sea snakes. *J. Med. Entomol.* 8(1):37–40.

10-37
Nicoll, W. 1918.
   The trematode parasites of North Queensland. IV. Parasites of reptiles and frogs. *Parasitology* 10(3).368–374.

10-37.5
Nilsson Cantell, C. A. 1921.
   Cirripeden-studien. Zur kenntnis der biologie, anatomie und systematik dieser gruppe. *Zool. Bidrag Uppsala* 7:75–395.

10-38
Nilsson Cantell, C. A. 1930.
   Cirripedes. *Mem. Mus. Hist. Nat. Belg.* 3(3):1–24.

10-38.5
Nilsson Cantell, C. A. 1934.
   Cirripeds from the Malay Archipelago in the Zoological Museum of Amsterdam. *Zool. Meded. (Leiden)* 17(1–2):31–63.

10-38.7
Nilsson Cantell, C. A. 1938.
   Cirripedes from the Indian Ocean in the collection of the Indian Museum, Calcutta. *Mem. Indian Mus.* 13(1):1–81.

10-39
Pilsbry, H. A. 1916.

The sessile barnacles (Cirripedia) contained in the collections of the U.S. National Museum; including a monograph of the American species. *Bull. U.S. Natl Mus.* No. 93:1–366.

10-39.4
Rageau, J.; Vervent, G. 1958.

Les tiques (Acariens, Ixodoidea) des Iles Francaises du Pacifique. *Bull. Soc. Pathol. Exot.* 52(6):819–835.

10-39.6
Rageau, J. 1960.

A propos d'*Amblyomma laticaudae* Warburton, 1933 (Acarien, Ixodidae) en Nouvelle-Caledonie. *Bull. Soc. Pathol. Exot.* 53(5):831–833.

10-40
Rageau, J. 1967.

Observations biologiques sur les tiques (Acari, Argasidae et Ixodidae) des iles Francaises d'oceanie. *Wiad. Parazytol.* 13(4–5):547–553.

10-41
Roskell, J. 1969.

A note on the ecology of *Conchoderma virgatum* (Spengler, 1790) (Cirripedia, Lepadomorpha). *Crustaceana* 16(1):103–104.

10-42
Sandars, D. F. 1960.

Hydrophitrema gigantica, n. gen., n. sp. (Trematoda, Digenea) from *Hydrophis elegans* (Gray, 1842) from Australia. *Libro homenaje al Dr. Eduardo Caballero y Caballero.* Mexico. Pp. 263–268.

10-43
Sandars, D. F. 1961.

Pulmoverminae n. subfam. (Trematoda: Digenea) for the genera *Pulmovermis* and *Hydrophitrema*. *J. Helminthol.* 35(3–4):305–308.

10-44
Schad, G. A. 1962.

Studies on the genus *Kalicephalus* (Nematoda: Diaphanocephalidae). II. A taxonomic revision of the genus *Kalicephalus* Molin, 1861. *Can. J. Zool.* 40:1035–1165.

10-45
Schmidt, G. D.; Kuntz, R. E. 1973.

Nematode parasites of Oceanica. XX. *Paraheterotyphlum ophiophagos* n. sp. (Heterocheilidae), from the banded yellow-lip sea snake, *Laticauda colubrina*. *Am. Midl. Nat.* 89(2):481–484.

10-46
Southcott, R. V. 1957.

On *Vatacarus ipoides* n. gen., n. sp. (Acarina: Trombidioidea) a new respiratory endoparasite from a Pacific sea-snake. *Trans. R. Soc. S. Aust.* 80:165–176.

10-47
Sprent, J. F. A. 1978.

Ascaridoid nematodes of amphibia and reptiles: *Paraheterotyphlum*. *J. Helminthol.* 52(2):163–170.

10-48
Sprent, J. F. A. 1978.

Ascaridoid nematodes of amphibians and reptiles: *Goezia*. *J. Helminthol.* 52(1):91–98.

10-49
Telford, S. R. 1967.

Studies on the parasites of oriental reptiles. I. Parasitology of the seasnake, *Laticauda semifasciata*, in the vicinity of Amami Island, Japan. *Jpn. J. Exp. Med.* 37(3):245–256.

10-50
Telford, S. R. 1970.

Laticaudatrema amamiensis Telford, 1967, a junior synonym of *Pulmovermis cyanovitellosus* Coil and Kuntz, 1960. *J. Parasitol.* 56(3):430.

10-51
Tubangui, M. A.; Masilungan, V. A. 1935.

Trematode parasites of Philippine vertebrates, VII. Additional records of new species. *Philipp. J. Sci.* 58(4):435–445.

10-51.5
Van der Meer Mohr, J. C. 1934.

Korte zoologische aanteekeningen II. *Trop. Natuur (Weltevreden)* 23(100):190–193.

10-52
Vercammen Grandjean, P. H. 1960.
Vatacarus ipoides Southcott 1957 est une larve de Trombiculidae apparentee a *Eutrombicula*. *Acarologia* 2(2):217–223.

10-53
Vercammen Grandjean, P. H. 1965.
Iguanacarus, a new subgenus of chigger mites from nasal fossae of the marine iguana in the Galapagos Islands, with a revision of the genus *Vatacarus* Southcott (Acarina, Trombiculidae). *Acarologia* 7(Suppl):266–274.

10-54
Vercammen Grandjean, P. H. 1966.
Evolutionary problems of mite-host specificity and its relevance to studies on Galapagos organisms. Bowman, R. I., ed. *The Galapagos*. Los Angeles: University of California Press. Pp. 236–239.

10-55
Vercammen Grandjean, P. H. 1969.
Le stade larvaire, sanctuaire de la phylogenie et de la taxonomie chez les Acariens. *Ann. Parasitol. Hum. Comp.* 44(2):205–210.

10-56
Vercammen Grandjean, P. H. 1971.
Eutrombicula (E.) poppi, a new chigger from a sea snake (Acarina, Trombiculidae). *Opusc. Zool.* No. 114:1–3.

10-57
Vercammen Grandjean, P. H.; Heyneman, D. 1964.
Pulmovermis and *Hydrophitrema*, hemiurid lung flukes of sea snakes: new host records with a corrigendum and reevaluation. *J. Helminthol.* 38(3–4):369–382.

10-57.5
Wang, P. Q. 1980.
Report on some trematodes from amphibians and reptiles in Fujian, South China. *Fujian Shida Xuebao* 1980(2):81–92. (In Chinese)

10-58
Warburton, C. 1932.
On five new species of ticks (Arachnida Ixodoidea). *Ixodes petauristae*, I. *Ampullaceus, Dermacentor imitans, Amblyomma laticaudae* and *Aponomma draconis*, with notes on three previously described species, *Ornithodorus franchinii*, Tonelli-Rondelli, *Haemaphysalis cooleyi* Bedford and *Rhipicephalus maculatus* Neumann. *Parasitology* 24(4):558–568.

10-59
Willey, A. 1910.
 Association of barnacles with snakes and worms. *Spolia Zeylan.* 6:180–181.

10-60
Wilson, N. 1970.
 New distributional records of ticks from Southeast Asia and the Pacific (Metastigmata: Argasidae, Ixodidae). *Orient. Insects* 4(1):37–46.

10-61
Yamaguti, S. 1933.
 Studies on the helminth fauna of Japan. 1. Trematodes of birds, reptiles, and mammals. *Jpn. J. Zool.* 5(1):1–134.

10-62
Yamaguti, S. 1935.
 Studies on the helminth fauna of Japan. Part 11. Reptilian nematodes. *Jpn. J. Zool.* 6:393–402.

10-63
Yamaguti, S. 1958.
 Systema helminthum. New York: Interscience Publishers. Volume 1.

10-63.5
Yamaguti, S. 1961.
 Systema helminthum. New York: Interscience Publishers. Volume 3.

10-64
Yamaguti, S. 1971.
 Synopsis of digenetic trematodes of vertebrates. Toyko: Keigaku Publ. Co. Vol. 1.

10-65
Zann, L. P. 1975.
 Biology of a barnacle (*Platylepas ophiophilus* Lanchester) symbiotic with sea snakes. Dunson, W. A., ed. *The biology of sea snakes.* Baltimore: University Park Press. Pp. 267–286.

10-66
Zann, L. P.; Cuffey, R. J.; Kropach, C. 1975.
 Fouling organisms and parasites associated with the skin of sea snakes. Dunson, W. A., ed. *The biology of sea snakes.* Baltimore: University Park Press. Pp. 251–265.

10-67
Zann, L. P.; Harker, B. M. 1978.
   Egg production of the barnacles *Platylepas ophiophilus* Lanchester, *Platylepas hexastylos* (O. Fabricius), *Octolasmis warwickii* Gray and *Lepas anatifera* Linnaeus. *Crustaceana* 35(2):206–214.

Genus-Species Index
for Chapter 10

Chapter 11

REPRODUCTION AND GROWTH

11-1
Andrews, R. M. 1982.
   Patterns of growth in reptiles. Gans, C.; Pough, F. H., eds. *Biology of the reptilia*. New York: Academic. Vol. 13, pp. 273–320.

11-2
Bacolod, P. T. 1983.
   Reproductive biology of two sea snakes of the genus *Laticauda* from Central Philippines. *Philipp. Sci.* 20:39–56.

11-2.5
Bauchot, R. 1965.
   La placentation chez les reptiles. *Ann. Biol.* 4(4):547–575.

11-2.7
Behler, J. L. 1979.
   Operation sea snakes. *Anim. Kingdom Rept.* 82(6):N2–N3.

11-3
Bergman, R. A. M. 1943
   The breeding habits of sea snakes. *Copeia* 1943(3):156–160.

11-4
Blackburn, D. G. 1985.
   Evolutionary origins of viviparity in the Reptilia. II. Serpentes, Amphisbaenia, and Ichthyosauria. *Amphib.-Reptilia* 6(3):259–291.

11-5

Burns, G. W. 1984.

Aspects of population movements and reproductive biology of Aipysurus laevis, the olive sea snake. Armidale,New South, Wales: University of New England. 171 pp. (Dissertation)

11-6

Chhapgar, B. F.; Kewalramani, H. G. 1967.

Mating and other observations on sea snakes in captivity. *J. Bombay Nat. Hist. Soc.* 64(3):563–564.

11-7

Choo, T. P. 1965–1966.

A preliminary study of the seasonal cycle of the gonads and growth rate of sea snake *Laticauda colubrina* (Hydrophiidae). Report for B.Sc., Univ. of Singapore.

11-7.3

Delsman, H. C. 1932.

Eierleggende zeeslangen? *Trop. Natuur (Weltevreden)* 21(6):98–99.

11-7.5

Dunham, A. E.; Miles, D. B.; Reznik, D. N. 1988.

Life history patterns in squamate reptiles. Gans, C.; Huey, R., eds. *Biology of the reptilia.* New York: Alan R. Liss. Vol. 16B, pp. 441–552.

11-8

Fitch, H. S. 1970.

Reproductive cycles of lizards and snakes. *Univ. Kans. Mus. Nat. Hist. Misc. Publ.* No. 52:1–247.

11-9

Fukada, H. 1965.

Breeding habits of some Japanese reptiles (critical review). *Bull. Kyoto Gakugei Univ. Ser. B Math. Nat. Sci.* No. 27:65–82.

11-10

Gorman, G. C.; Licht, P.; McCollum, F. 1981.

Annual reproductive patterns in three species of marine snakes from the central Phillippines. *J. Herpetol.* 15(3):335–354.

11-11
Guibe, J. 1970.
    La reproduction. Grasse, P. P. *Traite de zoologie*. Paris: Masson et Cie. Vol. 14, no. 3, pp. 859–892.

11-12
Harrison, R. J. 1956.
    Corpus luteum of pregnancy: vertebrates. Spector, W. S., ed. *Handbook of biological data*. Philadelphia: W. B. Saunders. Pp. 126–127.

11-13
Honma, Y.; Yoshie, S.; Ishiyama, M. 1982.
    Histological studies on some organs of a pregnant sea snake, *Pelamis platurus*, caught in summer off Northern coast of Sado Island in the Sea of Japan. *Ann. Rep. Sado Mar. Biol. Stn Niigata Univ.* No. 12:1–15.

11-14
Kasturirangan, L. R. 1951.
    Placentation in the sea-snake, *Enhydrina schistosa* (Daudin). *Proc. Indian Acad. Sci. Sect. B* 34:1–32.

11-15
Kasturirangan, L. R. 1951.
    The allantoplacenta of the sea snake *Hydrophis cyanocinctus* Daudin. *J. Zool. Soc. India* 3(2):277–290.

11-16
Lance, V. 1984.
    Endocrinology of reproduction in male reptiles. *Symp. Zool. Soc. Lond.* No. 52:357–383.

11-17
Lemen, C. A.; Voris, H. K. 1981.
    A comparison of reproductive strategies among marine snakes. *J. Anim. Ecol.* 50:89–101.

11-18
Lofts, B. 1978.
    Reptilian reproductive cycles and environmental regulators. Assenmacher, I.; Farner, D. S., eds. *Environmental endocrinology*. New York: Springer-Verlag. Pp. 37–43.

11-18.5
Matsui, T. 1972.

On an embryo of an oviparous sea-snake, *Laticauda laticaudata* (Linne). *Jpn. J. Herpetol.* 5(1):13–14.

11-18.7
Matthews, L. H. 1955.

The evolution of viviparity in vertebrates. Jones, I. C.; Eckstein, P. The comparative endocrinology of vertebrates, Part I. The comparative physiology of reproduction and the effects of sex hormones in vertebrates. *Mem. Soc. Endocrinol.* 4:129–148.

11-19
McCann, C. 1937.

Breeding season of the Jew's nosed seasnake (*Enhydrina valakadyen* (Boie)) in Bombay waters. *J. Bombay Nat. Hist. Soc.* 39(3–4):872–873.

11-19.5
Nagai, K. 1928.

Trimeresurus tokarensis and *Laticauda semifasciata. Kagoshimaken Hakubutsu Chosajhe* 1928(3):1–64. (In Japanese)

11-20
Nakamoto, E.; Toriba, M. 1986.

Successful artificial incubation of the eggs of erabu sea snake, *Pseudolaticauda semifasciata* (Reinwardt). *Snake* 18:55–56.

11-21
Neill, W. T. 1964.

Viviparity in snakes: some ecological and zoogeographical considerations. *Am. Nat.* 98(898):35–55.

11-22
Okada, Y. 1933.

Trimeresurus tokarensis and *Laticauda semifasciata* by K. Nagai. *Copeia* 1933(4):227. (Review)

11-23
Pernetta, J. C. 1977.

Observations on the habits and morphology of the sea snake *Laticauda colubrina* (Schneider) in Fiji. *Can. J. Zool.* 55(10):1612–1619.

11-24
Pimento, R. J. 1972.
   Some notes on the sea snake *Laticauda colubrina* (Schneider). *J. Bombay Nat. Hist. Soc.* 69(1):191–192.

11-25
Poyntz, A. R. 1927.
   The pairing of sea snakes. *J. Bombay Nat. Hist. Soc.* 31:1038–1039.

11-26
Raj, B. S. 1926.
   Parturitions of electric rays and a sea snake in the Marine Aquarium, Madras. *J. Bombay Nat. Hist. Soc.* 31:828.

11-26.5
Rasmussen, A. R. 1989.
   An analysis of *Hydrophis ornatus*(Gray), *H. lamberti* (Smith), and *H. inornatus* (Gray) (Hydrophiidae, Serpentes) based on samples from various localities, with remarks on feeding and breeding biology of *H. ornatus*. *Amphib.-Reptilia* 10:397–417.

11-27
Saint Girons, H. 1966.
   Le cycle sexuel des serpents venimeux. *Mem. Inst. Butantan* 33(1):105–114.

11-27.4
Saint Girons, H. 1975.
   Criteres d'age structure et dynamique des populations des reptiles. Lamotte, M.; Bourliere, F., eds. *Problems d'ecologie: la demographie des populations de vertebres*. Paris: Masson et C[ie]. Pp. 233–252.

11-27.6
Saint Girons, H. 1985.
   Comparative data on Lepidosaurian reproduction and some time tables. Gans, C.; Billett, F., eds. *Biology of the reptilia*. New York: Wiley. Vol. 158, pp. 35–58.

11-28
Samuel, M. 1944.
   Studies on the corpus luteum in *Enhydrina schistosa* (Daudin) and *Hydrophis cyanocinctus* (Daudin) of the Madras Coast. *Proc. Indian Acad. Sci. Sect. B.* 20(5):143–174.

11-28.4
Shine, R. 1985.
   The evolution of viviparity in reptiles: an ecological analysis. Gans, C.; Billett, F., eds. *Biology of the reptilia*. New York: Wiley. Vol. 15B, pp. 605–694.

11-28.6
Shine, R. 1988.
   Parental care in reptiles. Gans, C.; Huey, R., eds. *Biology of the reptilia*. New York: Alan R. Liss. Vol. 16B, pp. 275–329.

11-29
Smedley, N. 1930.
   Oviparity in a sea-snake (*Laticauda colubrina*). *Nature (Lond.)* 126(3174):312–313.

11-30
Smedley, N. 1931.
   Oviparity in a sea-snake, *Laticauda colubrina* (Schneid). *Bull. Raffles Mus.* 5:54–59.

11-31
Smedley, N. 1931.
   Ovoviviparity in sea-snakes. *Nature (Lond.)* 127(3192):13.

11-32
Smith, M. 1930.
   Ovoviviparity in sea-snakes. *Nature (Lond.)* 126(3180):568.

11-32.5
Tinkle, D. W.; Gibbons, J. W. 1977.
   The distribution and evolution of viviparity in reptiles. *Misc. Publ. Mus. Zool. Univ. Mich.* 154:1–55.

11-33
Tochimoto, T.; Uchida, I. 1973.
   On the artificial incubation and breeding of sea snake (*Laticauda semifasciata*). *Jpn. J. Herpetol.* 5(2):27. (In Japanese) (Abstract)

11-33.5
Toriba, M.; Nakamoto, E. 1987.
   Reproductive biology of erabu sea snake, *Laticauda semifasciata*. *Snake* 19:101–106.

11-34
Visser, J. 1967.

Color varieties, brood size, and food of South African *Pelamis platurus* (Ophidia: Hydrophiidae). *Copeia* 1967:219.

11-35
Voris, H. K.; Jayne, B.C. 1979.

Growth, reproduction and population structure of a marine snake, *Enhydrina schistosa* (Hydrophiidae). *Copeia* 1979(2):307–318.

11-36
Wall, F. 1911.

Notes on a brood of young sea-snakes (*Distira spiralis*, Shaw). *J. Bombay Nat. Hist. Soc.* 20:858–863.

11-37
Wall, F. 1918.

Notes on a gravid *Hydrophis cyanocinctus* and her brood. *J. Bombay Nat. Hist. Soc.* 25(4):754–756.

11-38
Yaron, Z. 1985.

Reptilian placentation and gestation: structure, function, and endocrine control. Gans, C.; Billett, F., eds. *Biology of the reptilia*. New York: Wiley. Vol. 15B, pp. 527–603.

Genus-Species Index
for Chapter 11

Chapter 12

ANATOMY, MORPHOLOGY, COLORS, AND PATTERNS

12-.05

Alexander, A. A.; Gans, C. 1966.

The pattern of dermal-vertebral correlation in snakes and amphibians. *Zool. Meded.* (*Leiden*) 41(11):171–190.

12-1

Baird, I. L. 1970.

The anatomy of the reptilian ear. Gans, C.; Parsons, T. S., eds. *Biology of the reptilia.* New York: Academic. Vol. 2, pp. 193–275.

12-2

Bal, D. V.; Nawathe, K. V. 1948.

The circulatory system of *Hydrophis caerulescens* (Shaw 1802). *34th Indian Sci. Cong. Assoc. Proc.* 3:186 (Abstract)

12-3

Bal, D. V.; Navathe, K. V. 1949.

The circulatory system of *Hydrophis caerulescans,* (Shaw). *J. Univ. Bombay,* New Ser. 17(5):1–14.

12-4

Becak, W. 1965.

Constituicao cromossomica e mecanismo de determinacao do sexo em ofidios sul-americanos. I. Aspectos cariotipicos. *Mem. Inst. Butantan* (*Sao Paulo*) 32:37–78.

12-4.5

Beddard, F. E. 1903.

On the trachea, lungs, and other points in the anatomy of the hamadryad snake (*Ophiophagus bungarus*). *Proc. Zool. Soc. Lond.* 1903:(2):319–328.

12-5
Beddard, F. E. 1904.
Contributions to the knowledge of the visceral anatomy of the pelagic serpents *Hydrus platyurus* and *Platyurus colubrinus*. *Proc. Zool. Soc. Lond.* 2:147–154.

12-6
Bergman, R. A. M. 1936.
Het uitvoerstelsel van pancreas en galblaas bij zeeslangen. *Ned. Tijdschr. Geneeskd.* 80(2):1679–1682.

12-7
Bergman, R. A. M. 1938.
Een nieuw orgaantje bij zeeslangen. *Geneesk. Tijdschr. Ned.-Indie* 78(34):2061–2070.

12-8
Bergman, R. A. M. 1949.
The anatomy of *Lapemis hardwickei* Gray I-II. *Proc. K. Ned. Akad. Wet.* 52(8):882–898.

12-9
Bergman, R. A. M. 1954.
Thalassophis anomalus Schmidt. Amsterdam: Koninklijk Instituut voor de Tropen. (In French)

12-10
Bergman, R. A. M. 1955.
L'anatomie de *Enhydrina schistosa* D. *Arch. Neerl. Zool.* 11(1):127–142.

12-11
Bergman, R. A. M. 1962.
The anatomy of *Hydrophis fasciatus atriceps*. *Biol. Jaarb. (Ghent)* 30:389–416.

12-12
Bhatnagar, H. M. 1957.
The fronto-parietal region in the skull of *Hydrophis spiralis* Shaw. *Sci. Cult.* 23(4):196.

12-13
Boettger, O. 1888.
Ueber aussere geschlechtscharactere bei den Seeschlangen. *Zool. Anz.* 11:395–398.

12-14
Boettger, O. 1899.
   Bau, lebensweise und unterscheidung der schlangen. *Ber. Senckenb. Nat. Ges. Frankfurt a. Main* 1899:75–88.

12-15
Bogert, C. M. 1943.
   Dentitional phenomena in cobras and other elapids with notes on adaptive modifications of fangs. *Bull. Am. Mus. Nat. Hist.* 81:285–360.

12-16
Bolanos, R.; Flores, A.; Taylor, R. T.; Cerdas, L. 1974.
   Color patterns and venom characteristics in *Pelamis platurus*. *Copeia* 1974(4):909–912.

12-17
Boulenger, G. A. 1888.
   Description of two new snakes from Hongkong, and note on the dentition of *Hydrophis viperina*. *Ann. Mag. Nat. Hist.* Ser. 6, 2:43–45.

12-18
Boulenger, G. A. 1890.
   Exhibition and remarks upon the skull of a large specimen of a sea-snake, *Distira cyanocinta*, and three skulls of the green turtle. *Proc. Zool. Soc. Lond.* 1890(6):617–618.

12-19
Boulenger, G. A. 1896.
   Remarks on the dentition of snakes and on the evolution of the poison-fangs. *Proc. Zool. Soc. Lond.* 1896:614–616.

12-20
Buffa, P. 1905.
   Ricerche sulla muscolatura cutanea dei serpenti e considerazioni sulla locomozione di questi animali. *Atti Accad. Veneto-Trent.-Istriana* New Ser. 1:145–228.

12-21
Burns, B. 1969.
   Oral sensory papillae in sea snakes. *Copeia* 1969(3):617–619.

12-22

Burns, B. 1969.

An oral sensory system in sea serpents (Hydrophiidae): histological and morphological considerations. Waterloo, Ontario, Canada: University of Waterloo. 22 pp. (Thesis)

12-23

Burns, B.; Pickwell, G. V. 1972.

Cephalic glands in sea snakes (*Pelamis, Hydrophis* and *Laticauda*). *Copeia* 1972(3):547–559.

12-24

Butler, G. W. 1892.

On the subdivision of the body-cavity in snakes. *Proc. Zool. Soc. Lond.* 1892:477–498.

12-25

Butler, G. W. 1895.

On the complete or partial suppression of the right lung in the *Amphisbaenidae* and of the left lung in snakes and snake-like lizards and amphibians. *Proc. Zool. Soc. Lond.* 1895:691–712.

12-26

Carus, C. G. 1835.

Traite elementaire d'anatomie comparee. Paris: J. B. Bailliere. Volume 2.

12-27

Chabanaud, P. 1923.

Description d'un chamaeleon nouveau d'Indo-Chine et d'un exemplaire monstrueux d'*Enhydris hardwicki* Gray. *Bull. Mus. Natl Hist. Nat.* Ser. 1, 29:209–210.

12-27.5

Clark, P. J.; Inger, R. F. 1942.

Scale reduction in snakes. *Copeia* 1942(3):163–170.

12-27.7

Cope, E. D. 1894.

The classification of snakes. *Am. Nat.* 28:831–844.

12-28

Cope, E. D. 1894.

On the lungs of the ophidia. *Proc. Am. Philos. Soc.* 33:217–223.

12-29
Cunningham, B. 1937.
   Axial bifurcation in serpents: an historical survey of serpent monsters having part of the axial skeleton duplicated. Durham, NC: Duke University Press.

12-29.5
Deraniyagala, P. E. P. 1949.
   Some vertebrate animals of Ceylon. Colombo.

12-30
Dunson, W. A.; Packer, R. K.; Dunson, M. K. 1971.
   Sea snakes: an unusual salt gland under the tongue. *Science (Wash. D.C.)* 173(3995):347–441.

12-31
Duvernoy, G. L. 1832.
   Memoire sur les caracteres tires de l'anatomie pour distinguer les serpens venimeux des serpens non venimeux. *Ann. Sci. Nat.* 26:113–160.

12-32
Duvernoy, G. L. 1833.
   Fragmens d'anatomie sur l'organisation des serpens. *Ann. Sci. Nat.* 30:5–32.

12-33
Edmund, A. G. 1960.
   Tooth replacement phenomena in the lower vertebrates. *Contrib. Life Sci. Div. R. Ont. Mus.* 52:3–190.

12-34
Edmund, A. G. 1970.
   Dentition. Gans, C.; Parsons, T. S., eds. *Biology of the reptilia.* New York: Academic. Vol. 1, pp. 117–200.

12-35
Emery, C. 1880.
   Intorno alle glandole del capo di alcuni serpenti proteroglifi. *Ann. Mus. Civ. Genova* 15:546–558.

12-35.5
Fahrenholz, C. 1937.
   Druesen der mundhoehle. Bolk, L.; Goeppert, E.; Kallius, E.; Lubosch, W. *Handbuch der vergleichenden anatomie der wirbeltiere.* Berlin: Urban & Schwarzenberg. Vol. 3, pp. 115–210.

12-36
Fitch, H. S. 1981.
   Sexual size differences in reptiles. *Univ. Kans. Mus. Nat. Hist. Misc. Publ.* No. 70:1–72.

12-37
Fox, H. 1977.
   The urinogenital system of reptiles. Gans, C.; Parsons, T. S., eds. *Biology of the reptilia.* New York: Academic. Vol. 6, pp. 1–157.

12-37.5
Friederich, U. 1978.
   Der pileus der squamata. *Stuttg. Beitr. Naturkd. Ser.A (Biol.)* 307:1–64.

12-38
Gabe, M. 1970.
   Donnees histologiques sur le pancreas endocrine des lepidosauriens (Reptiles). *Adv. Anat. Embryol. Cell Biol.* 42(2):1–63.

12-39
Gabe, M. 1970.
   Pancreas endocrine. Grasse, P. P. *Traite de zoologie.* Paris: Masson et C[ie]. Vol. 14, no. 3, pp. 1333–1399.

12-40
Gabe, M.; Saint Girons, H. 1969.
   Donnees histologiques sur les glandes salivaires des Lipidosauriens. *Mem. Mus. Natl Hist. Nat. Ser.A Zool.* 58(1):1–112.

12-41
Gabe, M.; Saint Girons, H. 1971.
   Histologie comparee des glandes salivaires du vestibule buccal chez les lezards et les serpents et evolution de la fonction venimeuse. DeVries, A.; Kochva, E. eds. *Toxins of animal, plant origins.* New York: Gordon and Breech. Vol. 1, pp. 65–69.

12-42
George, C. J.; Varde, M. R. 1941.
   A note on the modification of the lung and the trachea in some Indian snakes. *J. Univ. Bombay* New Ser. 108:70–73.

12-43
Gopalakrishnakone, P. 1985.

Structure of the spinous scales of *Lapemis hardwickii* (1). Light, transmission electron and scanning electron microscopic study. *Snake* 17(2):148–155.

12-43.5
Gopalakrishnakone, P. 1989.

Structure of the thyroid gland of the sea snake *Hydrophis cyanocinctus*. *Snake* 21(1):44–48.

12-44
Griffiths, I. 1962.

Skeletal lamellae as an index of age in heterothermous tetrapods. *Ann. Mag. Nat. Hist.* Ser. 13, 4(44):449–465.

12-45
Guerin Meneville, F. E. 1844.

Iconographie du regne animal de G. Cuvier, ou representation d'apres natur de l'une des especes les plus remarquables, et souvent non encore figurees, de chaque genre d'animaux. Paris: J. B. Bailliere. Volume 3.

12-45.5
Guibe, J. 1970.

L'appareil respiratoire. Grasse, P. P. *Traite de zoologie.* Paris: Masson et Cie. Vol. 14, no. 2, pp. 499–520.

12-46
Haas, G. 1930.

Ueber die schadelmechanik und die kiefermuskulatur einiger Proteroglypha. *Zool. Jahrb. Abt. Anat. Ontog. Tiere* 52(2):347–404.

12-47
Haas, G. 1931.

Die kiefermuskulatur und die schadelmechanik der schlangen in vergleichender darstellung. *Zool. Jahrb. Abt. Anat. Ontog. Tiere* 53:127–198.

12-48
Haas, G. 1973.

Muscles of the jaws and associated structures in the Rhynchocephalia and Squamata. Gans, C; Parsons, T. S., eds. *Biology of the reptilia.* New York: Academic. Vol. 4, pp. 285–490.

12-49

Hager, P. K. 1905.

Die kiefermuskeln der schlangen und ihre beziehungen zu den speicheldrusen. *Zool. Jahrb. Abt. Anat. Ontog. Tiere* 22:173–224.

12-49.5

Hallowell, E. 1845.

Remarks on the geographical distribution of reptiles, with descriptions of several species supposed to be new, and corrections of former papers. *Proc. Acad. Nat. Sci. Phil.* 7:98–105.

12-50

Harnack, M. von. 1953.

Die hautzeichnungen der schlangen. *Z. Morphol. Oekol. Tiere* 41:513–573.

12-51

Hattori, Z.; Inoue, T.; Okonogi, T. 1974.

In front of a pile of sea snake carcasses. *Snake* 6(2):94–98. (In Japanese with English abstract)

12-52

Heatwole, H.; Seymour, R. 1975

Diving physiology. Dunson, W. A., ed. *The biology of sea snakes*. Baltimore: University Park Press. Pp. 289–327.

12-52.3

Hediger, H. 1932.

Zum problem der "fliegenden" schlangen. *Rev. Suisse Zool.* 39(5):239–246.

12-52.6

Henle, J. 1839.

Vergleichend-anatomische beschreibung des kehlkopfs mit besonderer beruecksichligung des kehlkopfs der reptilien. Leipzig: Leopold Voss.

12-52.8

Hesse, R. 1910.

Der tierkoerper als selbstandiger organismus. Hesse, R.; Dolflein, F. *Tierbau und tierleben in ihrem zusammenhang*. Leipzig: G. B. Teubner. Volume 1.

12-53

Hibbard, E.; Lavergne, J. 1972.

Morphology of the retina of the sea-snake, *Pelamis platurus*. *J. Anat.* 112(1):125–136.

12-54
Hibbard, E. 1975.
Eyes and other sense organs of sea snakes. Dunson, W. A., ed. *The biology of sea snakes*. Baltimore: University Park Press. Pp. 355–382.

12-54.5
Hilzheimer, M.; Haempel, O. 1913.
Handbuch der biologie der wirbeltiere. Stuttgart: Ferdinand Enke.

12-55
Hoffstetter, R. 1939.
Contribution a l'etude des Elapidae actuels et fossiles et de l'osteologie des ophidiens. *Arch. Mus. Hist. Nat. Lyon* 15:1–78.

12-56
Hoffstetter, R.; Gasc, J. P. 1970.
Vertebrae and ribs of modern reptiles. Gans, C.; Parsons, T. S., eds. *Biology of the reptilia*. New York: Academic. Vol. 1, pp. 201–310.

12-57
Honma, Y.; Yoshie, S.; Ishiyama, M. 1982.
Histological studies on some organs of a pregnant sea snake, *Pelamis platurus*, caught in summer off northern coast of Sado Island in the Sea of Japan. *Ann. Rep. Sado Mar. Biol. Stn Niigata Univ.* No. 12:1–15.

12-58
Hunter, J. 1861.
Essays and observations on natural history, anatomy, physiology, psychology, and geology. London: John van Voorst. Volume 2.

12-59
Jacobeshagen, E. 1937.
Mittel und enddarm (Rumpfdarm). Bok, L.; Kallius, E.; Goppert, E.; Lubasch, W., eds. *Handbuch vergleichenden anatomie der wirbeltiere*. Berlin: Urban und Schwarzenberg. Vol. 3, pp. 638–654.

12-60
Jan, G. 1859.
Prodome d'une iconographie descriptive des ophidiens, et description sommaire de nouvelles especes de serpents venimeux. Paris.

12-61
Kathariner, L. 1897.
   Ueber bildung und ersatz der giftzahne bei giftschlangen. *Zool. Jahrb. Abt. Anat. Ontog. Tiere* 10:55–92.

12-62
Kathariner, L. 1899.
   Anatomische eigentumlichkeiten im bau der nase der im wasser lebenden schlangen. *Bull. Soc. Fribourg. Sci. Nat.* 7:186–195.

12-63
Kathariner, L. 1900.
   Die nase der im wasser lebenden schlangen als luftweg und geruchsorgan. *Zool. Jahrb. Abt. Syst. Oekol. Geogr. Tiere* 13:415–442.

12-64
Kiran, U. 1979.
   Evolutionary significance of the relative disposition of the bones in the jaw complex of snakes and lizards. *Ann. Zool. (Agra)* 15(2):59–78.

12-65
Kochva, E. 1978.
   Oral glands of the reptilia. Gans, C.; Gans, K. A., eds. *Biology of the reptilia.* New York: Academic Press. Vol. 8, pp. 43–161.

12-65.5
Kopstein, F. 1930.
   Herpetologische notizen. III. Reptilien des oestlichen Preanger (West Java). *Treubia* 12(3–4):273–276.

12-66
Kropach, C. 1971.
   Another color variety of the sea snake *Pelamis platurus* from Panama Bay. *Herpetologica* 27(3):326–327.

12-67
Kukihara, T. 1959.
   Studies on the nerve distribution in thymus. 2. On the nerve distribution in the thymus of hen, Japanese terrapin, and banded seasnake. *Med. J. Kagoshima Univ.* 32:1108–1120. (In Japanese with English abstract)

12-68
Lakjer, T. 1926.
Studien ueber die trigeminus-versorgte kaumuskulatur der sauropsiden. Kopenhagen: C. A. Reitzel.

12-69
Langebartel, D. A. 1968.
The hyoid and its associated muscles in snakes. Chicago, IL: University of Illinois Press.

12-70
Lindsey, C. C. 1975.
Pleomerism, the widespread tendency among related fish species for vertebral number to be correlated with maximum body length. *J. Fish. Res. Board Can.* 32(12):2453–2469.

12-70.5
Lockington, W. N. 1886.
The form of the pupil in snakes. *Proc. Acad. Nat. Sci. Phil.* 1866:300.

12-71
Lowne, B. T. 1872.
Descriptive catalogue of the teratological series in the Museum of the Royal College of Surgeons of England. London: Royal College of Surgeons.

12-72
Luedicke, M. 1962.
Ordnung der klasse reptilia: serpentes (5). *Handbuch der zoologie.* Berlin: Walter de Gruyter. Vol. 7, no. 1, pp. 1–128.

12-73
MacKinnon, M. R.; Heatwole, H. 1981.
Comparative cardiac anatomy of the reptilia. IV. The coronary arterial circulation. *J. Morphol.* 170:1–27.

12-74
Maderson, P. F. A. 1965.
Histological changes in the epidermis of snakes during the sloughing cycle. *J. Zool. (Lond.)* 146:98–113.

12-75
Mao, S. H.; Chen, B. Y. 1980.
Sea snakes of Taiwan: a natural history of sea snakes. Taipei, Taiwan: National Science Council.

12-76
Matthes, E. 1934.

Sinnesorgane. Gerushorgan. *Handbuch der vergleichenden anatomie der wirbeltiere.* Berlin: Urban & Schwarzenberg. Amsterdam: A. Asher; 1967. Vol. 2, pp. 879–948.

12-77
Mays, C. E. 1971.

Comparative morphology and histochemistry of the venom apparatus of some species of the proteroglyphous snakes *Laticauda* and *Dendroaspis. Wasmann J. Biol.* 29(1):81–96.

12-78
McCann, C. 1937.

Sexual dimorphism in the seasnake [*Distira cyanocincta* (Daud.)]. *J. Bombay Nat. Hist. Soc.* 39:872.

12-78.5
McCann, C. 1946.

The hemipenis in reptiles. *J. Bombay Nat. Hist. Soc.* 46(2):348–373.

12-79
McDowell, S. B. 1967.

Aspidomorphus, a genus of New Guinea snakes of the family Elapidae, with notes on related genera. *J. Zool. (Lond.)* 151:497–543.

12-80
McDowell, S. B. 1969.

Notes on the Australian sea-snake *Ephalophis greyi* M. Smith (Serpentes: Elapidae, Hydrophiinae) and the origin and classification of sea-snakes. *Zool. J. Linn. Soc.* 48:333–349.

12-81
Meckel, J. F. 1817.

Ueber den darmkanal der reptilien. *Dtsch. Arch. Physiol.* 3:199–232.

12-82
Mell, R. 1929.

Beitrage zur fauna sinica. IV. Grundzuege einer oekologie der Chinesischen reptilien und einer herpetologischen tiergeographie Chinas. Berlin: Walter de Gruyter.

12-83
Mell, R. 1929.
   Leitgedanken zu einer oekologie ostasiatischer Reptilien, insbesondere Schlan-
gen. *10th Congres International de Zoologie.* 1:1470–1477.

12-84
Mell, R. 1929.
   Preliminary contribution to an ecology of east Asiatic reptiles, especially
snakes. *Lingnan Sci. J.* 8:187–197.

12-85
Miller, M. R. 1966.
   The cochlear duct of lizards and snakes. *Am. Zool.* 6(3):421–429.

12-86
Miller, M. R. 1968.
   The cochlear duct of snakes. *Proc. Calif. Acad. Sci.* Ser. 4, 35(19):425–476.

12-87
Milne Edwards, H. 1858.
   Lecons sur la physiologie et l'anatomie comparee de l'homme et des animaux.
Paris: Librairie de Victor Masson. Volume 2.

12-88
More, N. K. 1977.
   Mucopolysaccharide heterogeneity of the reptilian kidney basement mem-
branes. *Acta Histochem.* 60:173–179.

12-89
Mosauer, W. 1935.
   The myology of the trunk region of snakes and its significance for ophidian
taxonomy and phylogeny. *Publ. Univ. Calif. Los Ang. Biol. Sci.* 1(6):81–120.

12-90
Nicholson, E. 1874.
   *Indian snakes. An elementary treatise on ophiology with a descriptive catalogue
of the snakes found in India and the adjoining countries.* Madras: Higginbotham.

12-90.5
Nishi, S. 1938.
   Muskeln des rumpfes. Bolk, L.; Goeppert, E.; Kallius, E.; Lubosch, W.
Handbuch der vergleichenden anatomie der wirbeltiere. Berlin: Urban & Schwar-
zenberg. Vol. 5, pp. 351–466.

12-91
Nopsca, F. 1923.
   Eidolosaurus und *Pachyophis*; zwei neue neocom-reptilien. *Palaeontographica* 65:97–154.

12-91.3
Ota, H.; Toriba, M.; Takahashi, H. 1986.
   The alteration pattern of dorsal scale rows in the yellow-lipped sea snake *Laticauda colubrina*, with special reference to sexual dimorphism. *Jpn. J. Herpetol.* 11(3):145–151.

12-91.6
Owen, R. 1840–1845.
   Odontography; or, a treatise on the comparative anatomy of the teeth; their physiological relations, mode of development and microscopic structure in the vertebrate animals. London: Hippolyte Bailliere. 2 vols.

12-92
Owen, R. 1866.
   On the anatomy of vertebrates. London: Longmans, Green and Co. Volume 1.

12-93
Owen, R. 1907.
   Descriptive and illustrated catalogue of the physiological series of comparative anatomy contained in the Museum of the Royal College of Surgeons of England. 2nd ed. London: Royal College of Surgeons. Volume 3.

12-93.5
Pagenstecher, H. A. 1878.
   Allgemeine zoologie oder grundgeste des thierischen baus und lebens. Berlin: Wiegandt, Hempel & Parey. Volume 3.

12-94
Parsons, T. S. 1959.
   Studies on the comparative embryology of the reptilian nose. *Bull. Mus. Comp. Zool.* 120(2):101–277.

12-95
Parsons, T. S. 1970.
   The nose and Jacobson's Organ. Gans, C.; Parsons, T. S., eds. *Biology of the reptilia.* New York: Academic. Vol. 2, pp. 99–191.

12-96
Pawlowsky, E. N. 1927.
   Gifttiere und ihre giftigkeit. Jena: Gustav Fischer.

12-97
Pernetta, J. C. 1977.
   Observations on the habits and morphology of the sea snake *Laticauda colubrina* (Schneider) in Fiji. *Can. J. Zool.* 55(10):1612–1619.

12-98
Perrier, E.; Perrier, R. 1928.
   Developpement embryogenique des vertebres allantoidiens les reptiles. Perrier, E. *Traite de zoologie*. Paris: Masson et Cie. Vol. 8, pp. 2285–3118.

12-99
Phisalix, M. 1912.
   Modifications que la fonction venimeuse imprime a la tete osseuse et aux dents chez les serpents. *Ann. Sci. Nat. Zool. Biol. Anim.* Ser. 9, 16(11):161–205.

12-100
Phisalix, M. 1914.
   Anatomie comparee de la tete et de l'appareil venimeux chez les serpents. *Ann. Sci. Nat. Zool. Biol. Anim.* Ser. 9, 19(1):1–114.

12-100.5
Picado Twight, C. 1931.
   Epidermal microornaments of the Crotalinae. *Bull. Antivenin Inst. Am.* 4(4):104–105.

12-101
Pockrandt, D. 1937.
   Beitrage zur histologie der schlangenhaut. *Zool. Jahrb. Abt. Anat. Ontog. Tiere* 62:275–322.

12-102
Pough, F. H.; Kwiecinski, G.; Bemis, W. 1978.
   Melanin deposits associated with the venom glands of snakes. *J. Morphol.* 155(1):63–71.

12-103
Prince, J. H. 1956.
   Comparative anatomy of the eye. Springfield, IL: Charles C. Thomas.

12-104
Radovanovic, M. 1935.
  Anatomische studien am schlangenkopf. *Jena. Z. Naturwiss.* 69(2):321–421.

12-105
Radovanovic, M. 1937.
  Osteologie des schlangenkopfes. *Jena. Z. Naturwiss.* 71(2):179–312.

12-106
Rochebrune, A. T. de. 1881.
  Memoire sur les vertebres des ophidiens. *J. Anat. Physiol.* 17:185–229.

12-107
Romer, A. S. 1956.
  Osteology of the reptiles. Chicago, IL: University of Chicago Press.

12-107.5
Rosen, N. 1904.
  Ueber die kaumuskeln der schlangen und ihre bedeutung bei der entleerung der giftdrusen. *Zool. Anz.* 28(1):1–7.

12-108
Rosenberg, H. I. 1967.
  Histology, histochemistry, and emptying mechanism of the venom glands of some elapid snakes. *J. Morphol.* 123(2):133–156.

12-109
Saint Girons, H.; Bassot, J. M.; Pfeffer, P. 1964.
  Donnees anatomiques et histologiques sur l'hypophyse de quelques Hydrophidae et Elapidae. *Cah. Pac.* 6:133–141.

12-110
Saint Girons, H. 1967.
  Morphologie comparee de l'hypophyse chez les squamata: donnees complementaires et apport a la phylogenie des reptiles. *Ann. Sci. Nat. Zool. Biol. Anim.* Ser. 12, 9:229–308.

12-111
Saint Girons, H. 1968.
  La morphologie comparee des glandes endocrines et la phylogenie des reptiles. *Bijdr. Dierkd.* 37:61–79.

12-112
Saint Girons, H. 1970.

Glandes endocrines. I. Hypophyse. Grasse, P. P. *Traite de zoologie*. Paris: Masson et Cie. Vol. 14, no. 3, pp. 681–726.

12-113
Saint Girons, H. 1970.

The pituitary gland. Gans, C.; Parsons, T. S., eds. *Biology of the reptilia*. New York: Academic. Vol. 3, pp. 135–199.

12-114
Saint Girons, M. C. 1970.

Morphology of the circulating blood cells. Gans, C.; Parsons, T. S., eds. *Biology of the reptilia*. New York: Academic. Vol. 3, pp. 73–91.

12-115
Samuel, M. 1944.

Studies on the corpus luteum in *Enhydrina schistosa* (Daudin) and *Hydrophis cyanocinctus* (Daudin) of the Madras Coast. *Proc. Indian Acad. Sci. Sect. B* 20(5):143–174.

12-115.5
Schlegel, H. 1828.

Untersuchung der speicheldruesen bei den schlangen mit gefurchten zaehnen, in vergleich mit denen der giftlosen und giftigen. *Nova Acta Phys.-Med. Acad. Natur. Curios.* 14(1):143–158.

12-116
Schlegel, H. 1837.

Essai sur la physionomie des serpens. Leide: Arnz. Volume 2.

12-117
Schlegel, H. 1843.

Essay on the physiognomy of serpents. Edinburgh: MacLachlan, Stewart.

12-118
Shine, R. 1978.

Sexual size dimorphism and male combat in snakes. *Oecologia (Berl.)* 33(3):269–277.

12-119
Shortt, J. 1868.

Notice of a double-headed water-snake. *J. Linn. Soc. Lond. Zool.* 9:49–50.

12-120
Smith, M. A. 1931.
Description of a new genus of sea-snake from the coast of Australia, with a note on the structures providing for complete closure of the mouth in aquatic snakes. *Proc. Zool. Soc. Lond.* 1931:397–398.

12-121
Smith, M. A. 1942.
Remarks on the nasal pit in snakes. *Copeia* 1942(4):256.

12-122
Smith, M.; Bellairs, A. d'A. 1947.
The head glands of snakes, with remarks on the evolution of the parotid gland and teeth of the Opisthoglypha. *J. Linn. Soc. Lond. Zool.* 41(279):351–368.

12-123
Smith, T. 1818.
On the structure of the poisonous fangs of serpents. *Philos. Trans. R. Soc.* 108(2):471–476.

12-124
Snoo, K. de. 1947
Het probleem menschwording gezien in het licht der vergelijkende verloskunde. 2nd ed. Haarlem: Bohn.

12-125
Stannius, H. 1856.
Handbuch der zootomie. Berlin: Veit. Volume 2

12-125.5
Sternfeld, R. 1913.
Die erschunungen der mimicry bei den schlangen. *Sitz. Ges. Nat. Freunde Berlin* 1913:98–117.

12-126
Strohl, J. 1925.
Les serpents a deux tetes et les serpents doubles a propos d'un cas de bicephalie chez un Hydrophyide *"Hydrophis spiralis"* (Shaw) et d'un cas de bifurcation axiale posterieure (Deradelphie) chez une *"Vipera berus"* L. *Ann. Sci. Nat. Zool. Biol. Anim.* Ser. 10, 8:105–132.

12-127
Taub, A. M. 1966.
Ophidian cephalic glands. *J. Morphol.* 118:529–542.

12-127.5
Taub, A. M. 1967.
    Comparative histological studies on Duvernoy's gland of colubrid snakes. *Bull. Am. Mus. Nat. Hist.* 138(1):1–50.

12-127.7
Taub, A. M.; Dunson, W. A. 1966.
    A new gland in sea snakes (Squamata, Reptilia). *Am. Zool.* 6(4):565. (Abstract)

12-128
Taub, A. M.; Dunson, W. A. 1967.
    The salt gland in a sea snake (*Laticauda*). *Nature (Lond.)* 215:995–996.

12-129
Thireau, M. 1968.
    Analyse descriptive et biometrique de la colonne vertebrale du serpent marin *Enhydrina schistosa* Daudin (Hydrophiines). Characteres vertebraux des Hydrophiides. *Bull. Mus. Natl Hist. Nat.* Ser. 2, 39(6):1044–1056.

12-130
Thompson, J. C. 1913.
    Contributions to the anatomy of the Ophidia. *Proc. Zool. Soc. Lond.* 1913:414–425.

12-131
Tomes, C. S. 1876.
    On the development and succession of the poison-fangs of snakes. *Philos. Trans. R. Soc.* 166:377–385.

12-132
Tu, A. T. 1976.
    Investigation of the sea snake, *Pelamis platurus* (Reptilia, Serpentes, Hydrophiidae), on the Pacific coast of Costa Rica, Central America. *J. Herpetol.* 10(1):13–18.

12-133
Underwood, G. 1970.
    The eye. Gans, C.; Parsons, T. S., eds. *Biology of the reptilia.* New York: Academic. Vol. 2, pp. 1–97.

12-133.5
Vanderhorst, A. 1940.
    Photographs of snake sloughs. *Herpetologica* 2(2):44, plate 4.

12-134
Varde, M. R. 1951.
   The morphology and histology of the lung in snakes. *J. Univ. Bombay* 19(5):79–89.

12-135
Voris, H. K. 1975.
   Dermal scale-vertebra relationships in sea snakes (Hydrophiidae). *Copeia* 1975(4):746–757.

12-136
Voris, H. K.; Jayne, B. C. 1976.
   The costocutaneous muscles in some sea snakes (Reptilia, Serpentes). *J. Herpetol.* 10(3):175–180.

12-137
Wall, F. 1902.
   Aids to the differentiation of snakes. *J. Bombay Nat. Hist. Soc.* 14:337–345.

12-137.5
Wall, F. 1905.
   Double-headed snakes. *J. Bombay Nat. Hist. Soc.* 16(2):386–388.

12-138
Wall, F. 1918.
   Notes on a gravid *Hydrophis cyanocinctus* and her brood. *J. Bombay Nat. Hist. Soc.* 25(4):754–756.

12-138.5
Wall, F. 1921.
   Notes on some Ceylon snakes recently acquired by the Colombo Museum. *Spolia Zeylan.* 11(43–44):405–406.

12-139
Walls, G. L. 1942.
   The vertebrate eye and its adaptive radiation. *Cranbrook Inst. Sci. Bull.* No. 19:1–785.

12-140
West, G. S. 1895.
   On the buccal glands and teeth of certain poisonous snakes. *Proc. Zool. Soc. Lond.* 1895:812–826.

12-141
West, G. S. 1895.

On the poison apparatus of certain snakes. *Rep. 65th Meet. Br. Assoc. Adv. Sci.* 1895:737. (Abstract)

12-142
West, G. S. 1896–1898.

On the histology of the salivary, buccal, and Harderian glands of the *Colubridae*, with notes on their tooth-succession and the relationships of the poison-duct. *J. Linn. Soc. Lond. Zool.* 26:517–526.

12-143
Wiedemann, E. 1932.

Zur ortsbewegung der schlangen und schleichen. *Zool. Jahrb. Abt. Allg. Zool. Physiol. Tiere* 50:557–596.

12-143.5
Wolf, S. 1933.

Zur kenntnis von bau und funktion der reptilienlunge. *Zool. Jahrb. Abt. Allg. Zool. Physiol. Tiere.* 57(1):139–190.

12-144
Yoshihara, M.; Uchiyama, M.; Murakami, T. 1979.

Notes on the ultimobranchial glands of some Japanese snakes. *Zool. Mag. (Tokyo)* 88(2):180–184.

12-145
Yoshizawa, H.; Suzuki, K. 1984.

Ultrastructural studies on the ultimobranchial and parathyroid glands of the sea snake, *Laticauda semifasciata. Acta Anat. Nippon* 59(4):538. (In Japanese)

12-146
Young, B. A. 1986.

The cephalic vascular anataomy of three species of sea snakes. *J. Morphol.* 196:195–204.

Genus-Species Index
for Chapter 12

Chapter 13

PHYSIOLOGY—SALT GLANDS
AND OTHER ASPECTS
(EXCEPT AS IN CHAPTER 14)

13-1

Barrett, R.; Maderson, P. F. A.; Meszler, R. M. 1970.

The pit organs of snakes. Gans, C.; Parsons, T. S., eds. *Biology of the reptilia.*
New York: Academic. Vol. 2, pp. 277–300.

13-2

Bentley, P. J. 1976.

Osmoregulation. Gans, C.; Dawson, W. R., eds. *Biology of the reptilia.* New
York: Academic. Vol. 5, pp. 365–412.

13-3

Benyajati, S.; Yokota, S. D.; Dantzler, W. H. 1982.

Secretion and reabsorption of Mg and K by the kidney of the sea snake,
Aipysurus laevis. Fed. Proc. 41(4):1005. (Abstract)

13-4

Benyajati, S.; Yokota, S. D.; Dantzler, W. H. 1985.

Renal function in sea snakes. II. Sodium, potassium, and magnesium excretion.
Am. J. Physiol. 249(2 Pt. 2):R237–R245.

13-5

Benyajati, S.; Dantzler, W. H. 1986.

Renal secretion of amino acids in ophidian reptiles. *Am. J. Physiol.* 250:R712–
R720.

13-6

Brattstrom, B. H. 1965.

Body temperatures of reptiles. *Am. Midl. Nat.* 73(2):376–422.

13-6.5
Brown, J. N. B. 1987.
   Sea-snakes in the waters of the United Arab Emirates. *Bull. Emirates Nat. Hist. Group (Abu Dhabi)* No. 32:18–20.

13-7
Bullock, T. H.; Barrett, R. 1968.
   Radiant heat reception in snakes. *Commun. Behav. Biol. Part A Orig. Artic.* 1:19–29.

13-8
Chiu, K. W.; Wong, C. C. 1977.
   Peripheral iodotyrosine de-iodinating activity of the non-mammalian species. *Acta Endocrinol.* (Suppl.) 212:169. (Abstract)

13-9
Conaway, C. H.; Fleming, W. R. 1960.
   Placental transmission of Na^{22} and I^{131} in *Natrix. Copeia* 1960(1):53–55.

13-10
Cowles, R. B. 1962.
   Semantics in biothermal studies. *Science (Wash. D.C.)* 135:670.

13-11
Dantzler, W. H.; Holmes, W. N. 1974.
   Water and mineral metabolism in reptilia. Florkin, M.; Scheer, B. T., eds. *Chemical zoology.* New York: Academic. Vol. 9, pp. 277–336.

13-12
Duggan, R. T.; Lofts, B. 1978.
   Adaptation to fresh water in the sea snake *Hydrophis cyanocinctus*: tissue electrolytes and peripheral corticosteroids. *Gen. Comp. Endocrinol.* 36(4):510–520.

13-13
Duggan, R. T.; Lofts, B. 1978.
   Steroid synthesis in the adrenal gland of the sea snake *Hydrophis cyanocinctus*: the metabolism of exogenous precursors. *Gen. Comp. Endocrinol.* 36(3):415–426.

13-14
Duggan, R. T.; Lofts, B. 1979.
   The pituitary-adrenal axis in the sea snake, *Hydrophis cyanocinctus* Daudin. *Gen. Comp. Endocrinol.* 38(3):374–383.

13-14.5
Dunson, M. K.; Dunson, W. A. 1975.
The relation beteen plasma Na concentration and salt gland Na-K ATPase content in the diamondback terrapin and the yellow-bellied sea snake. *J. Comp. Physiol.* 101:89–97.

13-15
Dunson, W. A. 1968.
Salt gland secretion in the pelagic sea snake *Pelamis. Am. J. Physiol.* 215(6):1512–1517.

13-16
Dunson, W. A. 1969.
Reptilian salt glands. Botelho, S. Y.; Brooks, F. P.; Shelley, W. B., eds. *Exocrine glands.* Philadelphia: University of Pennsylvania Press. Pp. 83–101.

13-17
Dunson, W. A. 1970.
Physiological ecology of the sea snake *Pelamis platurus.* San Diego, CA: *Alpha Helix Research Program 1969–1970.* Page 34.

13-18
Dunson, W. A. 1975.
Salt and water balance in sea snakes. Dunson, W. A., ed. *The biology of sea snakes.* Baltimore: University Park Press. Pp. 329–353.

13-19
Dunson, W. A. 1976.
Salt glands in reptiles. Gans, C.; Dawson, W. R., eds. *Biology of the reptilia.* San Francisco: Academic. Vol. 5, p. 413–445.

13-20
Dunson, W. A. 1977.
Permeability of the skin of aquatic snakes to water and sodium. *Physiologist* 20(4):23. (Abstract)

13-21
Dunson, W. A. 1978.
Role of the skin in sodium and water exchange of aquatic snakes placed in seawater. *Am. J. Physiol.* 235(3):R151–R159.

13-22
Dunson, W. A. 1979.
Control mechanisms in reptiles. Gilles, R. ed. *Mechanisms of osmoregulation in animals: maintenance of cell volume.* New York: Wiley. Pp. 273–322.

13-22.3
Dunson, W. A. 1981.
Control of secretion in reptilian salt glands. *Adv. Physiol. Sci.* 18:31–41.

13-22.6
Dunson, W. A. 1984.
The contrasting roles of salt glands, the integument and behavior in osmoregulation of marine reptiles. Pequeux, A.; Gilles, R.; Bolis, L., eds. *Osmoregulation in estuarine and marine animals.* New York: Springer-Verlag. Pp. 107–129.

13-23
Dunson, W. A.; Taub, A. M. 1967.
Extrarenal salt excretion in sea snakes (*Laticauda*). *Am. J. Physiol.* 213(4):975–982.

13-24
Dunson, W. A.; Murphy, D. M. 1970.
Physiology of the sea snake salt gland. San Diego, CA: *Alpha Helix Research Program 1969–1970.* Page 34.

13-25
Dunson, W. A.; Packer, R. K.; Dunson, M. K. 1971.
Sea snakes: an unusual salt gland under the tongue. *Science (Wash. D.C.)* 173(3995):347–441.

13-26
Dunson, W. A.; Dunson, M. K. 1973.
Convergent evolution of sublingual salt glands in the marine file snake and the true sea snakes. *J. Comp. Physiol.* 86(3):193–208.

13-27
Dunson, W. A.; Dunson, M. K. 1974.
Interspecific differences in fluid concentration and secretion rate of sea snake salt glands. *Am. J. Physiol.* 227(2):430–438.

13-29
Dunson, W. A.; Robinson, G. D. 1976.
Sea snake skin: permeable to water but not to sodium. *J. Comp. Physiol. B Biochem. Syst. Environ. Physiol.* 108:303–311.

13-30

Dunson, W. A.; Stokes, G. D. 1983.

Asymmetrical diffusion of sodium and water through the skin of sea snakes. *Physiol. Zool.* 56(1):106–111.

13-30.5

Dunson, W. A.; Mazzotti, F. J. 1989.

Salinity as a limiting factor in the distribution of reptiles in Florida Bay: a theory for the estuarine origin of marine snakes and turtles. *Bull. Mar. Sci.* 44(1):229–244.

13-31

Ehlert, G. W. 1970.

Temperature tolerance in *Pelamis*. San Diego, CA: *Alpha Helix Research Program, 1969–1970*. Pp. 35–36.

13-32

Graham, J. B. 1974.

Body temperatures of the sea snake *Pelamis platurus*. *Copeia* 1974(2):531–533.

13-33

Graham, J. B.; Rubinoff, I.; Hecht, M. K. 1971.

Temperature physiology of the sea snake *Pelamis platurus*: an index of its colonization potential in the Atlantic Ocean. *Proc. Nat. Acad. Sci. U.S.A.* 68(6):1360–1363.

13-33.5

Heatwole, H. 1976.

Reptile ecology. St. Lucia, Queensland: University of Queensland Press.

13-34

Heatwole, H. 1977.

Adaptation of sea snakes. Messel, H.; Butler, S. T., eds. *Australian animals and their environment*. Sydney: Shakespeare Head Press. Pp. 293–204.

13-35

Heatwole, H. 1978.

Adaptations of marine snakes. *Am. Sci.* 66(5):594–604.

13-36

Heatwole, H. 1981.

Temperature relations of some sea snakes. *Snake* 13:53–57.

13-36.5
Heatwole, H. 1987.
  Sea snakes. Kensington, N.S.W.: New South Wales University Press.

13-37
McDonald, H. S. 1976.
  Methods for the physiological study of reptiles. Gans, C.; Dawson, W. R., eds. *Biology of the reptilia*. New York: Academic. Vol. 5, pp. 19–126.

13-38
Murphy, D. M. 1970.
  Vein cannulation in *Pelamis*. San Diego, CA: *Alpha Helix Research Program, 1969–1970*. Pp. 34–35.

13-39
Peaker, M.; Linzell, J. L. 1975.
  Salt glands in birds and reptiles. New York: Cambridge University Press.

13-40
Potts, W. T. W.; Parry, G. 1964.
  Osmotic and ionic regulation in animals. New York: Macmillan.

13-41
Pough, F. H.; Lillywhite, H. B. 1984.
  Blood volume and blood oxygen capacity of sea snakes. *Physiol. Zool.* 57(1):32–39.

13-41.5
Sasayama, Y.; Oguro, C.; Yui, R.; Kambegawa, A. 1984.
  Immunohistochemical demonstration of calcitonin in ultimobranchial glands of some lower vertebrates. *Zool. Sci. (Tokyo)* 1:755–758.

13-42
Schmidt Nielsen, B.; Davis, L. E. 1968.
  Fluid transport and tubular intercellular spaces in reptilian kidneys. *Science (Wash. D.C.)* 159:1105–1108.

13-43
Schmidt Nielsen, K.; Fange, R. 1958.
  Salt glands in marine reptiles. *Nature (Lond.)* 182(4638):783–785.

13-44
Seymour, R. S. 1982.
   Physiological adaptations to aquatic life. Gans, C.; Pough, F. H., eds. *Biology of the reptilia*. New York: Academic. Vol. 13, pp. 1–51.

13-45
Skoczylas, R. 1978.
   Physiology of the digestive tract. Gans, C.; Gans, K. A., eds. *Biology of the reptilia*. New York: Academic. Vol. 8, pp. 589–717.

13-46
Sood, M. S. 1947.
   Some new lymph-glands in the serpents. *Indian Sci. Cong. Assoc. Proc.* 34:187. (Abstract)

13-47
Stokes, G. D.; Dunson, W. A. 1982.
   Passage of water and electrolytes through natural and artificial keratin. *Desalination* 42(3):321–328.

13-48
Stokes, G. D.; Dunson, W. A. 1982.
   Permeability and channel structure of reptilian skin. *Am. J. Physiol.* 242(16):F681–F689.

13-49
Tammar, A. R. 1974.
   Bile salts in reptilia. *Chem. Zool.* 9:337–351.

13-50
Wallach, J. D.; Hoessle, C. 1970.
   M-99 as an immobilizing agent in poikilothermes. *Vet. Med. Small Anim. Clin.* 65:163–167.

13-51
Wong, C. C.; Chiu, K. W.; Choy, Y. M. 1983.
   Metabolism of monoiodotyrosine in lower vertebrates: with emphasis on snakes. *Comp. Biochem. Physiol. B Comp. Biochem.* 74(4):739–742.

13-52
Yokota, S. D.; Benyajati, S.; Dantzler, W. H. 1982.
   Renal function in the sea snake, *Aipysurus laevis*. *Fed. Proc.* 41(4):1005 (Abstract)

13-53

Yokota, S. D.; Benyajati, S.; Dantzler, W. H. 1985.

Renal function in sea snakes. I. Glomerular filtration rate and water handlng. *Am. J. Physiol.* 249(2 Pt. 2):R228–R236.

13-53.5

Zimmerman, K.; Heatwole, H. 1990.

Cutaneous photoreception: a new sensory mechanism for reptiles. *Copeia* 1990:856–858.

13-54

Zug, D. A.; Dunson, W. A. 1979.

Salinity preference in fresh water and estuarine snakes (*Nerodia sipedon* and *N. fasciata*). *Fla. Sci.* 42(1):1–8.

Genus-Species Index
for Chapter 13

Chapter 14

PHYSIOLOGY—BLOOD, RESPIRATORY, CARDIAC, AND DIVING

14-1
Anonymous. 1975.
How sea snakes may avoid the bends. *Sea Front.* 21(6):358.

14-1.5
Anonymous. 1987.
Sea snakes keep their heads in deep water. *New Sci.* 1987(1562):34.

14-2
Baldwin, J.; Seymour, R. S. 1977.
Adaptation to anoxia in snakes: levels of glycolytic enzymes in skeletal muscle. *Aust. J. Zool.* 25:9–13.

14-3
Benyajati, S.; Dantzler, W. H. 1986.
Plasma levels and renal handling of endogenous amino acids in snakes: a comparative study. *J. Exp. Zool.* 238(1):17–28.

14-4
Feder, M. E. 1980.
Blood oxygen stores in the file snake, *Acrochordus granulatus*, and in other marine snakes. *Physiol. Zool.* 53(4):394–401.

14-5
Feder, M. E.; Burggren, W. W. 1985.
Cutaneous gas exchange in vertebrates: design, patterns, control and implications. *Biol. Rev. Camb. Philos. Soc.* 60:1–45.

14-6
Feder, M. E.; Burggren, W. W. 1985.
   Skin breathing in vertebrates. *Sci. Am.* 253(5):126–142.

14-7
Graham, J. B. 1973.
   Aquatic respiration and the physiological responses to submersion of the sea snake *Pelamis platurus. Am. Zool.* 13(4):1296. (Abstract)

14-8
Graham, J. B. 1974.
   Aquatic respiration in the sea snake *Pelamis platurus. Respir. Physiol.* 21:1–7.

14-9
Graham, J. B.; Gee, J. H.; Robison, F. S. 1975.
   Hydrostatic and gas exchange functions of the lung of the sea snake *Pelamis platurus. Comp. Biochem. Physiol. A Comp. Physiol.* 50:477–482.

14-9.3
Graham, J. B.; Lowell, W. R.; Rubinoff, I.; Motta, J. 1987.
   Surface and subsurface swimming of the sea snake *Pelamis platurus. J. Exp. Biol.* 127:27–44.

14-9.5
Graham, J. B.; Gee, J. H.; Motta, J.; Rubinoff, I. 1987.
   Subsurface buoyancy regulation by the sea snake *Pelamis platurus. Physiol. Zool.* 60(2):251–261.

14-10
Heatwole, H. 1975.
   Voluntary submergence times of marine snakes. *Mar. Biol. (Berl.)* 32:205–213.

14-11
Heatwole, H. 1977.
   Adaptation of sea snakes. Messel, H.; Butler, S. T., eds. *Australian animals and their environment.* Sydney: Shakespeare Head Press. Pp. 193–204.

14-12
Heatwole, H. 1977.
   Ecological studies on marine snakes. *Mar. Res. Indones.* No. 19:147 (Abstract)

14-13
Heatwole, H. 1977.
Heart rate during breathing and apnea in marine snakes (Reptilia, Serpentes). *J. Herpetol.* 11(1):67–76.

14-13.5
Heatwole, H. 1977.
Sea snakes: a contrast to other vertebrate divers. *J. South Pac. Underwater Med. Soc.* 1977:35–38

14-14
Heatwole, H. 1978.
Adaptations of marine snakes. *Am. Sci.* 66(5):594–604.

14-15
Heatwole, H. 1981.
Role of the saccular lung in the diving of the sea krait, *Laticauda colubrina* (Serpentes: Laticaudidae). *Aust. J. Herpetol.* 1(1):11–16.

14-15.5
Heatwole, H. 1987.
Sea snakes. Kensington, N.S.W.: New South Wales University Press.

14-16
Heatwole, H.; Seymour, R. 1975.
Diving physiology. Dunson, W. A., ed. *The biology of sea snakes.* Baltimore: University Park Press. Pp. 289–327.

14-17
Heatwole, H.; Seymour, R. 1975.
Pulmonary and cutaneous oxygen uptake in sea snakes and a file snake. *Comp. Biochem. Physiol. A Comp. Physiol.* 51:399–405.

14-18
Heatwole, H.; Seymour, R. 1976.
Respiration of marine snakes. Hughes, G. M., ed. *Respiration of amphibious vertebrates.* New York: Academic. Pp. 375–389.

14-19
Heatwole, H.; Seymour, R. S. 1978.
Cutaneous oxygen uptake in three groups of aquatic snakes. *Aust. J. Zool.* 26:481–486.

14-20
Heatwole, H.; Seymour, R. S.; Webster, M. E. D. 1978.
    Heart rates of sea snakes diving in the sea. *Comp. Biochem. Physiol. A Comp. Physiol.* 62:453–455.

14-21
Jacobs, M. 1985.
    Sea snake research sheds light on diving biology of marine reptiles. *Smithson. Inst. Res. Rep.* No. 45:5.

14-22
Lillywhite, H. B. 1985.
    Behavioral control of arterial pressure in snakes. *Physiol. Zool.* 58(2):159–165.

14-22.3
Lillywhite, H. B. 1985.
    Postural edema and blood pooling in snakes. *Physiol. Zool.* 58(6):759–766.

14-22.6
Lillywhite, H. B. 1988.
    Snakes, blood circulation and gravity. *Sci. Am.* 259(6):92–98.

14-23
Lillywhite, H. B.; Seymour, R. S. 1978.
    Regulation of arterial blood pressure in Australian tiger snakes. *J. Exp. Biol.* 75:65–80.

14-24
Lillywhite, H. B.; Pough, F. H. 1981.
    Arterial pressure and hemodynamics in sea snakes. *Am. Zool.* 21(4):963.

14-25
Lillywhite, H. B.; Pough, F. H. 1983.
    Control of arterial pressure in aquatic sea snakes. *Am. J. Physiol.* 244(1):R66–R73.

14-26
McDonald, H. S. 1959.
    Respiratory functions of the ophidian air sac. *Herpetologica* 15:193–198.

14-27
Nair, S. G. 1955.
    The oxyphoric capacity of the blood of some reptiles and mammals. *J. Anim. Morphol. Physiol.* 1(2):48–54.

14-28
Nicol, J. A. C. 1961.
   The biology of marine animals. London: Sir Isaac Pitman & Sons.

14-28.5
Pough, F. H. 1979.
   Summary of oxygen transport characteristics of reptilian blood. *Smithson. Herpetol. Inf. Serv.* No. 45:1–18.

14-29
Pough, F. H.; Lillywhite, H. B. 1982.
   Oxygen transport properties of blood of sea snakes. *Am. Zool.* 22(4):934. (Abstract)

14-30
Rosenberg, H. I.; Voris, H. K. 1980.
   Cutaneous capillaries of sea snakes and their possible role in gas exchange. *Am. Zool.* 20(4):758. (Abstract)

14-30.5
Rubinoff, I.; Graham, J. B.; Motta, J. 1986.
   Diving of the sea snake *Pelamis platurus* in the Gulf of Panama. I. Dive depth and duration. *Mar. Biol. (Berl.)* 91:181–191.

14-31
Seymour, R. S. 1974.
   How sea snakes may avoid the bends. *Nature (Lond.)* 250:489–490.

14-32
Seymour, R. S. 1976.
   Blood respiratory properties in a sea snake and a land snake. *Aust. J. Zool.* 24:313–320.

14-33
Seymour, R. S. 1978.
   Gas tensions and blood distribution in sea snakes at surface pressure and at simulated depth. *Physiol. Zool.* 51(4):388–407.

14-34
Seymour, R. S. 1979.
   Blood lactate in free-diving sea snakes. *Copeia* 1979(3):494–497.

14-35
Seymour, R. S.; Webster, M. E. D. 1975.
Gas transport and blood acid-base balance in diving sea snakes. *J. Exp. Zool.* 191(2):169–181.

14-36
Seymour, R. S.; Lillywhite, H. B. 1976.
Blood pressure in snakes from different habitats. *Nature (Lond.)* 264:664–666.

14-37
Seymour, R. S.; Spragg, R. G.; Hartman, M. T. 1981.
Distribution of ventilation and perfusion in the sea snake, *Pelamis platurus.* *J. Comp. Physiol. B. Biochem. Syst. Environ. Physiol.* 145:109–115.

14-38
Wood, S. C.; Lenfant, C. J. M. 1976.
Respiration: mechanics, control and gas exchange. Gans, C.; Dawson, W. R., eds. *Biology of the reptilia.* New York: Academic. Vol. 5, pp. 255–274.

Genus-Species Index
for Chapter 14

Chapter 15

BIOCHEMISTRY—GENERAL

15-1

Alexander, K. M. 1955.

A comparative study of the ash and iron content in the skeletal muscles of some representative vertebrates. *J. Zool. Soc. India* 7(2):163–169.

15-2

Alexander, K. M. 1956.

A comparative study of the total protein content in the skeletal muscles of some representative vertebrates. *J. Zool. Soc. India* 8(2):149–156.

15-3

Benyajati, S.; Dantzler, W. H. 1986.

Plasma levels and renal handlng of endogenous amino acids in snakes: a comparative study. *J. Exp. Zool.* 238(1):17–28.

15-4

Cohen, E. 1954.

A comparison of the total protein and albumin content of the blood sera of some reptiles. *Science (Wash. D.C.)* 119:98–99.

15-5

Dessauer, H. C. 1970.

Blood chemistry of reptiles: physiological and evolutionary aspects. Gans, C.; Parsons, T. S., eds. *Biology of the reptilia.* New York: Academic. Vol. 3, pp. 1–72.

15-6

Duggan, R. T. 1981.

Plasma corticostcroids in marinc, tcrrcstrial and freshwater snakes. *Comp. Biochem. Physiol. A Comp. Physiol.* 68(1):115–118.

15-7
Dunson, M. K.; Dunson, W. A. 1975.
The relation between plasma Na concentration and salt gland Na-K ATPase content in the diamondback terrapin and the yellow-bellied sea snake. *J. Comp. Physiol.* 101:89–97.

15-8
Haslewood, G. A. D. 1967.
Bile salts. London: Methuen.

15-9
Holmes, W. N.; Pearce, R. B. 1979.
Hormones and osmoregulation in the vertebrates. Gilles, R., ed. *Mechanisms of osmoregulation in animals: maintenance of cell volume.* New York: John Wiley. Pp. 413–533.

15-9.5
Jiang, L. F.; Chen, Y. C. 1981.
Preliminary studies on the relationship between the snake classification and its venom and serum proteins. I. Electrophoretic analysis of the venoms and serum proteins of snakes. *Zool. Res. (China)* 2 (Suppl 4):161–164. (In Chinese with English abstract)

15-10
Kamat, D. N.; Pore, A. A. 1972.
Membrane esterases from vertebrate kidney. *Acta Histochem.* 44:308–312.

15-11
Liu, C. S. 1975.
Preparation and chemical characterization of the three chains of the major hemoglobin of the sea snake, *Pelamis platurus. J. Chin. Biochem. Soc.* 4(1):6P. (Abstract)

15-12
Liu, C. S. 1975.
Preparation and chemical characterization of the three chains of the major hemoglobin of the sea snake, *Pelamis platurus. J. Biochem. (Tokyo)* 78:19–29.

15-13
Mao, S. H.; Dessauer, H. C.; Chen, B. Y. 1978.
Fingerprint correspondence of hemoglobins and the relationships of sea snakes. *Comp. Biochem. Physiol. B Comp. Biochem.* 59(4):353–361.

15-14
More, N. K. 1977.
Histochemistry of mucopolysaccharides from snake kidney and their possible role in excretion. *Cell. Mol. Biol.* 22(2):151–161.

15-15
Nair, S. G. 1955.
The non-protein nitrogen in the blood of some reptiles and mammals. *J. Anim. Morphol. Physiol.* 2:96–100.

15-16
Nair, S. G. 1958.
A study of the plasma proteins of some reptiles and mammals. *J. Anim. Morphol. Physiol.* 5:95–100.

15-17
Nair, S. G. 1960.
Free amino acids in the blood of some reptiles and mammals. *J. Anim. Morphol. Physiol.* 7(1):98–100.

15-18
Nakajima, H.; Hasekura, H.; Shinonaga, S. 1965.
Blood group antigen tests on the sea-snakes. *Jpn. J. Leg. Med.* 19(4):280–283. (In Japanese with English abstract)

15-19
Pore, A. A.; Kamat, D. N. 1973.
Histochemistry of lipases in the kidney of reptiles from different habitats. *Indian Sci. Cong. Assoc. Proc.* 60:446. (Abstract)

15-20
Sasayama, Y.; Oguro, C.; Yui, R.; Kambegawa, A. 1984.
Immunohistochemical demonstration of calcitonin in ultimobranchial glands of some lower vertebrates. *Zool. Sci. (Tokyo)* 1:755–758.

15-21
Tsunoo, S.; Horisaka, K.; Motonishi, K.; Takeda, J. 1964.
Ueber das ophidin in den muskeln von den seeschlangen *Laticauda semifasciata* und *laticaudata. J. Biochem. (Tokyo)* 55(6):604–606.

Genus-Species Index
for Chapter 15

Chapter 16

VENOM—GENERAL, TOXICITY STUDIES, EVOLUTIONARY ASPECTS, REVIEWS, MORPHOLOGY OF VENOM GLANDS, AND ASSOCIATED STRUCTURES

16-1
Amaral, A. do. 1925.
A general consideration of snake poisoning and observations on neotropical pit-vipers. Cambridge: Harvard University Press.

16-1.1
Anonymous. 1902.
Les serpents de mer et leur venin. *Revue Scient.* Ser. 4, 18(7):219.

16-1.2
Anonymous. 1957.
Serpents marins venimeux (Hydrophiidae) et leurs venins. *Inst. Pasteur Viet-Nam Rapp. Annu. Fonct. Tech.* 1958:30–36.

16-1.3
Anonymous. 1958.
Serpents marins venimeux (Hydrophiidae) et leurs venins. *Inst. Pasteur Viet-Nam Rapp. Annu. Fonct. Tech.* 1958:53–56.

16-1.6
Anonymous. 1960.
Serpents marins venimeux (Hydrophiidae) et leur venins. *Inst. Pasteur Viet-Nam Rapp. Annu. Fonct. Tech.* 1960:52–61.

16-2
Anthony, J. 1955.
Essai sur l'evolution anatomique de l'appareil venimeux des ophidiens. *Ann. Sci. Nat. Zool. Biol. Anim.* Ser. 11, 17:7–53.

16-3
Anthony, J. 1970.
Anatomie de l'appareil venimeux des reptiles. Grasse, P. P., ed. *Traite de zoologie*. Paris: Masson et Cie. Vol. 14, no. 2, pp. 549–598.

16-4
Barme, M. 1958.
Contribution a l'etude des serpents marins venimeux Hydrophiidae du Viet-Nam. *Bull. Soc. Pathol. Exot.* 51:258–265.

16-5
Barme, M. 1963.
Venomous sea snakes of Viet Nam and their venoms. Keegan, H. L.; McFarlane, M. V., eds. *Venomous and poisonous animals and noxious plants of the Pacific region*. New York: Macmillan. Pp. 373–378.

16-6
Barme, M. 1964.
Les serpents marins venimeux du Viet-Nam et leurs venins. *Cah. Pac.* 6:192.

16-7
Barme, M. 1968.
Venomous sea snakes (Hydrophiidae). Buecherl, W.; Buckley, E. E.; Deulofeu, V., eds. *Venomous animals and their venoms*. New York: Academic. Vol. 1, pp. 285–308.

16-8
Barme, M.; Detrait, J. 1959.
Etude de la composition des venins des Hydrophiides. *C. R. Acad. Sci. Paris* 248(2):312–315.

16-9
Baslow, M. H. 1977.
Marine pharmacology: a study of toxins and other biologically active substances of marine origin. 2nd ed. Huntington, NY: Robert E. Krieger.

16-10
Berman, D. M. 1983.
The toxicities of snake venoms to goldfish (*Carassius auratus*) and the susceptibilities of reef fish and crabs to olive sea snake (*Aipysurus laevis*) venom. *Toxicon* 1983(Suppl.3):37–40.

16-11
Berncastle, J. 1870.
On the distinction between the harmless and venomous snakes of Australia. *Aust. Med. Gaz.* 1870:21–22.

16-12
Bolanos, R. 1972.
Toxicity of Costa Rican snake venoms for the white mouse. *Am. J. Trop. Med. Hyg.* 21(3):360–363.

16-13
Bolanos, R.; Flores, A.; Taylor, R. T.; Cerdas, L. 1974.
Color patterns and venom characteristics in *Pelamis platurus*. *Copeia* 1974(4):909–912.

16-14
Boquet, P. 1964.
Venins de serpents (1ere partie): physio-pathologie de l'envenimation et proprietes biologiques des venins. *Toxicon* 2:5–41.

16-15
Boquet, P. 1970.
Venins de serpents. Grasse, P. P., ed. *Truile de zoologie*. Paris: Masson et Cie. Vol. 14, no. 2, pp. 599–675.

16-16
Botes, D. P.; Carlsson, F. H. H.; Joubert, F. J.; Louw, A. I.; Strydom, A. J. C; Strydom, D. J.; Viljoen, C. C. 1974.
Nomenclature of snake venom toxins. *Toxicon* 12:99–101.

16-17
Broad, A. J.; Sutherland, S. K.; Coulter, A. R. 1979.
The lethality in mice of dangerous Australian and other snake venom. *Toxicon* 17:661–664.

16-18
Brown, J. H. 1973.
Toxicology and pharmacology of venoms from poisonous snakes. Springfield, IL: Charles C. Thomas.

16-18.5
Calmette, A. 1905.
Intoxikationskrankheiten. 2. Vergiftungen durch tierische gifte. Mense, C. A. *Handbuch der tropenkrankheiten.* Leipzig: J. A. Barth. Pp. 291–337.

16-19

Calmette, A. 1907.

Les venins: les animaux venimeux et la serotherapie antivenimeuse. Paris: Masson et C[ie].

16-20

Calmette, A. 1908.

Venoms: venomous animals and antivenomous serum—therapeutics. London: John Bale, Sons & Danielsson, Ltd.

16-21

Cantor, T. 1841.

Observations upon pelagic serpents. *Trans. Zool. Soc. Lond.* 2(4):303–313.

16-22

Carey, J. E.; Wright, E. A. 1960.

The toxicity and immunological properties of some sea-snake venoms with a particular reference to that of *Enhydrina schistosa. Trans. R. Soc. Trop. Med. Hyg.* 54(1):50–67.

16-23

Chang, T. W.; Tang, N. 1972.

Selection pressures on homologous proteins of varied activities. *Nat. New Biol.* 239:207

16-24

Chang, W. Y.; Lu, C. C.; Tu, T. C. 1959.

Effect of the venom of a sea snake, *Laticauda semifasciata* (Reinwardt) on the capillary permeability in rabbits. *J. Formosan Med. Assoc.* 58:740. (In Chinese) (Abstract)

16-25

Cheymol, J.; Bourillet, F.; Roch Arveiller, M. 1972.

Venins et toxines de serpents effets neuromusculaires. *Actual. Pharmacol.* 25:179–240.

16-26

Christensen, P. A. 1955.

South African snake venoms and antivenoms. Johannesburg: South African Institute for Medical Research.

16-27
Courville, D. A. 1970.
Venomous sea snakes. Chemistry. Halstead, B. W. *Poisonous and venomous marine animals of the world*. Washington, DC: Government Printing Office. Vol. 3, pp. 671–673.

16-28
Courville, D. A. 1978.
Venomous sea snakes. Chemistry. Halstead, B. W. *Poisonous and venomous marine animals of the world*. Rev. ed. Princeton, NJ: Darwin Press. Pp. 970–975.

16-29
Durkin, J. P.; Pickwell, G. V.; Trotter, J. T.; Shier, W. T. 1981.
Phospholipase A_2 electrophoretic variants in reptile venoms. *Toxicon* 19(4):535–546.

16-30
Duvernoy, G. L. 1832.
Memoire sur les caracteres tires de l'anatomie pour distinguer les serpens venimeux des serpens non venimeux. *Ann. Sci. Nat.* 26:113–160.

16-31
Endean, R. 1979.
Neurotoxins occurring in marine animals from Australian waters. Chubb, I. W.; Geffen, L. B., eds. *Neurotoxins: fundamental & clinical advances*. Adelaide, South Australia: Adelaide University Union Press. Pp. 57–73.

16-32
Faust, E. S. 1906.
Die tierischen gifte. Braunschweig: F. Vieweg und Sohn.

16-33
Fayrer, J. 1872.
The thanatophidia of India, being a description of the venomous snakes of the Indian Peninsula, with an account of the influence of their poison on life; and a series of experiments. London: Churchill.

16-34
Fayrer, J. 1874.
The thanatophidia of India, being a description of the venomous snakes of the Indian Peninsula: with an account of the influence of their poison on life; and a series of experiments. 2nd ed. rev'd and enlarged. London: Churchill.

16-35
Gail, R.; Rageau, J. 1958.
Introduction a l'etude des serpents marins (Ophidiens, Hydrophiidae) en Nouvelle-Caledonie. *Bull. Soc. Pathol. Exot.* 51(3):448–459.

16-36
Gans, C.; Elliott, W. B. 1968.
Snake venoms: production, injection, action. Staple, P. A., ed. *Advances in oral biology.* Orlando, FL: Academic. Vol. 3, pp. 45–81.

16-37
Ganthavorn, S. 1969.
Toxicities of Thailand snake venoms and neutralization capacity of antivenin. *Toxicon* 7:239–241.

16-38
Gawade, S. P.; Bhide, M. B. 1978.
Chromatographic separation of the venom of *Lapemis curtus* and pharmacological characterization of its toxic components: preliminary studies. *Toxicon* 16(2):134. (Abstract)

16-39
Gawade, S. P.; Tare, T. G.; Gaitonde, B. B. 1981.
Simple method of milking of sea snakes. *Snake* 13:154–157.

16-40
Gawade, S. P.; Gaitonde, B. B. 1982.
Isolation and characterisation of toxic components from the venom of the common Indian sea snake (*Enhydrina schistosa*). *Toxicon* 20(4):797–801.

16-41
Gopalakrishnakone, P. 1985.
Structure of the venom glands of sea snakes (Hydrophidae). *Toxicon* 23(4):571. (Abstract)

16-42
Habermehl, G. 1976.
Gift-Tiere und ihre waffen: eine einfuhrung fuer biologen, chemiker und mediziner. New York: Springer Verlag.

16-43
Habermehl, G. G. 1981.
Venomous animals and their toxins. New York: Springer-Verlag.

16-44
Halstead, B. W. 1970.
Poisonous and venomous marine animals of the world. Washington, DC: Government Printing Office. Volume 3.

16-45
Halstead, B. W. 1978.
Poisonous and venomous marine animals of the world. Rev. ed. Princeton, NJ: Darwin Press.

16-46
Halstead, B. W. 1981.
Current status of marine biotoxicology—an overview. *Clin. Toxicol.* 18(1):1–24.

16-47
Halstead, B. W.; Engen, P. C.; Tu, A. T. 1978.
The venom and venom apparatus of the sea snake *Lapemis hardwicki* (Gray). *Zool. J. Linn. Soc.* 63(4):371–396.

16-47.5
Heatwole, H. 1987.
Sea snakes. Kensington, N.S.W.: New South Wales University Press.

16-48
Homma, M.; Okonogi, T.; Mishima, S. 1964.
Studies on sea snake venom. 1. Biological toxicities of venoms possessed by three species of sea snakes captured in coastal waters of Amami Oshima. *Gunma J. Med. Sci.* 13:283–296.

16-49
Homma, M.; Abe, R.; Okonogi, T.; Kosuge, T.; Mishima, S. 1965.
Studies on habu snake and erabu sea snake venoms: outlines of biological toxicities of the two snake venoms, and inhibitory actions of tannic acid on them. *Jpn. J. Bacteriol.* 20(6):281–289. (In Japanese with English abstract)

16-50
Hong, B. S. 1970.
Studies on the toxic principles in the venoms of sea snakes. Fort Collins, CO: Colorado State University. 179 pp. (Dissertation)

16-50.5
Hseu, T. H.; Jou, E. D.; Wang, C.; Yang, C. C. 1977.
Molecular evolution of snake venom toxins. *J. Mol. Evol.* 10(2):167–182.

16-51
Institut Pasteur du Viet-Nam. 1960.
  Serpents marins venimeux (Hydrophiidae) et leurs venins. *Inst. Pasteur Viet-Nam Rapp. Annu. Fonct. Tech.* 1960:52–61.

16-52
Ivanov, O. C.; Ivanov, C. P. 1976.
  Sequence analogy of proinsulin and some neurotoxins. Fox, J. L.; Deyl, Z.; Blazej, A., eds. *Protein structure and evolution.* New York: Marcel Dekker. Pp. 413–428.

16-53
Ivanov, C. P.; Ivanov, O. C. 1979.
  The evolution and ancestors of toxic proteins. *Toxicon* 17(3):205–220.

16-53.5
Jiang, L. F.; Chen, Y. C. 1981.
  Preliminary studies on the relationship between the snake classification and its venom and serum proteins. I. Electrophoretic analysis of the venoms and serum proteins of snakes. *Zool. Res. (China)* 2 (Suppl.4):161–164. (In Chinese wtih English abstract)

16-54
Kaiser, E.; Michl, H. 1958.
  Die biochemie der tierischen gifte. Wein: Franz Deuticke.

16-55
Kao, C. S.; Liu, C. S.; Blackwell, R. Q. 1973.
  Presence of erabutoxins a and b in venom of the sea snake *Laticauda semifasciata* from Taiwan. *Toxicon* 11:383–385.

16-56
Kent, C. G.; Tu, A. T.; Geren, C. R. 1984.
  Isotachophoretic and immunological analysis of venoms from sea snakes (*Laticauda semifasciata*) and brown recluse spiders (*Loxosceles reclusa*) of different morphology, locality, sex, and developmental stages. *Comp. Biochem. Physiol. B Comp. Biochem.* 77(2):303–311.

16-57
Kochva, E. 1978.
  Oral glands of the reptilia. Gans, C.; Gans, K. A., eds. *Biology of the reptilia.* New York: Academic. Vol. 8, pp. 43–161.

16-58
Kochva, E.; Gans, C. 1971.
   Salivary glands of snakes. Minton, S. A., ed. *Snake venoms and envenomation.* New York: Marcel Dekker. Pp. 17–42.

16-59
Lee, C. Y.. 1972.
   Chemistry and pharmacology of polypeptide toxins in snake venoms. *Annu. Rev. Pharmacol.* 12:265–286.

16-60
Lee, C. Y. 1979.
   Recent advances in chemistry and pharmacology of snake toxins. *Adv. Cytopharmacol.* 3:1–16.

16-60.5
Lee, C. Y. 1979.
   Snake venoms. Lee, C. Y., ed. *Handbuch der experimentallen pharmakologie.* New York: Springer Verlag. Volume 52.

16-61
Levey, H. A. 1969.
   Toxicity of the venom of the sea-snake, *Laticauda colubrina*, with observations on a Malay 'folk cure'. *Toxicon* 6:269–276.

16-62
Levey, H. H.; Soh, N. M.; Leong, L. T.; Koh, K. H. 1968.
   Toxicity of the venom of the sea-snake, *Laticauda colubrina* (with observations on a Malay "folk-cure"). *Malay. Nat. J.* 21(Suppl):xxii–xxiii. (Abstract)

16-63
Limpus, C. J. 1975.
   A study of the ecology and toxicology of subtropical Queensland sea snakes (*Hydrophiinae*). Brisbane, Australia: University of Queensland. 197 pp. (Thesis)

16-64
Limpus, C. J. 1978.
   Toxicology of the venom of subtropical Queensland Hydrophiidae. Rosenberg, P., ed. *Toxins: animal, plant and microbial.* New York: Pergamon. Pp. 341–363.

16-65
Limpus, C. J. 1978.
   The venom apparatus and venom yields of sub-tropical Queensland Hydrophiidae. Rosenberg, P., ed. *Toxins: animal, plant and microbial.* New York: Pergamon. Pp. 39–70.

16-65.5
Limpus, C. J. 1987.
   Sea snakes. Coracevich, J.; Davie, P.; Pearn, J., eds. *Toxic plants and animals.* Brisbane: Queensland Museum Press. Pp. 195–203.

16-66
Lin, M. W. 1975.
   Venomous snakes of southeast Asia and some physiological and biochemical aspects of their venoms. Chicago, IL: Northeastern Illinois University. 65 pp. (Thesis)

16-67
Maegraith, B. G. 1958.
   Serpents marins venimeux (Hydrophiidae) et leurs venins. *Inst. Pasteur Viet-Nam Rapp. Annu. Fonct. Tech.* 1958:53–56.

16-68
Martin, C. J.; Lamb, G. 1907.
   Snake-poison and snake-bite. Allbutt, T. C.; Rolleston, H. D., eds. *A system of medicine.* New York: Macmillan. Vol. 2, no. 2, pp. 783–821.

16-69
Mays, C. E. 1971.
   Comparative morphology and histochemistry of the venom apparatus of some species of the proteroglyphous snakes *Laticauda* and *Dendroaspis. Wasmann J. Biol.* 29(1):81–96.

16-69.5
Meier, J. 1990.
   Venomous snakes. Stocker, K. F., ed. *Medical use of snake venom proteins.* Boca Raton, FL: CRC Press. Pp. 1–132.

16-71
Minton, S. A. 1974.
   Venom diseases. Springfield, IL: Charles C. Thomas.

16-72

Minton, S. A. 1983.

Lethal toxicity of venoms of snakes from the Coral Sea. *Toxicon* 21(6):901–902.

16-73

Minton, S. A.; Schoettler, W. H. A.; Slotta, K. H.; Graydon, J. J.; Morgan, F. G.; Christensen, P. A.; Amaral, A. do. 1964.

Animal toxins: Part I. Reptiles. Altman, P. L.; Dittmer, D. S., eds. *Biology data book.* Washington DC: Federation of American Societies for Experimental Biology. Pp. 328–335.

16-74

Minton, S. A.; Tamiya, N.; Takasaki, C. 1978.

Toxicity of sea snake venoms for amphibians and reptiles. Rosenberg, P., ed. *Toxins: animal, plant and microbial.* Oxford: Pergamon. (Abstract)

16-75

Mishima, S.; Okonogi, T. 1962.

Studies on the effect of sea snake venom. On the toxicity of *Hydrophis cyanocinctus* Daudin venom (Report 2). *Jpn. J. Sanit. Zool.* 13(2):153. (In Japanese) (Abstract)

16-76

Mishima, S.; Okonogi, T.; Honma, M. 1964.

Studies on the effect of sea snakes venom (Report 4). *Jpn. J. Sanit. Zool.* 14(2):100. (In Japanese) (Abstract)

16-77

M'Kenzie, A. 1820.

An account of venomous sea snakes, on the coast of Madras. *Asiat. Res.* 13:329–336.

16-77.5

Mori, N.; Tu, A. T. 1988.

Isolation and primary structure of the major toxin from sea snakes, *Acalyptophis peronii*, venom. *Arch. Biochem. Biophys.* 260(1):10–17.

16-78

Nauck, E. G. 1929.

Untersuchungen ueber das gift einer Seeschlange (*Hydrus platurus*) des Pazifischen Ozeans. *Arch. Schiffs- Tropen- Hyg.* 33:167–170.

16-79
Nishida, S.; Tamiya, N. 1986.
   Phospholipases A$_2$ from snake and bee venoms. *Tanpakushitsu Kakusan Koso* 31:158–165.

16-80
Noguchi, H. 1909.
   Snake venoms: an investigation of venomous snakes with special reference to the phenomena of their venoms. Washington, DC: Carnegie Institution of Washington.

16-80.5
Oppel, A. 1900
   Lehrbuch des vergleichenden mikroskopischen anatomie der wirbeltiere. Jena: Gustav Fischer. Volume 3.

16-81
Pawlowsky, E. N. 1927.
   Gifttiere und ihre giftigkeit. Jena: Gustav Fischer.

16-82
Phisalix, M. 1922.
   Animaux venimeux et venins. Paris: Masson & Cie. Volume 2.

16-84
Pickwell, G. V.; Vick, J. A.; Shipman, W. H.; Grenan, M. M. 1973.
   Production, toxicity and preliminary pharmacology of venom from the sea snake, *Pelamis platurus* with observations on its probable threat to man along middle America. Worthen, L. R., ed. *Proc. 3rd Food-Drugs from the Sea Conference, 1972.* Washington, DC: Marine Technology Society. Pp. 247–265.

16-84.5
Rapp, W. von. 1843.
   Untersuchungen ueber die giftwerkzeuge der schlangen. Eine inaugural-dissertation. Tuebingen: Ludwig Friedrich Fues.

16-85
Reid, H. A. 1956.
   Sea-snake bite research. *Trans. R. Soc. Trop. Med. Hyg.* 50(6):517–542.

16-85.5
Reid, H. A. 1980.
   Snake bite and other venomous bites and stings. Maegraith, B. *Adams & Maegraith: Clinical tropical diseases.* London: Blackwell Scientific. Pp. 411–420.

16-86
Richards, V. 1885.
   The land-marks of snake-poison literature: being a review of the more important researches into the nature of snake-poisons. Calcutta: D. M. Traill.

16-87
Rogers, L. 1903.
   Demonstration of *Enhydrina* poisoning. *Proc. Physiol. Soc. (London)* 30:iv–v.

16-88
Rosenberg, H. I. 1967.
   Histology, histochemistry, and emptying mechanism of the venom glands of some elapid snakes. *J. Morphol.* 123(2):133–156.

16-88.5
Russell, F. E. 1979.
   Snake venom poisoning. Kelley, V. C., ed. *Practice of pediatrics.* Hagerstown, MD: Harper & Row. Vol. 1, no. 19, pp. 1–14.

16-89
Russell, F. E. 1980.
   Snake venom poisoning. Philadelphia: J. P. Lippincott.

16-89.5
Russell, F. E. 1983.
   Snake venom poisoning. Reprint with corrections. Great Neck, NY: Scholium International.

16-90
Russell, F. E.; Brodie, A. F. 1974.
   Venoms of reptiles. *Chem. Zool.* 9:449–478.

16-91
Russell, P. 1796.
   An account of Indian serpents, collected on the coast of Coromandel; containing descriptions and drawings of each species; together with experiments and remarks on their several poisons. London: George Nicol.

16-92
Sakai, A.; Tseng, C. S.; Toriba, M.; Miyata, H.; Moriguchi, H.; Sawai, Y. 1982.
   Note on the extraction of snake venoms at The Japan Snake Institute. *Snake* 14:61–67. (In Japanese with English abstract)

16-93
Savoor, S. R. 1957.
   Sea snakes. *Annu. Rep. Inst. Med. Res. (Kuala Lumpur)* 1945:85–88.

16-93.2
Schlegel, H. 1828.
   Untersuchung der speicheldruesen bei den schlangen mit gefurchten zaehnen, in vergleich mit denen der giftlosen und giftigen. *Nova Acta Phys.-Med. Acad. Natur. Curios.* 14(1):143–158.

16-93.3
Singh, Y. N. 1984.
   Effect of freeze drying and storage on the toxicity of the venom of the common Fijian sea snake, *Laticauda colubrina* (Schneider). *South Pac. J. Nat. Sci.* 5:93–98.

16-93.7
Singh, Y. N.; Guinea, M. L. 1984.
   Yield and toxicity studies on the venom of the common Fijian sea snake, *Laticauda colubrina* (Schneider). *South Pac. J. Nat. Sci.* 5:71–29.

16-93.8
Slowinski, J. B. 1989.
   The interrelationships of laticaudine sea snakes based on the animo acid sequences of short-chain neurotoxins. *Copeia* 1989(3):783–788.

16-94
Smith, M.; Hindle, E. 1931.
   Experiments with the venom of *Laticauda, Pseudechis* and *Trimeresurus* species. *Trans. R. Soc. Trop. Med. Hyg.* 25(2):115–120.

16-95
Smith, M.; Bellairs, A. d'A. 1947.
   The head glands of snakes, with remarks on the evolution of the parotid gland and teeth of the Opisthoglypha. *J. Linn. Soc. Lond. Zool.* 41(279):351–368.

16-95.1
Stocker, K. 1990.
   Composition of snake venoms. Stocker, K. F., ed. *Medical use of snake venom proteins.* Boca Raton, FL: CRC Press. Pp. 35–56.

16-95.2
Strong, R. P. 1942.
Stitt's diagnosis, prevention and treatment of tropical diseases. 6th ed. Philadelphia: Blakiston. (reprinted in the 7th ed., 1944)

16-95.3
Strydom, D. J. 1972.
Phylogenetic relationships of proteroglyphae toxins. *Toxicon* 10:39–45.

16-95.7
Strydom, D. J. 1973.
Snake venom toxins: structure-function relationships and phylogenetics. *Comp. Biochem. Physiol. B Comp. Biochem.* 44:269–281.

16-96
Strydom, D. J. 1973.
Snake venom toxins: the evolution of some of the toxins found in snake venoms. *Syst. Zool.* 22:596–608.

16-97
Strydom, D. J. 1979.
The evolution of toxins found in snake venoms. Lee, C. Y., ed. *Snake venoms.* New York: Springer-Verlag. Pp. 258–275.

16-98
Sutherland, S. K. 1979.
Australian venoms and care of the envenomed patient. Parkville, Victoria, Australia: University of Melbourne. 323 pp. (Medical Dissertation)

16-99
Sutherland, S. K. 1983.
Australian animal toxins: the creatures, their toxins and care of the poisoned patient. New York: Oxford University Press.

16-99.5
Takahashi, H. 1985.
Venom of fish egg feeding sea snake *Emydocephalus ijimae. Jpn. J. Herpetol.* 11(2):63. (In Japanese) (Abstract)

16-100
Tamiya, N. 1972.
Contents of toxic components in *Laticauda semifasciata* venom. *J. Formosan Med. Assoc.* 71:389–393.

16-101
Tamiya, N. 1975.
Sea snake venoms and toxins. Dunson, W. A., ed. *The biology of sea snakes.* Baltimore: University Park Press. Pp. 385–415.

16-102
Tamiya, N. 1978.
Introduction to sea snake toxinology—a round table discussion. *Toxicon* 1978(Suppl.1):314–315. (Abstract)

16-102.3
Tamiya, N. 1985.
A comparison of amino acid sequences of neurotoxins and of phospholipases of some Australian elapid snakes with those of other proteroglyphous snakes. Grigg, G.; Shine, R.; Ehmann, H., eds. *Biology of Australasian frogs and reptiles.* Chipping Norton, N.S.W.: Surrey Beatty & Sons. Pp. 209–219.

16-102.4
Tamiya, N. 1987.
Non-divergence theory of evolution: amino acid sequence comparison of proteroglyphous snake venom components. Gopalakrishnakone, P.; Tan, C. K., eds. *Progress in venom and toxin research.* Singapore: National University of Singapore. Pp. 137–145.

16-102.5
Tamiya, N. 1988.
Non-divergence theory of evolution: amino acid sequence comparison of proteroglyphous snake venom components. *Toxicon* 26(1):42. (Abstract)

16-103
Tamiya, N.; Puffer, H. 1974.
Lethality of sea snake venoms. *Toxicon* 12:85–87.

16-104
Tamiya, N.; Maeda, N. 1978.
Chemical taxonomy of snake neurotoxins. Matsubara, H.; Yamanaka, T., eds. *Evolution of protein molecules.* Tokyo, Japan: Scientific Societies Press. Pp. 297–310.

16-105
Tamiya, N.; Yagi, T. 1985.
Non-divergence theory of evolution: sequence comparison of some proteins from snakes and bacteria. *J. Biochem. (Tokyo)* 98(2):289–303.

16-105.5
Tan, N. H.; Tan, C. S. 1988.

Thermal stability of snake venom enzymatic activities. *Toxicon* 26(1):42. (Abstract)

16-106
Trethewie, E. R. 1971.

The pharmacology and toxicology of the venoms of the snakes of Australia and Oceania. Buecherl, W.; Buckley, E., eds. *Venomous animals and their venoms.* New York: Academic. Vol. 2, pp. 79–101.

16-107
Tu, A. T. 1969.

Immunological and biochemical studies of sea snakes. *Abstr. West. Soc. Nat. Annu. Meeting, 50th, Long Beach, Calif., Dec. 27–30, 1969.* Page 27. (Abstract)

16-107.5
Tu, A. T. 1970.

Sea snake (Hydrophiidae) collection and venoms. *Herpetol. Rev.* 2:7. (Abstract)

16-108
Tu, A. T. 1977.

Venoms: chemistry and molecular biology. New York: Wiley.

16-109
Tu, A. T. 1982.

Venom toxicology and composition. Tu, A. T., ed. *Survey of contemporary toxicology.* New York: Wiley. Vol. 2, pp. 145–201.

16-109.5
Tu, A. T. 1987.

Biotoxicology of sea snake venoms. *Ann. Emerg. Med.* 16(9):1023–1028.

16-110
Tu, A. T.; Tu, T. 1970.

Sea snakes from Southeast Asia and Far East and their venoms. Halstead, B. W. *Poisonous and venomous marine animals of the world.* Washington, DC: Government Printing Office. Vol. 3, pp. 885–903.

16-111
Tu, T. 1959.

Toxicological studies on the venom of a sea snake, *Laticauda semifasciata* (Reinwardt) in Formosan waters. *J. Formosan Med. Assoc.* 58(4):182–203.

16-112
Tu, T. 1961.
Toxicological studies on the venom of a sea snake, *Laticauda semifasciata* (Reinwardt). *Biochem. Pharmacol.* 8:75–76. (Abstract)

16-113
Tu, T. 1967.
Toxicological studies on the venom of the sea snake *Laticauda laticaudata affinis. Toxicon* 4(4):296. (Abstract)

16-114
Tu, T. 1967.
Toxicological studies on the venom of the sea snake *Laticauda laticaudata affinis.* Russell, F. E.; Saunders, P. R., eds. *Animal toxins.* Oxford: Pergamon. Pp. 245–248.

16-115
Tu, T. C. 1957.
Toxicological studies on the venom of *Laticauda semifasciata* (Reinwardt) in Formosan waters. 1st report: Toxicity and toxicological action on the respiration and the circuratory system. *J. Formosan Med. Assoc.* 56:609–610. (In Chinese) (Abstract)

16-116
Tu, T. C. 1959.
Toxicological studies on the venom of a sea snake, *Laticauda semifasciata* (Reinwardt) in Formosan waters. Fifth Report. *J. Formosan Med. Assoc.* 58(4):740. (Abstract)

16-116.5
Tu, T. C.; Lin, M. J.; Yang, H. M.; Lin, H. J.; Chen, C. N. 1962.
Toxicological studies on the venom of a sea snake (*Laticauda colubrina*) (Schneider) (First report). *J. Formosan Med. Assoc.* 61:1296. (also cited as 61:122) (Abstract)

16-116.6
Van der Walle, N. 1930.
Een en ander over gifslangen en slangengif. *Natuurw. Tijdschr. Ned.-Indie* 90(2):183–233.

16-116.7
Viaud Grand Marais, A. 1880.
L'envenomation ophidienne etudiee dans les differents groupes de serpents. *Gaz. Hosp. Civil Mil.* 53(113):901–902.

16-117
Vick, J. A. 1970.
   Venomous sea snakes. Halstead, B. W. *Poisonous and venomous marine animals of the world.* Washington, DC: Government Printing Office. Vol. 3, pp. 974–975.

16-118
Yang, C. C. 1974.
   Chemistry and evolution of toxins in snake venoms. *Toxicon* 12:1–43.

16-118.5
Zhou, D.; Liang, X.; Wu, X.; Wei, Q. 1986.
   The gamma-glutamyl transpeptidase in snake venoms. *Zool. Res. (China)* 7(3):273–279. (In Chinese with English abstract)

16-119
Zimmerman, K. D. 1989.
   The effects of the venom of Aipysurus laevis on its prey species. Armidale, N.S.W.: University of New England. 211 pp. (Dissertation)

Genus-Species Index
for Chapter 16

Astrotia stokesii 16-7, 16-10, 16-47.5, 16-64, 16-65, 16-77.5, 16-89, 16-89.5, 16-99, 16-101, 16-102.3, 16-104, 16-105, 16-108, 16-110

Disteira kingii 16-64, 16-65.5, 16-99

Disteira major 16-47.5, 16-64, 16-65, 16-65.5, 16-99

Disteira melanocephala 16-111

Disteira melanosoma 16-82

Distira annandalei 16-82

Distira belcheri 16-82

Distira bituberculata 16-82

Distira brungmansii 16-82

Distira cincinatii 16-82

Distira cyanocincta 16-19, 16-20, 16-45, 16-80, 16-81, 16-82

Distira cyanosoma 16-82

Distira gillespiae 16-82

Distira godeftovi 16-82

Distira grandis 16-82

Distira hendorsoni 16-82

Distira jerdonii 16-19, 16-20, 16-80, 16-82

Distira lapimoides 16-82

Distira macfarlani 16-82

Distira mertoni 16-82

Distira orientalis 16-82

Distira ornata 16-18.5, 16-19, 16-20, 16-80, 16-82

Distira pachycercus 16-82

Distira saravacensis 16-82

Distira semperi 16-32, 16-82

Distira stokesi 16-82

Distira subcincta 16-19, 16-20, 16-80, 16-82

Distira tuberculata 16-82

Distira viperina 16-82

Distira wrayi 16-82

Emydocephalus 16-14, 16-102.4, 16-109.5

Emydocephalus annulatus 16-7, 16-44, 16-47.5, 16-63, 16-64, 16-65.5, 16-66, 16-72, 16-99

Emydocephalus ijimae 16-7, 16-44, 16-99.5, 16-110

Enhydrina 16-18.5, 16-77.5, 16-95.2

Enhydrina bengalensis 16-32, 16-33, 16-34, 16-45, 16-87, 16-111

Enhydrina schistosa 16-1.2, 16-1.3, 16-1.6, 16-4, 16-5, 16-7, 16-8, 16-9, 16-10, 16-14, 16-15, 16-16, 16-17, 16-18, 16-22, 16-23, 16-25, 16-28, 16-36, 16-37, 16-39, 16-40, 16-42, 16-43, 16-44, 16-45, 16-47.5, 16-50.5, 16-51, 16-53, 16-54, 16-57, 16-59, 16-60, 16-63, 16-64, 16-65, 16-65.5, 16-66, 16-67, 16-71, 17-73, 16-79, 16-85, 16-88.5, 16-89, 16-89.5, 16-93, 16-95.1, 16-96, 16-97, 16-98, 16-99, 16-101, 16-102.3, 16-105, 16-105.5, 16-106, 16-107.5, 16-108, 16-109.5, 16-110, 16-111, 16-118

Enhydrina valakadien 16-1, 16-20, 16-32, 16-45, 16-68, 16-111

Enhydrina valakadien (bengalensis) 16-19, 16-80

Enhydris curtus 16-19, 16-20, 16-80, 16-82

Enhydris hardwickii 16-82

Enhydris valakadien 16-82

Enhydrus curtus 16-45

Enhydrus hardwickii 16-80.5

Ephalophis 16-47.5

Ephalophis greyi 16-65.5, 16-99

Ephalophis mertoni 16-65.5, 16-99

Hydrelaps 16-47.5, 16-109.5

Hydrelaps darwiniensis 16-7, 16-20, 16-44, 16-65.5, 16-80, 16-99

Chapter 17

VENOM—BASIC AND STRUCTURAL BIOCHEMISTRY

17-1
Abe, T.; Tamiya, N. 1979.
Amino acid residues associated with antigenicity of erabutoxin. *Koen Yoshishu-Tanpakushitsu Kozo Toronkai, 30th. (Symposium on protein structures, Tokyo, 30)*. Tokyo: Tokyo Daigaku Rigakubu. Pp. 41–44. (In Japanese)

17-2
Allen, M.; Tu, A. T. 1985.
The effect of tryptophan modification on the structure and function of a sea snake neurotoxin. *Mol. Pharmacol.* 27:79–85.

17-3
Anonymous. 1971.
Sea snake venom structure probed. *Chem. Engr. News* 49(1):25–26.

17-4
Arai, H.; Tamiya, N.; Toshioka, S. I.; Shinonaga, S.; Kano, R. 1964.
Studies on sea-snake venoms I. Protein nature of the neurotoxic component. *J. Biochem. (Tokyo)* 56(6):568–571.

17-5
Arseniev, A. S.; Pashkov, V. S.; Pluzhnikov, K. A.; Rochat, H.; Bystrov, V. F. 1981.
The ^{1}H nuclear-magnetic-resonance spectra and spatial structure of the *Naja mossambica mossambica* neurotoxin III. *Eur. J. Biochem.* 118:453–462.

17-6
Banerjee, S.; Devi, A.; Copley, A. L. 1973.
Studies of actions of snake venoms on blood coagulation II. Electrophoretic analysis of venoms of Viperidae, Crotalidae, Elapidae, and Hydrophidae. *Thromb. Res.* 3:451–464.

17-7
Bouet, F.; Menez, A.; Fromageot, P. 1977.
Etude des vitesses de renaturation structurale des polypeptides curarisants des venins des cobras. *C. R. Acad. Sci. Paris* 285D:1527–1530. (In French with English abstract)

17-8
Bouet, F.; Menez, A.; Hider, R.C.; Fromageot, P. 1982.
Separation of intermediates in the refolding of reduced erabutoxin b by analytical isoelectric focusing in layers of polyacrylamide gel. *Biochem. J.* 201:495–499.

17-9
Bourne, P. E.; Sato, A.; Corfield, P. W. R.; Rosen, L. S.; Birken, S.; Low, B. W. 1985.
Erabutoxin b: initial protein refinement and sequence analysis at 0.140-nm resolution. *Eur. J. Biochem.* 153:521–527.

17-10
Bystrov, V. F.; Ivanov, V. T.; Okanov, V. V.; Miroshnikov, A. I.; Arseniev, A. S.; Tsetlin, V. I.; Pashkov, V. S.; Karlsson, E. 1981.
The composite physical and chemical approach to the solution spatial structure of polypeptide neurotoxins. Bertini, I.; Lunazzi, L.; Dei, A., eds. *Advances in solution chemistry*. New York: Plenum. Pp. 231–251.

17-11
Carey, J. E.; Wright, E. A. 1960.
Isolation of the neurotoxic component of the venom of the sea snake, *Enhydrina schistosa. Nature (Lond.)* 185(4706):103–104.

17-12
Carey, J. E.; Wright, E. A. 1962.
Studies on the fractions of the venom of the sea snake *Enhydrina schistosa. Aust. J. Exp. Biol. Med. Sci.* 40:427–435.

17-13
Chen, Y. H.; Lu, H. S.; Lo, T. B. 1975.
Conformation of snake toxins: prediction and comparison. *J. Chin. Biochem. Soc.* 4(2):69–82.

17-13.5
Corfield, P. W. R.; Lee, T. J.; Low, B. W. 1989.
The crystal structure of erabutoxin a at 2.0-A resolution. *J. Biol. Chem.* 264(16):9239–9242.

17-14
Dayhoff, M. O.; Barker, W. C.; Hunt, L. T. 1976.
   Protein superfamilies. Dayhoff, M. O., ed. *Atlas of protein sequence and structure.* Silver Spring, MD: National Biomedical Research Foundation. Vol. 5, pp. 9–19.

17-15
Drake, A. F.; Dufton, M. J.; Hider, R. C. 1977.
   The flexible nature of a critical peptide region common to all Elapidae 'short' neurotoxins. *FEBS (Fed. Eur. Biochem. Soc.) Lett.* 83(2):202–206.

17-16
Drenth, J.; Low, B. W.; Richardson, J. S.; Wright, C. S. 1980.
   The toxin-agglutinin fold. A new group of small protein structures organized around a four-disulfide core. *J. Biol. Chem.* 255(7):2652–2655.

17-16.5
Ducancel, F.; Guignery-Frelat, G.; Boulain, J. C.; Menez, A. 1990.
   Nucleotide sequence and structure analysis of cDNAs encoding short-chain neurotoxins from venom glands of a sea snake (*Aipysurus laevis*). *Toxicon* 28(1):119–123.

17-17
Dufton, M. J.; Hider, R. C. 1977.
   Snake toxin secondary structure predictions: structure activity relationships. *J. Mol. Biol.* 115:177–193.

17-18
Dufton, M. J.; Hider, R. C. 1980.
   Lethal protein conformations. *Trends Biochem. Sci.* 5(2):53–57.

17-19
Dufton, M. J.; Hider, R. C. 1982.
   Conformational properties of the neurotoxins and cytotoxins isolated from elapid snake venoms. *CRC Crit. Rev. Biochem.* 14(2):113–171.

17-20
Dufton, M. J.; Eaker, D.; Hider, R. C. 1983.
   Conformational properties of phospholipases A_2: secondary-structure prediction, circular dichroism and relative interface hydrophobicity. *Eur. J. Biochem.* 137:537–544.

17-21
Durkin, J. P.; Pickwell, G. V.; Trotter, J. T.; Shier, W. T. 1981.
Phospholipase A_2 electrophoretic variants in reptile venoms. *Toxicon* 19(4):535–546.

17-22
Eaker, D. 1975.
Structure and function of snake venom toxins. Walter, R.; Meierhofer, J., eds. *Peptides: chemistry, structure and biology.* Ann Arbor, MI: Ann Arbor Science. Pp. 17–30.

17-23
Eaker, D. 1978.
Studies of presynaptically neurotoxic and myotoxic phospholipases A_2. Li, C. H., ed. *Versatility of proteins.* New York: Academic. Pp. 413–431.

17-24
Elliott, W. B. 1978.
Chemistry and immunology of reptilian venoms. Gans, C.; Gans, K. A., eds. *Biology of the reptilia.* New York: Academic. Vol. 8, pp. 163–436.

17-25
Endo, T.; Kim, H. S.; Maeda, N.; Tamiya, N.; Miyazawa, T. 1986.
Proton nuclear magnetic resonance analyses of the molecular conformations of unique long neurotoxins bearing Phe-25 astrotia stokesii b, astrotia stokesii c, and acanthophis antarcticas b. *J. Biochem. (Tokyo)* 99:681–691.

17-26.3
Endo, T.; Tamiya, N. 1987.
Current view on the structure-function relationship of postsynaptic neurotoxins from snake venoms. *Pharmacol. Ther.* 34(3):403–451.

17-26.4
Endo, T.; Oya, M.; Tamiya, N.; Miyazawa, T. 1987.
Proton nuclear magnetic resonance characterization of phospholipase A_2 from *Laticauda semifasciata. J. Biochem. (Tokyo)* 101(3):795–804.

17-26.5
Endo, T.; Oya, M.; Tamiya, N.; Hayashi, K. 1987.
Role of c-terminal tail of long neurotoxins from snake venoms in molecular conformation and acetylcholine receptor binding: proton nuclear magnetic resonance and competition binding studies. *Biochemistry* 26(14):4592–4598.

17-26.7
Endo, Y.; Sato, S.; Ishii, S.; Tamiya, N. 1971.
The disulphide bonds of erabutoxin a, a neurotoxic protein of a sea-snake (*Laticauda semifasciata*) venom. *Biochem. J.* 122:463–467.

17-27
Eterovic, V. A.; Ferchmin, P. A. 1977.
Predicted secondary structure of snake venom toxins from their primary structures. *Int. J. Pept. Protein Res.* 10:245–251.

17-28
Faure, G.; Menez, A.; Boulain, J. C.; Sato, A.; Tamiya, N.; Fromageot, P. 1982.
The role of the amino groups in the curare-like activity of toxin (alpha) from *Naja nigricollis*. *Toxicon* 20(1):115–116. (Abstract)

17-29
Fohlman, J.; Eaker, D. 1977.
Isolation and characterization of a lethal myotoxic phospholipase a from the venom of the common sea snake *Enhydrina schistosa* causing myoglobinuria in mice. *Toxicon* 15(3):385–393.

17-31
Fox, J. W. 1978.
Studies on the structures of two snake venom toxins. Fort Collins, CO: Colorado State University. 164 pp. (Dissertation).

17-32
Fox, J. W. 1979.
Studies on the structures of two snake venom toxins. *Diss. Abstr. Int. B Sci. Eng.* 39(12B):5889. (Abstract)

17-33
Fox, J. W.; Elzinga, M.; Tu, A. T. 1977.
Amino acid sequence of a snake neurotoxin from the venom of *Lapemis hardwickii* and the detection of a sulfhydryl group by laser Raman spectroscopy. *FEBS (Fed. Eur. Biochem. Soc.) Lett.* 80(1):217–220.

17-34
Fox, J. W.; Elzinga, M.; Tu, A. T. 1979.
Amino acid sequence and disulfide bond assignment of myotoxin a isolated from the venom of prairie rattlesnake (*Crotalus viridis viridis*). *Biochemistry* 18(4):678–684.

17-34.5
Fox, J.; Tu, A. T. 1979.
Conformational analysis of a snake neurotoxin by prediction from sequence, circular dichroism, and Raman spectroscopy. *Arch. Biochem. Biophys.* 193(2):407–414.

17-35
Fryklund, L. 1973.
Sequence homology in snake venom proteins. *Abstr. Uppsal. Diss. Sci.* 233:1–13.

17-36
Fryklund, L.; Eaker, D.; Karlsson, E. 1972.
Amino acid sequences of the two principal neurotoxins of *Enhydrina schistosa* venom. *Biochemistry* 11(24):4633–4640.

17-37
Fung, C. H.; Chang, C. C.; Gupta, R. K. 1979.
Proton nuclear magnetic resonance study of cobrotoxin. *Biochemistry* 18(3):457–460.

17-38
Galaktionov, S. G.; Rodionov, M. A. 1979.
Calculation of the system of intramolecular residue-residue contacts in the globule of neurotoxin I from the venom of the central asian cobra. *Sov. J. Bioorg. Chem.* 4(11):1132–1134.

17-39
Garnet, J. R. 1970.
Snake venoms and antivenenes. *Australas. J. Pharm. Sci. Suppl.* 91:S65–S68.

17-40
Gawade, S. P.; Bhide, M. B. 1977.
Enzymatic activities of the venom of *Enhydrina schistosa* (common sea snake) from the western coast of India. *Bull. Haffkine Inst.* 5(2):48–51.

17-41
Gawade, S. P.; Bhide, M. B. 1978.
Chromatographic separation of the venom of *Enhydrina schistosa* (common sea snake) and characterisation of its principal toxic component. *Indian J. Med. Res.* 67:854–861.

17-42
Guinea, M. L.; Tamiya, N.; Cogger, H. G. 1983.
The neurotoxins of the sea snake *Laticauda schistorhynchus*. *Biochem. J.* 213:39–41.

17-43
Harada, I. 1979.
Laser Raman spectroscopy of protein molecules. *Tanpakushitsu Kakusan Koso* 24(13):1441–1451. (In Japanese)

17-44
Harada, I.; Shimanouchi, T. 1976.
Raman spectra of some neurotoxins and denatured neurotoxins in relation to the structures and toxicities. *Proc. Int. Conf. Raman Spectrosc.*, 5th, 1976. Pp. 44–45.

17-45
Harada, I.; Takamatsu, T.; Shimanouchi, T.; Miyazawa, T.; Tamiya, N. 1976.
Raman spectra of some neurotoxins and denatured neurotoxins in relation to structures and toxicities. *J. Phys. Chem.* 80(11):1153–1156.

17-46
Hayashi, K.; Ohta, M. 1975.
Neurotoxins in snake venoms. *Protein Nucleic Acid Enzyme (Tokyo)* 20:53–69. (In Japanese)

17-47
Hider, R. C. 1979.
An alternative view of the structure of the reactive site in the short-series snake venom postsynaptic neurotoxins. *Adv. Cytopharmacol.* 3:149.

17-47.5
Hider, R. C.; Drake, A. F.; Tamiya, N. 1988.
An analysis of the 225–230-nm CD band of elapid toxins. *Biopolymers* 27(1):113–1212.

17-48
Hong, B. S. 1970.
Studies on the toxic principles in the venoms of sea snakes. Fort Collins, CO: Colorado State University. 179 pp. (Dissertation)

17-49
Hong, B. S. 1971.
Studies on the toxic principles in the venoms of sea snakes. *Diss. Abstr. Int. B Sci. Eng.* 32B(3):1353. (Abstract)

17-50
Hong, B. S.; Tu, A. T. 1970.
Importance of tryptophane residue for toxicity in sea snake venom toxins. *Fed. Proc.* 29:888. (Abstract)

17-51
Hori, H.; Tamiya, N. 1976.
Preparation and activity of guanidinated or acetylated erabutoxins. *Biochem. J.* 153:217–222.

17-52
Hseu, T. H.; Chang, H.; Hwang, D. M.; Yang, C. C. 1978.
Laser Raman studies on cobrotoxin. *Biochim. Biophys. Acta* 537(2):284–292.

17-53
Huang, J. S.; Liu, S. S.; Ling, K. H.; Chang, C. C.; Yang, C. C. 1973.
Iodination of cobrotoxin. *Toxicon* 11:39–45.

17-54
Hunt, L. T.; Dayhoff, M. O. 1976.
Toxins. Dayhoff, M. O., ed. *Atlas of protein sequence and structure.* Silver Spring, MD: National Biomedical Research Foundation. Vol. 5, pt. 2, pp. 147–164.

17-55
Ibrahim, S. A. 1970.
A study on sea-snake venom phospholipase A. *Toxicon* 8:221–224.

17-56
Inagaki, F.; Miyazawa, T.; Hori, H.; Tamiya, N. 1978.
Conformation of erabutoxins a and b in aqueous solution as studied by nuclear magnetic resonance and circular dichroism. *Eur. J. Biochem.* 89:433–442.

17-57
Inagaki, F.; Tamiya, N.; Miyazawa, T. 1980.
Molecular conformation and function of erabutoxins as studied by nuclear magnetic resonance. *Eur. J. Biochem.* 109:129–138.

17-58
Inagaki, F.; Miyazawa, T.; Williams, R. J. P. 1981.
The dynamic structures of proteins: short and long neurotoxins as examples. *Biosci. Rep.* 1:743–755.

17-59
Inagaki, F.; Clayden, N. J.; Tamiya, N.; Williams, R. J. P. 1981.
A proton-magnetic-resonance study on the molecular conformation and structure-function relationship of a long neurotoxin, laticauda semifasciata III from *Laticauda semifasciata*. *Eur. J. Biochem.* 120(2):313–322.

17-60
Inagaki, F.; Tamiya, N.; Miyazawa, T.; Williams, R. J. P. 1981.
Structural differences between erabutoxins in aqueous solution and in crystalline states. *Eur. J. Biochem.* 118(3):621–625.

17-61
Inagaki, F.; Boyd, J.; Campbell, I. D.; Clayden, N. J.; Hull, W. E.; Tamiya, N.; Williams, R. J. P. 1982.
Dynamics of erabutoxin b as studied by nuclear magnetic resonance relaxation studies of methyl proton resonances. *Eur. J. Biochem.* 121(3):609–616.

17-62
Inagaki, F.; Clayden, N. J.; Tamiya, N.; Williams, R. J. P. 1982.
Individual assignments of the amide proton resonances involved in the triple-stranded antiparallel pleated (beta)-sheet structure of a long neurotoxin, laticauda semifasciata III from *Laticauda semifasciata*. *Eur. J. Biochem.* 123(1):99–104.

17-63
Inagaki, F.; Miyazawa, T.; Tamiya, N.; Williams, R. J. P. 1982.
Structural dynamics of erabutoxin b: a ^{13}C nuclear magnetic resonance relaxation study of methyl groups. *Eur. J. Biochem.* 123(2):275–282.

17-64
Ishikawa, Y.; Menez, A.; Hori, H.; Yoshida, H.; Tamiya, N. 1977.
Structure of snake toxins and their affinity to the acetylcholine receptor of fish electric organ. *Toxicon* 15:477–488.

17-65
Ishikawa, Y.; Yoshida, H.; Tamiya, N. 1980.
Purification and properties of the acetylcholine receptor protein from *Narke japonica*. *J. Biochem. (Tokyo)* 87:313–321.

17-66
Ishizaki, H.; Allen, M.; Tu, A. T. 1984.
   Effect of sulfhydryl group modification on the neurotoxic action of a sea snake toxin. *J. Pharm. Pharmacol.* 36:36–41.

17-67
Iwanaga, S.; Suzuki, T. 1979.
   Enzymes in snake venom. Lee, C. Y., ed. *Snake venoms.* New York: Springer-Verlag. Pp. 61–158.

17-68
Jimenez Porras, J. M. 1971.
   Biochemistry of snake venoms. Minton, S. A., ed. *Snake venoms and envenomation.* New York: Marcel Dekker. Pp. 43–85.

17-69
Jones, G. P. 1979.
   Curare-like structure of snake neurotoxin. Chubb, I. W.; Geffen, L. B., eds. *Neurotoxins: fundamental & clinical advances.* Adelaide: Adelaide University Union Press. Page 264. (Abstract)

17-70
Kao, C. S.; Liu, C. S.; Blackwell, R. Q. 1973.
   Presence of erabutoxins a and b in venom of the sea snake *Laticauda semifasciata* from Taiwan. *Toxicon* 11:383–385.

17-71
Karlsson, E. 1973.
   Chemistry of some potent animal toxins. *Experientia (Basel)* 29:1319–1327.

17-72
Karlsson, E. 1979.
   Chemistry of protein toxins in snake venoms. Lee, C. Y., ed. *Snake venoms.* New York: Springer-Verlag. Pp. 159–212.

17-73
Karlsson, E.; Eaker, D.; Fryklund, L.; Kadin, S. 1972.
   Chromatographic separation of *Enhydrina schistosa* (common sea snake) venom and the characterization of two principal neurotoxins. *Biochemistry* 11(24):4628–4633.

17-74
Kim, H. S.; Abe, T.; Tamiya, N. 1980.
   The acetylation of the amino groups of laticauda semifasciata III, a sea snake venom component. *J. Biochem. (Tokyo)* 88(3):889–893.

17-75
Kim, H. S.; Tamiya, N. 1981.
   The amino acid sequence and position of the free thiol group of a short-chain neurotoxin from common-death-adder (*Acanthophis antarcticus*) venom. *Biochem. J.* 199(1):211–218.

17-76
Kim, H. S.; Tamiya, N. 1981.
   Isolation, properties and amino acid sequence of a long-chain neurotoxin, Acanthophis antarcticus b, from the venom of an Australian snake (the common death adder, *Acanthophis antarcticus*). *Biochem J.* 193(3):899–906.

17-77
Kim, H. S.; Tamiya, N. 1982.
   Amino acid sequences of two novel long-chain neurotoxins from the venom of the sea snake *Laticauda colubrina*. *Biochem J.* 207(2):215–223.

17-78
Kimball, M. R.; Sato, A.; Richardson, J. S.; Rosen, L. S.; Low, B. W. 1979.
   Molecular conformation of erabutoxin b; atomic coordinates at 2.5 (angstrom) resolution. *Biochem. Biophys. Res. Commun.* 88(3):950–959.

17-79
Kini, R. M.; Iwanaga, S. 1986.
   Structure-function relationships of phospholipases I: prediction of presynaptic neurotoxicity. *Toxicon* 24(6):527–541.

17-80
Lauterwein, J.; Wuethrich, K. 1978.
   A possible structural basis for the different modes of action of neurotoxins and cardiotoxins from snake venoms. *FEBS (Fed. Eur. Biochem. Soc.) Lett.* 93(2):181–184.

17-81
Lauterwein, J.; Lazdunski, M.; Wuethrich, K. 1978.
   The ^{1}H nuclear-magnetic-resonance spectra of neurotoxin I and cardiotoxin $V^{II}4$ from *Naja mossambica mossambica*. *Eur. J. Biochem.* 92:361–371.

17-82
Lee, C. Y. 1971.
Elapid neurotoxins and their mode of action. Minton, S. A., ed. *Snake venoms and envenomation*. New York: Marcel Dekker. Pp. 111–126.

17-83
Lee, C. Y. 1971.
Mode of action of cobra venom and its purified toxins. Simpson, L. L., ed. *Neuropoisons*. New York: Plenum. Vol. 1, pp. 21–70.

17-84
Lee, C. Y. 1972.
Classification of polypeptide toxins from elapid and sea snake venoms according to their pharmacological properties and chemical structures. *J. Formosan Med. Assoc.* 71:311–317.

17-85
Lee, C. Y. 1972.
Recent advances in chemistry of polypeptide toxins from snake venoms. *J. Chin. Biochem. Soc.* 1(1):47P–56P.

17-86
Lind, P.; Eaker, D. 1981.
Amino acid sequence of a lethal myotoxic phospholipase A_2 from the venom of the common sea snake (*Enhydrina schistosa*). *Toxicon* 19(1):11–24.

17-87
Liu, C. S.; Huber, G. S.; Lin, C. S.; Blackwell, R. Q. 1973.
Fractionation of toxins from *Hydrophis cyanocinctus* venom and determination of amino acid composition and end groups of hydrophitoxin a. *Toxicon* 11(1):73–79.

17-88
Liu, C. S.; Blackwell, R. Q. 1974.
Hydrophitoxin b from *Hydrophis cyanocinctus* venom. *Toxicon* 12:543–546.

17-89
Liu, C. S.; Wang, C. L.; Blackwell, R. Q. 1975.
Isolation and partial characterization of pelamitoxin a from *Pelamis platurus* venom. *Toxicon* 13:31–35.

17-90
Low, B. W. 1976.
Structure studies of a sea snake neurotoxin "erabutoxin b". *J. Biochem. (Tokyo)* 79(4):27p.

17-91
Low, B. W. 1979.
   The three-dimensional structure of postsynaptic snake neurotoxins: considera-
tion of structure and function. Lee, C. Y., ed. *Snake venoms*. New York: Springer-
Verlag. Pp. 213–257.

17-92
Low, B. W. 1979.
   Three-dimensional structure of erabutoxin b, prototype structure of the snake
venom postsynaptic neurotoxins: consideration of structure and function; description
of the reactive site. *Adv. Cytopharmacol.* 3:141–147.

17-93
Low, B. W.; Potter, R.; Jackson, R. B.; Tamiya, N.; Sato, S. 1971.
   X-ray crystallographic study of the erabutoxins and of a diiodo derivative. *J.
Biol. Chem.* 246(13):4366–4368.

17-94
Low, B. W.; Preston, H. S.; Sato, A.; Rosen, L. S.; Searl, J. E.; Rudko, A. D.;
Richardson, J. S. 1976.
   Three dimensional structure of erabutoxin b neurotoxic protein: inhibitor of
acetylcholine receptor. *Proc. Nat. Acad. Sci. U.S.A.* 73(9):2991–2994.

17-95
Low, B. W.; Preston, H. S.; Sato, A.; Rosen, L. S.; Searl, J. E.; Rudko, A. D.;
Richardson, J. S. 1978.
   Three dimensional structure of the sea snake toxin—erabutoxin b. *Toxicon* 16
(Suppl.1):312. (Abstract)

17-96
Low, B. W.; Corfield, P. W. R. 1985.
   Binding site of neurotoxins to acetylcholine receptor. *Toxicon* 23(4):590. (Ab-
stract)

17-97
Maeda, N.; Takagi, K.; Tamiya, N.; Chen, Y. M.; Lee, C. Y. 1974.
   The isolation of an easily reversible post-synaptic toxin from the venom of a
sea snake, *Laticauda semifasciata*. *Biochem. J.* 141:383–387.

17-98
Maeda, N.; Tamiya, N. 1974.
   The primary structure of the toxin Laticauda semifasciata III, a weak and
reversibly acting neurotoxin from the venom of a sea snake, *Laticauda semifasciata*.
Biochem. J. 141:389–400.

17-99
Maeda, N.; Tamiya, N.; Chen, Y. M.; Lee, C. Y. 1976.
The isolation, properties, and amino acid sequence of Laticauda semifasciata III, a weak and reversible neurotoxin of a sea snake, *Laticauda semifasciata*, venom. Ohsaka, A.; Hayashi, K.; Sawai, Y., eds. *Animal, plant, and microbial toxins*. New York: Plenum. Vol. 2, pp. 1–14.

17-100
Maeda, N.; Tamiya, N. 1976.
Isolation, properties and amino acid sequences of three neurotoxins from the venom of a sea snake, *Aipysurus laevis*. *Biochem. J.* 153:79–87.

17-101
Maeda, N.; Tamiya, N. 1977.
Correction of partial amino acid sequence of erabutoxins. *Biochem. J.* 167(1):289–291.

17-102
Maeda, N.; Tamiya, N. 1978.
Three neurotoxins from the venom of a sea snake *Astrotia stokesii*, including two long-chain neurotoxic proteins with amidated C-termini. *Biochem. J.* 175(2):507–517.

17-103
Mebs, D. 1973.
Chemistry of animal venoms, poisons and toxins. *Experientia (Basel)* 29(11):1328–1334.

17-104
Mebs, D.; Samejima, Y. 1980.
Myotoxic phospholipases A from snake venom, *Pseudechis colletti*, producing myoglobinuria in mice. *Experientia (Basel)* 36(7):868–869.

17-104.5
Mebs, D. 1990.
Use of toxins in neurobiology and muscle research. Stocker, K. F., ed. *Medical use of snake venom proteins*. Boca Raton, FL: CRC Press. Pp. 57–78.

17-105
Menez, A. 1977.
Marquages a l'iode et au tritium d'agents curarisants et cardiotoxiques. Etude de la conformation en solution des polypeptides toxiques extraits de venin de serpents. Paris: University of Paris. 198 pp. (Dissertation)

17-105.5
Menez, A.; Bouet, F.; Fromageot, P.; Tamiya, N. 1976.
On the role of tyrosyl and tryptophanyl residues in the conformation of two snake neurotoxins. *Bull. Inst. Pasteur* 74:57–64.

17-106
Menez, A.; Bouet, F.; Tamiya, N.; Fromageot, P. 1976.
Conformational changes in two neurotoxic proteins from snake venoms. *Biochim. Biophys. Acta* 453:121–132.

17-107
Menez, A.; Morgat, J. L.; Fromageot, P. 1976.
Some biological properties of two titrated snake neurotoxins. Ohsaka, A.; Hayashi, K.; Sawai, Y., eds. *Animal, plant, and microbial toxins.* New York: Plenum. Vol. 1, pp. 39–47.

17-108
Menez, A.; Langlet, G.; Tamiya, N.; Fromageot, P. 1978.
Conformation of snake toxic polypeptides studied by a method of prediction and circular dichroism. *Biochimie (Paris)* 60(5):505–516.

17-109
Menez, A.; Montenay Garestier, T.; Helene, C.; Fromageot, P. 1978.
Solvent effects on neurotoxic protein polymorphism. *Toxicon* 16 (Suppl.1):313. (Abstract)

17-110
Menez, A.; Montenay Garestier, T.; Fromageot, P.; Helene, C. 1980.
Conformation of two homologous neurotoxins. Fluorescence and circular dichroism studies. *Biochemistry* 19(23):5202–5208.

17-111
Menez, A.; Bouet, F.; Guschlbauer, W.; Fromageot, P. 1980.
Refolding of reduced short neurotoxins: circular dichroism analysis. *Biochemistry* 19:4166–4172.

17-112
Menez, A.; Tamiya, N. 1982.
Role des lysines dans la stabilite d'une neurotoxine de serpent. *C.R. Acad. Sci. Paris* Ser. 3, 294(19):957–962. (In French with English abstract)

17-113
Miyazawa, T. 1981.
   Microenvironments and roles of protein side chains as elucidated by NMR analyses. *Indian J. Biochem. Biophys.* 18(4):27. (Abstract)

17-114
Miyazawa, T.; Endo, T.; Inagaki, F.; Hayashi, K.; Tamiya, N. 1983.
   NMR analysis of structure and function of snake neurotoxins. *Biopolymers* 22(1):139–145.

17-114.5
Mori, N.; Tu, A. T. 1988.
   Isolation and primary structure of the major toxin from sea snake, *Acalyptophis peronii* venom. *Arch. Biochem. Biophys.* 260(1):10–17.

17-114.7
Mori, N.; Tu, A. T. 1989.
   Isolation and characterization of *Pelamis platurus* (yellow-bellied sea snake) postsynaptic isoneurotoxin. *J. Pharm. Pharmacol.* 41(5):331–334.

17-115
Muszkat, K. A.; Khait, I.; Hayashi, K.; Tamiya, N. 1984.
   Photochemically induced nuclear polarization study of exposed tryosines, tryptophans, and histidines in postsynaptic neurotoxins and in membranotoxins of Elapid and Hydrophid snake venoms. *Biochemistry* 23(21):4913–4920.

17-116
Nabedryk Viala, E.; Thiery, C.; Menez, A.; Tamiya, N.; Thiery, J. M. 1980.
   Molecular dynamics of two homologous neurotoxins revealed by ^{1}H-^{2}H exchange: an infrared spectrometry study. *Biochim. Biophys. Acta* 626:321–331.

17-117
Nakanishi, M.; Kobayashi, M.; Tsuboi, M.; Takasaki, C.; Tamiya, N. 1980.
   Electronic spectroscopy and deuteration kinetics of tyrosine and tryptophan residues: an application to the study of erabutoxin b. *Biochemistry* 19(14):3204–3208.

17-118
Nishida, S.; Kim, H. S.; Tamiya, N. 1982.
   Amino acid sequences of three phospholipases A I, III, and IV from the venom of the sea snake *Laticauda semifasciata*. *Biochem. J.* 207(3):589–594.

17-119

Nishida, S.; Kokubun, Y.; Tamiya, N. 1985.

Correction of the amino acid sequences of erabutoxins from the venom of the sea snake *Laticauda semifasciata*. *Biochem. J.* 226(3):879–880.

17-120

Pashkov, V. S.; Pluzhnikov, K. A.; Utkin, Y. N.; Hintsche, R.; Arseniev, A. S.; Tsetlin, V. I.; Ivanov, V. T.; Bystrov, V. F. 1983.

^{1}H NMR study of neurotoxin II *Naja naja oxiana* and its spin labeled derivatives: conformation of "short" neurotoxins. *Bioorg. Khim.* 9(9):1181–1219. (In Russian with English abstract)

17-121

Passey, R. B. 1969.

Snake venom enzymes. Fort Collins, CO: Colorado State University. 125 pp. (Dissertation)

17-122

Passey, R. B. 1970.

Snake venom enzymes. *Diss. Abstr. Int. B Sci. Eng.* 30:4494. (Abstract)

17-123

Porath, J. 1965.

Some separation methods based on molecular size and charge and their application to purification of polypeptides and proteins in snake venoms. *Mem. Inst. Butantan (Sao Paulo)* 33(2):379–387.

17-124

Preston, H. S.; Kay, J.; Sato, A.; Low, B. W.; Tamiya, N. 1975.

Crystalline erabutoxin C. *Toxicon* 13:273–275.

17-125

Ramachandran, L. K.; Achyuthan, K. E.; Agarwal, O. P.; Chaudhury, L.; Vedasiromani, J. R.; Ganguli, D. K. 1984.

Toxic proteins of snakes and scorpions. *Proc. Indian Acad. Sci. Chem. Sci.* 93(7):1117–1136.

17-126

Raymond, M. L.; Tu, A. T. 1972.

Role of tyrosine in sea snake neurotoxin. *Biochim. Biophys. Acta* 285:498–502.

17-127
Rosenfeld, G.; Nahas L.; Kelen, E. M. A. 1968.
Coagulant, proteolytic, and hemolytic properties of some snake venoms. Buecherl, W.; Buckley, E. E.; Deulofeu, V., eds. *Venomous animals and their venoms*. New York: Academic. Vol. 1, pp. 229–273.

17-128
Ryden, L.; Gabel, D.; Eaker, D. 1973.
A model of the three-dimensional structure of snake venom neurotoxins based on chemical evidence. *Int. J. Pept. Protein Res.* 5:261–273.

17-129
Sarkar, N. K.; Devi, A. 1968.
Enzymes in snake venoms. Buecherl, W.; Buckley, E. E.; Deulofeu, V., eds. *Venomous animals and their venoms*. New York: Academic. Vol. 1, pp. 167–216.

17-130
Sato, S. 1971.
The structure of neurotoxins from sea snakes. *Protein Nucleic Acid Enzyme (Tokyo)* 16(2):86–95. (In Japanese)

17-131
Sato, S.; Yoshida, H.; Abe, H.; Tamiya, N. 1969.
Properties and biosynthesis of a neurotoxic protein of the venoms of sea snakes *Laticauda laticaudata* and *Laticauda colubrina*. *Biochem. J.* 115:85–90.

17-132
Sato, S.; Tamiya, N. 1970.
Iodination of erabutoxin b: diiodohistidine formation. *J. Biochem. (Tokyo)* 68(6):867–872.

17-133
Sato, S.; Tamiya, N. 1971.
The amino acid sequences of erabutoxins, neurotoxic proteins of sea-snake (*Laticauda semifasciata*) venom. *Biochem. J.* 122:453–461.

17-134
Searl, J. E.; Fullerton, W. W.; Low, B. W. 1973.
X-ray crystallographic study of laticotoxin a. *J. Biol. Chem.* 248(17):6057–6058.

17-135
Seto, A.; Sato, S.; Tamiya, N. 1970.
The properties and modification of tryptophan in a sea snake toxin, erabutoxin a. *Biochim. Biophys. Acta* 214:483–489.

17-136
Setoguchi, Y.; Morisawa, S.; Obo, F. 1968.
Investigation of sea snake venom (III): Acid and alkaline phosphatases (phosphodiesterase, phosphomonoesterase, 5'-nucleotidase and ATPase) in sea snake (*Laticauda semifasciata*) venom. *Acta Med. Univ. Kagoshima.* 10:53–60.

17-137
Shier, W. R.; Durkin, J. P.; Trotter, J. R.; Pickwell, G. V. 1979.
Phospholipase A2 electrophoretic variants in reptile venoms. *Toxicon* 17 (Suppl.1):167.

17-138
Shipman, W. H.; Pickwell, G. V. 1973.
Venom of the yellow-bellied sea snake (*Pelamis platurus*): some physical and chemical properties. *Toxicon* 11:375–377.

17-139
Smythies, J. R.; Benington, F.; Bradley, R. J.; Bridgers, W. F.; Morin, R. D.; Romine, W. O. 1975.
The molecular structure of the receptor-ionophore complex at the neuromuscular junction. *J. Theor. Biol.* 51:111–126.

17-140
Stekol'nikov, L. I. 1972.
Venoms from sea snakes. *Priroda (Mosc.)* 1972(2):106–107. (In Russian)

17-141
Takahashi, H.; Iwanaga, S.; Suzuki, T. 1974.
Distribution of proteinase inhibitors in snake venoms. *Toxicon* 12:193–197.

17-142
Takamatsu, T.; Harada, I.; Shimanouchi, T.; Ohta, M.; Hayashi, K. 1976.
Raman spectrum of toxin B in relation to structure and toxicity. *FEBS (Fed. Eur. Biochem. Soc.) Lett.* 72(2):291–294.

17-143
Takamatsu, T.; Harada, I.; Hayashi, K. 1980.
Raman spectra of some snake venom components. *Biochim. Biophys. Acta* 622:189–200.

17-144
Takasaki, C.; Tamiya, N. 1982.
Isolation and properties of lysophospholipases from the venom of an Australian elapid snake, *Pseudechis australis. Biochem. J.* 203(1):269 276.

17-144.5
Takasaki, C.; Kuramochi, H.; Shimazu, T.; Tamiya, N. 1988.
Correction of amino acid sequence of phospholipase A_2 I from the venom of *Laticauda semifasciata* (Erabu sea snake). *Toxicon* 26(8):747–749.

17-144.7
Takasaki, C.; Kimura, S.; Kokubun, Y.; Tamiya, N. 1988.
Isolation, properties and amino acid sequences of a phospholipase A_2 and its homologue without activity from the venom of a sea snake, *Laticauda colubrina*, from the Solomon Islands. *Biochem. J.* 253(3):869–875.

17-145
Takeda, M.; Yoshida, H.; Tamiya, N. 1974.
Biosynthesis of erabutoxins in the sea snake, *Laticauda semifasciata*. *Toxicon* 12:633–641.

17-146
Takeda, M.; Yoshida, H.; Tamiya, N. 1976.
Biosynthesis of erabutoxins in the sea snake, *Laticauda semifasciata*. Ohsaka, A.; Hayashi, K.; Sawai, Y., eds. *Animal, plant, and microbial toxins*. New York: Plenum. Vol. 1, pp. 1–15.

17-147
Tamiya, N. 1972.
Contents of toxic components in *Laticauda semifasciata* venom. *J. Formosan Med. Assoc.* 71:389–393.

17-148
Tamiya, N. 1972.
Studies on sea snake toxins: structure of erabutoxin c and its content, as well as those of erabutoxins a and b in *Laticauda semifasciata* venom. *Toxicon* 10(5):535. (Abstract)

17-149
Tamiya, N. 1973.
Erabutoxins a, b and c in sea snake *Laticauda semifasciata* venom. *Toxicon* 11(1):95–97.

17-150
Tamiya, N. 1973.
Untersuchungen ueber toxine von seeschlangen: Die struktur von erabutoxin c und der gehalt an erabutoxin a, b, und c im gift von *Laticauda semifasciata*. Studies on sea snake toxins: the structure of erabutoxin c and the content of

erabutoxins a, b, and c in the venom of *Laticauda semifasciata. Tier- und plan-zengifte: Animal and plant toxins.* Muenchen: Verlag. Pp. 41–44. (Abstract)

17-151
Tamiya, N. 1974.
  Toxic proteins of sea snake venoms. *High Polymers (Kobun Shi)* 23(10):728–732. (In Japanese)

17-152
Tamiya, N. 1975.
  Sea snake venoms and toxins. Dunson, W. A., ed. *Biology of sea snakes.* Baltimore: University Park Press. Pp. 385–415.

17-153
Tamiya, N. 1977.
  Neurotoxic proteins: their binding to the nicotinic acetylcholine receptor. *Tanpakushitsu Kakusan Koso* 22(6):554–565. (In Japanese)

17-154
Tamiya, N. 1980.
  Active sites of the snake neurotoxin molecules. *Snake* 12(1–2):147–148.

17-155
Tamiya, N. 1980.
  Snake venom neurotoxins acting on postsynaptic membrane. *Shinkei Kenkyu No Shinpo* 24(5):870–876. (In Japanese)

17-156
Tamiya, N.; Arai, H. 1966.
  Studies on sea-snake venoms: crystallization of erabutoxins a and b from *Laticauda semifasciata* venom. *Biochem. J.* 99:624–630.

17-157
Tamiya, N.; Sato, S. 1966.
  Studies on sea-snake venoms: structure and function of crystalline toxins from sea-snakes Laticaudinae. *Proc. Int. Congr. Biochem.*, 7th, Tokyo, 1966. Page 497.

17-158
Tamiya, N.; Arai, H.; Sato, S. 1967.
  Crystalline sea snake toxins. *Toxicon* 4:298. (Abstract)

17-159
Tamiya, N.; Arai, H.; Sato, S. 1967.
   Studies on sea snake venoms: crystallization of erabutoxins "a" and "b" from *Laticauda semifasciata*, venom and of laticotoxin "a" from *Laticauda laticaudata* venom. Russell, F. E.; Saunders, P. R., eds. *Animal toxins*. Oxford: Pergamon. Pp. 249–258.

17-160
Tamiya, N.; Sato, S.; Seto, A. 1971.
   Structure and action of the sea snake neurotoxins, erabutoxins a and b. DeVries, A.; Kochva, E., eds. *Toxins of animal and plant origins*. New York: Gordon and Breach 1971–1973. Vol. 1, pp. 237–249.

17-161
Tamiya, N.; Abe, H. 1972.
   The isolation, properties and amino acid sequence of erabutoxin c, a minor neurotoxic component of the venom of a sea snake, *Laticauda semifasciata*. *Biochem. J.* 130:547–555.

17-162
Tamiya, N.; Maeda, N. 1978.
   Chemistry of sea snake neurotoxins. *Toxicon* 16 (Suppl.1): 315. (Abstract)

17-163
Tamiya, N.; Takasaki, C. 1978.
   Detection of erabutoxins in the venom of sea snake *Laticauda semifasciata* from the Philippines. *Biochim. Biophys. Acta* 532:199–201.

17-164
Tamiya, N.; Ishikawa, Y.; Menez, A.; Hori, H.; Yoshida, A. 1978.
   The structure of snake neurotoxins and their affinity for the acetylcholine receptor. *Toxicon* 16 (Suppl.1):243–253.

17-165
Tamiya, N.; Abe, T. 1979.
   Antigenicity determining amino acid residues of erabutoxin b. *Toxicon* 17 (Suppl.1):186. (Abstract)

17-166
Tamiya, N.; Maeda, N. 1979.
   Neurotoxins of Australian sea snakes. Chubb, I. W.; Geffen, L. B., eds. *Neurotoxins: fundamental & clinical advances*. Adelaide: Adelaide University Union Press. Pp. 95–101.

17-167
Tamiya, N.; Abe, T. 1980.
   Antigenicity-determining amino acid residues of erabutoxin b. Eaker, D.; Wadstrom, T., eds. *Natural toxins*. New York: Pergamon. Pp. 91–98.

17-168
Tamiya, N.; Nishida, S.; Kim, H. S.; Yoshida, H. 1980.
   Phospholipases A_2 of sea snake *Laticauda semifasciata*: role of tryptophan in their reaction kinetics. *Kagaku, Zokan (Kyoto)* 85:137–143. (In Japanese)

17-169
Tamiya, N.; Nishida, S.; Kim, H. S.; Yoshida, H. 1980.
   Phospholipases A_2 of sea snake *Laticauda semifasciata*: role of tryptophan in their reaction kinetics. *Dev. Biochem.* 10:287–296.

17-170
Tamiya, N.; Takasaki, C.; Sato, A.; Menez, A.; Inagaki, F.; Miyazawa, T. 1980.
   Structure and function of erabutoxins and related neurotoxins from sea snakes and cobras. *Biochem. Soc. Trans.* 8(6):753–755.

17-171
Tamiya, N.; Maeda, N.; Cogger, H. G. 1983.
   Neurotoxins from the venoms of the sea snakes *Hydrophis ornatus* and *Hydrophis lapemoides*. *Biochem. J.* 213(1):31–38.

17-172
Tamiya, N.; Sato, A.; Kim, H. S.; Teruuchi, T.; Takasaki, C.; Ishikawa, Y.; Guinea, M. L.; McCoy, M.; Heatwole, H.; Cogger, H. G. 1983.
   Neurotoxins of sea snakes genus *Laticauda*. *Toxicon* (Suppl.3):445–447.

17-173
Tamiya, T.; Lamouroux, A.; Julien, J. F.; Grima, B.; Mallet, J.; Fromageot, P.; Menez, A. 1985.
   Cloning and sequence analysis of the cDNA encoding a snake neurotoxin precursor. *Biochimie (Paris)* 67(2):185–189.

17-174
Tamiya, T.; Guignery Frelat, G.; Mariat, D.; Boulain, J. C.; Menez, A. 1986.
   Translation of poly(A^+) mRNAs and expression of cDNAs encoding sea snake neurotoxins. *Proc. 7th European Symposium on Animal, Plant & Microbial Toxins, Prague, Aug.18–22*. Page 137. (Abstract)

17-175
Tan, N. H. 1982.
Acidic phospholipases A_2 from the venom of common sea snake *Enhydrina schistosa*. *Biochim. Biophys. Acta* 717:503–508.

17-176
Tanaka, S.; Scheraga, H. A. 1977.
Statistical mechanical treatment of protein conformation. 6. Elimination of empirical rules for prediction by use of a high-order probability. Correlation between the amino acid sequences and conformations for homologous neurotoxin proteins. *Macromolecules* 10(2):305–316.

17-177
Taylor, L. P. 1983.
Synthetic studies on the mode of action of erabutoxin b. Detroit: MI: Wayne State University. 125 pp. (Dissertation)

17-178
Taylor, L. P.; Hudson, R. A.; Doscher, M. S. 1981.
Synthetic studies of the mode of action of erabutoxin b. Rich, D. H.; Gross, E., eds. *Peptides: synthesis-structure, function*. Rockford, IL: Pierce Chemical Co. Pp. 241–244.

17-179
Thiery, C.; Nabedryk Viala, E.; Menez, A.; Fromageot, P.; Thiery, J. M. 1980.
Hydrogen exchange kinetics and dynamic structure of erabutoxin b from 1H NMR and infrared spectrometry. *Biochem. Biophys. Res. Commun.* 93(3):889–897.

17-180
Toom, P. M.; Squire, P. G.; Tu, A. T. 1969.
Characterization of the enzymatic and biological activities of snake venoms by isoelectric focusing. *Biochim. Biophys. Acta* 181:339–341.

17-181
Tsernoglou, D.; Petsko, G. A. 1976.
The crystal structure of a post-synaptic neurotoxin from sea snake at 2.2 A resolution. *FEBS (Fed. Eur. Biochem. Soc.) Lett.* 68(1):1–4.

17-182
Tsernoglou, D.; Petsko, G. A.; Tu, A. T. 1977.
Protein sequencing by computer graphics. *Biochim. Biophys. Acta* 491:605–608.

17-183
Tsernoglou, D.; Petsko, G. A. 1977.
Structure-function relationships in snake venom neurotoxins. *Fed. Proc.* 36(3):802. (Abstract)

17-184
Tsernoglou, D.; Petsko, G. A. 1977.
The three dimensional structure of a snake venom neurotoxin. *Biophys. J.* 17(2):224. (Abstract)

17-185
Tsernoglou, D.; Petsko, G. A. 1977.
Three-dimensional structure of neurotoxin a from venom of the Philippines sea snake. *Proc. Nat. Acad. Sci. U.S.A.* 74(3):971–974.

17-186
Tsernoglou, D.; Petsko, G. A.; Hudson, R. A. 1978.
Structure and function of snake venom curarimimetic neurotoxins. *Mol. Pharmacol.* 14:710–716.

17-187
Tsetlin, V. I.; Arseniev, A. S.; Utkin, Y. N.; Gurevich, A. Z.; Senyavina, L. B.; Bystrov, V. F.; Ivanov, V. T.; Ovchinnikov, Y. A. 1979.
Conformational studies of neurotoxin II from *Naja naja oxiana*: selective N-acylation, circular dichroism and nuclear-magnetic-resonance study of acylation products. *Eur. J. Biochem.* 94:337–346.

17-188
Tu, A. T. 1973.
Neurotoxins of animal venoms: snakes. *Annu. Rev. Biochem.* 42:235–258.

17-189
Tu, A. T. 1974.
Distribution of sea snakes in Southeast Asia and the far east and chemistry of venoms of three species. Humm, H. J.; Lane, C. E., eds. *Bioactive compounds from the sea.* New York: Marcel Dekker. Vol. 1, pp. 207–230.

17-190
Tu, A. T. 1974.
Sea snake venoms and neurotoxins. *Agric. Food Chem.* 22(1):36–43.

17-191
Tu, A. T. 1976.
Structure analysis of sea snake and rattlesnake toxins. *Toxicon* 14(6):419–420.

17-192
Tu, A. T. 1977.
   Venoms: chemistry and molecular biology. New York: Wiley.

17-193
Tu, A. T. 1979.
   Conformational analysis of snake toxins by laser Raman spectroscopy. Theophanides, T. M., ed. *Infrared and raman spectroscopy of biological molecules*. Boston: D. Reidel. Pp. 139–152.

17-193.5
Tu, A. T. 1988.
   Chemistry. Halstead, B. W. *Poisonous and venomous marine animals of the world*. 2nd rev. ed. Princeton: Darwin Press. Pp. 1089–1097.

17-194
Tu, A. T.; Passey, R. B.; Tu, T. 1966.
   Proteolytic enzyme activities of snake venoms. *Toxicon* 4:59–60.

17-195
Tu, A. T.; Toom, P. M. 1967.
   The presence of L-leucyl-(beta)-napthylamide hydrolyzing enzyme in snake venoms. *Experientia (Basel)* 23:439–443.

17-196
Tu, A. T.; Passey, R. B.; Toom, P. M. 1970.
   Isolation and characterization of phospholipase A from sea snake, *Laticauda semifasciata* venom. *Arch. Biochem. Biophys.* 140:96–106.

17-197
Tu, A. T.; Hong, B. S.; Solie, T. N. 1971.
   Characterization and chemical modifications of toxins isolated from the venoms of the sea snake, *Laticauda semifasciata* from Philippines. *Biochemistry* 10(8):1295–1304.

17-198
Tu, A. T.; Toom, P. M. 1971.
   Isolation and characterization of the toxic component of *Enhydrina schistosa* (common sea snake) venom. *J. Biol. Chem.* 246(4):1012–1016.

17-199

Tu, A. T.; Passey, R. B. 1971.

Phospholipase A from sea snake venom and its biological properties. DeVries, A.; Kochva, F., eds. *Toxins of animal and plant origin.* New York: Gordon and Breach. Vol. 1, pp. 419–436.

17-200

Tu, A. T.; Hong, B. S. 1971.

Purification and chemical studies of a toxin from the venom of *Lapemis hardwickii* (Hardwick's sea snake). *J. Biol. Chem.* 246(9):2772–2779.

17-201

Tu, A. T.; Hong, B.; Toom, P. M.; Tsernoglou, D. 1972.

Chemical study of sea snake venom toxins from three species in Southeast Asia. *Toxicon* 10(5):535–536. (Abstract)

17-202

Tu, A. T.; Hong, B.; Toom, P. M.; Tsernoglou, D. 1973.

Chemische untersuchungen ueber die toxine der gifte von drei sudostasiatischen seeschlangen. Chemical study of sea snake venom toxins from three species in Southeast Asia. *Tier- und pflanzengifte; Animal and plant toxins.* Muenchen: Verlag. Pp. 45–50.

17-203

Tu, A. T.; Lin, T. S. 1974.

Laser Raman studies on sea snake neurotoxins. *Fed. Proc.* 2(5):1564. (Abstract)

17-204

Tu, A. T.; Lin, T. S.; Bieber, A. L. 1975.

Purification and chemical characterization of the major neurotoxin from the venom of *Pelamis platurus. Biochemistry* 14(15):3408–3413.

17-205

Tu, A. T.; Jo, B. H.; Yu, N. T. 1976.

Laser Raman spectroscopy of snake venom neurotoxins: conformation. *Int. J. Pept. Protein Res.* 8:337–343.

17-205.3

Tu, A. T.; Allen, M.; Ishizaki, H. 1986.

Role of sulfhydryl group and the tryptophan residue in sea snake postsynaptic neurotoxin-acetylcholine receptor binding. *J. Toxicol. Toxin Rev.* 5(2):139. (Abstract)

17-205.6
Tu, A. T.; Mori, N. 1987.
Primary structure of neurotoxins from sea snakes *Acalyptophis peronii* and *Pelamis platurus* venoms. *FASEB (Fed. Am. Soc. Exp. Biol.) J.* 2(6):abstract 8415. (Abstract)

17-206
Ui, N.; Takasaki, C.; Tamiya, N. 1982.
Isoelectric points of erabutoxins and monoacyl derivatives of erabutoxin b. *Biochem. J.* 203(2):427–433.

17-207
Uwatoko, Y.; Nomura, Y.; Kojima, K.; Obo, F. 1966.
Investigation of sea snake venom (I). Fractionation of sea snake (*Laticauda semifasciata*) venom. *Acta Med. Univ. Kagoshima.* 8:141–150.

17-208
Uwatoko, Y.; Nomura, Y.; Kojima, K.; Obo, F. 1966.
Investigation of sea snake venom (II). Crystallization of toxic compounds in sea snake (*Laticauda semifasciata*) venom. *Acta Med. Univ. Kagoshima.* 8:151–156.

17-209
Uwatoko Setogushi, Y; Minamishima, Y.; Obo, F. 1968.
Studies on sea snake venom (IV): Purification of phospholipase A in *Laticauda semifasciata* venom. *Acta Med. Univ. Kagoshima.* 10:219–225.

17-210
Uwatoko Setoguchi, Y.; Ohbo, F. 1969.
Studies on sea snake venom (V). Some properties of phospholipase A in *Laticauda semifasciata* venom. *Acta Med. Univ. Kagoshima.* 11:139–144.

17-211
Uwatoko Setoguchi, Y. 1970.
Studies on sea snake venom (VI). Pharmacological properties of *Laticauda semifasciata* venom and purification of toxic components, acid phosphomonoesterase and phospholipase A in the venom. *Acta Med. Univ. Kagoshima.* 12(1):73–96.

17-212
Verheij, H. M.; Boffa, M. C.; Rothen, C.; Bryckaert, M. C.; Verger, R.; De Haas, G. H. 1980.
Correlation of enzymatic activity and anticoagulant properties of phospholipase A_2. *Eur. J. Biochem.* 112:25–32.

17-213
Wang, C. L.; Liu, C. S.; Hung, Y. O.; Blackwell, R. Q. 1976.
Amino acid sequence of pelamitoxin a, the main neurotoxin of the sea snake, *Pelamis platurus. Toxicon* 14:459–466.

17-213.5
Wu, X.; Zhou, D.; Liang, X.; Wei, Q. 1985.
Studies on three enzymes in three snake venoms. *Shengwu Huaxue Zazhi* 1:57–63.

17-214
Yang, C. C. 1974.
Chemistry and evolution of toxins in snake venoms. *Toxicon* 12:1–43.

17-215
Yang, C. C. 1978.
Chemistry and biochemistry of snake venom neurotoxins. *Toxicon* 16 (Suppl.1):261–292.

17-216
Yang, C. C. 1984.
Snake neurotoxins. *Snake* 16:90–103.

17-217
Yang, C. C.; Chang, C. C.; Liou, I. F. 1974.
Studies on the status of arginine residues in cobrotoxin. *Biochim. Biophys. Acta* 365:1–14.

17-218
Yang, C. C.; Chang, C. C. 1976.
Cationic groups and biological activity of cobrotoxin. Ohsaka, A.; Hayashi, K.; Sawai, Y., eds. *Animal, plant, microbial toxins.* New York: Plenum. Vol. 1, pp. 49–65.

17-219
Yang, C. C.; Chang, L. C. 1978.
Carbamylation of cobrotoxin with cyanate. *J. Chin. Biochem. Soc.* 7(2):63–77.

17-220
Yoshida, H.; Kudo, T.; Shinkai, W.; Tamiya, N. 1979.
Phospholipase A of sea snake *Laticauda semifasciata* venom: isolation and properties of novel forms lacking tryptophan. *J. Biochem. (Tokyo)* 85(2):379–388.

17-221

Yu, N. T.; Lin, T. S.; Tu, A. T. 1975.

Laser Raman scattering of neurotoxins isolated from the venoms of sea snakes *Lapemis hardwickii* and *Enhydrina schistosa. J. Biol. Chem.* 250(5):1782–1785.

17-222

Zaval'nyi, A. A.; Popov, E. M. 1982.

A priori computations of the spatial structure of neurotoxin fragment leu 1-Cys 23. *Mol. Biol.* 16(1 pt. 2):105–115.

17-223

Zeller, E. A. 1948.

Enzymes of snake venoms and their biological significance. *Adv. Enzymol.* 8:459–495.

17-224

Zeller, E. A. 1949.

Ueber die cholinesterase der schlangengifte 5. Mitteilung ueber die biochemie der tierischen gifte. *Helv. Chim. Acta* 32(1):94–105.

Genus-Species Index
for Chapter 17

Chapter 18

VENOM—PHARMACOLOGY,
NEUROPHYSIOLOGY,
MODE OF ACTION

18-1

Arenson, M. S.; Nistri, A. 1975.

The effect of erabutoxin B on the rat diaphragm and the frog spinal cord. *Exp. Brain Res.* 23 (Suppl):9. (Abstract)

18-2

Ariyoshi, M.; Hasuo, H.; Koketsu, K.; Ohta, Y.; Tokimasa, T. 1985.

Histamine is an antagonist of the acetylcholine receptor at the frog endplate. *Br. J. Pharmacol.* 85(1):65–73.

18-3

Barber, D. W.; Puffer, H. W.; Tamiya, N.; Shynkar, T. P. 1974.

Aspects of the neuromuscular activity of sea snake venom. *Proc. West. Pharmacol. Soc.* 17:235–238.

18-4

Bhise, S. B.; Bhide, M. B. 1978.

Presynaptic action of '*Hydrophis cyanocinctus*' venom. *Bull. Haffkine Inst.* 6(3):92–95.

18-4.5

Bradley, R. J.; Edge, M. T.; Chau, W. C. 1990.

The A-neurotoxin erabutoxin b causes fade at the rat end-plate. *Eur. J. Pharmacol.* 176(1):11–22.

18-5

Brook, G. A.; Duchen, L. W.; Gopalakrishnakone, P.; Torres, L. F. 1986.

Selective myonecrosis induced by phospholipase of *Enhydrina schistosa* venom. *Proc. 2nd Annual Symposium on Animal, Plant & Microbial Toxins, Tempe, Arizona.* Page 45. (Abstract)

18-5.5
Brook, G. A.; Torres, L. F.; Gopalakrishnakone, P.; Duchen, L. W. 1987.
Effects of phospholipase of *Enhydrina schistosa* venom on nerve, motor end-plate and muscle of the mouse. *Q. J. Exp. Physiol.* 72:571–591.

18-6
Brown, J. H. 1973.
Toxicology and pharmacology of venoms from poisonous snakes. Springfield, IL: Charles C. Thomas.

18-7
Burne, J. A. 1978.
The action of erabutoxin b in the cerebral cortex: evidence for short latency nicotinic responses. *Proc. Aust. Physiol. Pharmacol. Soc.* 9(1):97P.

18-8
Burne, J. A. 1978.
The action of erabutoxin 'b' in the cerebral cortex: evidence for short latency potentially nicotinic neurones. *Life Sci.* 23(8):775–781.

18-9
Burne, J. A.; Webster, M. E. D. 1973.
Effects of erabutoxin 'b' on the electrocorticogram of anaesthetised albino rats. *Proc. Aust. Physiol. Pharmacol. Soc.* 5(1):49–50. (Abstract)

18-10
Burne, J. A.; Webster, M. E. D. 1976.
Some preliminary observations on the effects of the neurotoxins, erabutoxins 'a' and 'b' on central neurones in the somatosensory cortex of the rat. *Proc. Aust. Physiol. Pharmacol. Soc.* 7(1):30P. (Abstract)

18-11
Burne, J. A.; Webster, M. E. D. 1977.
The action of erabutoxin 'b' on spontaneous and glutamate-induced cortical activity. *Life Sci.* 20:2023–2028.

18-12
Campbell, C. W. B. 1978.
Effects of post- and pre-synaptic snake venom neurotoxins on chloride and choline association with synaptic plasma membranes in vitro. *Neurosci. Lett.* 10:71–75.

18-13
Carey, J. E.; Wright, E. A. 1961.
   The site of action of the venom of the sea snake *Enhydrina schistosa*. *Trans. R. Soc. Trop. Med. Hyg.* 55(1):153–160.

18-14
Chan, K. E.; Geh, S. L. 1967.
   Antagonism of intra-arterial acetylcholine induced contraction of skeletal muscle by sea snake venom. *Nature (Lond.)* 213:1147–1148.

18-15
Chan, K. E.; Chang, P. 1971.
   Normal ganglionic transmission in the presence of acetylcholine block by sea snake (*Enhydrina schistosa*) venom. *Eur. J. Pharmacol.* 13:277–279.

18-16
Chang, C. C. 1979.
   The action of snake venoms on nerve and muscle. Lee, C. Y., ed. *Snake venoms*. New York: Springer-Verlag. Vol. 52, pp. 309–376.

18-16.3
Cheah, L. S.; Gwee, M. C. E.; Gopalakrishnakone, P. 1987.
   Possible involvement of a presynaptic mechanism in the neuromuscular blocking action of the venom of a local sea snake (*Laticauda semifasciata*). *Asia Pac. J. Pharmacol.* 2:155–60.

18-16.5
Cheah, L. S.; Gopalakrishnakone, P.; Gwee, M. C. E. 1987.
   Train-of-four fade during recovery from neuromuscular blockade by the venom of a local sea snake (*Laticauda semifasciata*). Gopalakrishnakone, P.; Tan, C. K., eds. *Progress in venom and toxin research*. Singapore: National University of Singapore. Pp. 275–279.

18-17
Cheymol, J.; Bourillet, F.; Roch, M. 1966.
   Action neuromusculaire des venins de quelques Crotalidae, Elapidae et Hydrophiidae. *Mem. Inst. Butantan (Sao Paulo)* 33(2):541–554.

18-18
Cheymol, J.; Barme, M.; Bourillet, F.; Roch Arveiller, M. 1967.
   Action neuromusculaire de trois venins d'Hydrophiides. *Toxicon* 5:111–119.

18-19
Cheymol, J.; Tamiya, N.; Bourillet, F.; Roch Arveiller, M. 1972.
Action neuromusculaire du venin de serpent marin 'erabu' (*Laticauda semifasciata*) et des erabutoxines a et b. *Toxicon* 10:125–131.

18-20
Cheymol, J.; Bourillet, F.; Roch Arveiller, M. 1972.
Actions neuromusculaires de fractions isolees de venins de serpents. DeVries, A.; Kochva, E., eds. *Toxins of animal and plant origin*. New York: Gordon and Breach. Vol. 2, pp. 655–665.

18-20.5
Chiappinelli, V. A. 1985.
Actions of snake venom toxins on neuronal nicotinic receptors and other neuronal receptors. *Pharmacol. Ther.* 31(1–2):1–32.

18-21
Chiappinelli, V. A.; Cohen, J. B.; Zigmond, R. E. 1981.
The effects of (alpha)- and (beta)-neurotoxins from the venoms of various snakes on transmission in autonomic ganglia. *Brain Res.* 211(1):107–126.

18-22
Copley, A. L.; Banerjee, S.; Devi, A. 1973.
Studies of snake venoms on blood coagulation. I. The thromboserpentin (thrombin-like) enzyme in the venoms. *Thromb. Res.* 2:487–508.

18-23
Dowdall, M. J.; Fohlman, J. P.; Eaker, D. 1977.
Inhibition of high-affinity choline transport in peripheral cholinergic endings by presynaptic snake venom neurotoxins. *Nature (Lond.)* 269(5630):700–702.

18-24
Dowdall, M. J.; Fohlman, J. P.; Watts, A. 1979.
Presynaptic action of snake venom neurotoxins on cholinergic systems. *Adv. Cytopharmacol.* 3:63–76.

18-25
Dowdall, M. J.; Fretten, P. 1981.
Inhibition of membrane transport systems in synaptosomes from *Torpedo* electric organ by snake neurotoxins. *Adv. Behav. Biol.* 25:249–259.

18-26

Elliott, W. B.; Gans, C. 1967.

Production by snake venoms of uncoupling activity and reverse acceptor control in rat liver mitochrondrial preparations, Part I. Previous work. Russell, F. E.; Saunders, P. R., eds. *Animal toxins.* New York: Pergamon. Pp. 235–243.

18-27

Endo, T.; Nakanishi, M.; Furukawa, S.; Joubert, F. J.; Tamiya, N.; Hayashi, K. 1986.

Stopped-flow fluorescence studies on binding kinetics of neurotoxins with acetylcholine receptor. *Biochemistry* 25(2):395–404.

18-27.3

Endo, T.; Tamiya, N. 1987.

Current view on the structure-function relationship of postsynaptic neurotoxins from snake venoms. *Pharmacol. Ther.* 34(3):403–451.

18-27.5

Endo, T.; Oya, M.; Tamiya, N.; Hayashi, K. 1987.

Role of c-terminal tail of long neurotoxins from snake venoms in molecular conformation and acetylcholine receptor binding: proton nuclear magnetic resonance and competition binding studies. *Biochemistry* 26(14):4592–4598.

18-27.7

Endo, T.; Nakanishi, M.; Furukawa, S.; Joubert, F. J.; Tamiya, N.; Hayashi, K. 1987.

Stopped-flow fluorescence studies on binding of snake neurotoxins to acetylcholine receptor. *J. Protein Chem.* 6:227–236.

18-28

Ferry, C. B.; Geh, S. L. 1977.

An electrophysiological study of the action of crude *Enhydrina schistosa* venom. *Br. J. Pharmacol.* 60:233–238.

18-29

Festoff, B. W. 1975.

Mechanism of action of neurotoxins. *Ann. Clin. Lab. Sci.* 5(5):377–382.

18-29.5

Fohlman, J. 1977.

Studies on taipan and sea snake venoms with special reference to neuro- and myotoxic mechanisms. Uppsala: Institute of Biochemistry, University of Uppsala.

18-30
Fraser, T. R.; Elliot, R. H. 1904.
Contributions to the study of the action of sea-snake venoms, Part I. *Proc. R. Soc.* 74:104–108. (Also appears in *Scott. Med. Surg. J.* 1904:26–31 and *Lancet* 2:141–142)

18-33
Fraser, T. R.; Elliot, R. H. 1905.
Contributions to the study of the action of sea-snake venoms, Part I: Venoms of *Enhydrina valakadien* and *Enhydris curtus*. *Philos. Trans. R. Soc.* 197B:249–279.

18-34
Fretten, P.; Dowdall, M. J.; Culliford, P. G. 1981.
Action of presynaptic phospholipase neurotoxins on synaptosomes from *Torpedo* electric organ. *Biochem. Soc. Trans.* 9(5):411–412.

18-35
Gawade, S. P.; Gaitonde, B. B. 1982.
Presynaptic and postsynaptic sites of action of enhydrotoxin-a (EsNTx-a) isolated from *Enhydrina schistosa* venom. *J. Pharm. Pharmacol.* 34:782–787.

18-35.5
Geh, J. S. L. 1968.
The pharmacological actions of the <u>Enhydrina schistosa</u> (Daudin) venom. Kuala Lumpur: University of Malaya. 138 pp. (Thesis)

18-36
Geh, S. L.; Chan, K. E. 1973.
The pre-junctional site of action of *Enhydrina schistosa* venom at the neuromuscular junction. *Eur. J. Pharmacol.* 21:115–120.

18-37
Geh, S. L.; Toh, H. T. 1978.
Ultrastructural changes in skeletal muscle caused by a phospholipase A_2 fraction isolated from the venom of a sea snake, *Enhydrina schistosa*. *Toxicon* 16:633–643.

18-38
Gerencser, G. A.; Loo, S. Y. 1982.
Effect of *Laticauda semifasciata* (sea snake) venom on sodium transport across the frog skin. *Comp. Biochem. Physiol. A Comp. Physiol.* 72(4):727–730.

18-38.5
Gibb, A. J.; Marshall, I. G. 1984.
Pre- and post-junctional effects of tubocurarine and other nicotinic antagonists during repetitive stimulation in the rat. *J. Physiol. (Lond.)* 351:275–297.

18-39
Gopalakrishnakone, P. 1984.
Histopathological changes induced by the sea snake *Enhydrina schistosa* venom on murine muscle and neuromuscular junction. Meier, J.; Stocker, K.; Freyvogel, T. A., eds. *Proc. 6th European Symposium on Animal, Plant and Microbial Toxins.* Basel. Page 89. (Abstract)

18-40
Gul, S.; Khara, J. S.; Smith, A. D. 1974.
Hemolysis of washed human red cells by various snake venoms in the presence of albumin and Ca^{2+}. *Toxicon* 12:311–315.

18-40.5
Gwee, M. C. E.; Gopalakrishnakone, P.; Cheah, L. S. 1987.
Tetanic fade during recovery from neuromuscular blockade by the venom of a local sea snake *Laticauda semifasciata*. Gopalakrishnakone, P.; Tan, C. K., eds. *Progress in venom and toxin research.* Singapore: National University of Singapore. Pp. 270–274.

18-41
Harris, J. B.; Johnson, M. A.; MacDonell, C. A. 1980.
Muscle necrosis induced by some presynaptically active neurotoxins. Eaker, D.; Wadstrom, T., eds. *Natural toxins.* Oxford: Pergamon. Pp. 569–578.

18-42
Harvey, A. L.; Rodger, I. W.; Tamiya, N. 1978.
Neuromuscular blocking activity of two fractions isolated from the venom of the seasnake, *Laticauda semifasciata. Toxicon* 16:45–50.

18-43
Harvey, A. L.; Rodger, I. W. 1978.
Reversibility of neuromuscular blockade produced by toxins isolated from the venom of the seasnake *Laticauda semifasciata. Toxicon* 16:219–226.

18-44
Harvey, A. L.; Tamiya, N. 1980.
Phospholipase A and neuromuscular blocking activities of components from the venom of the seasnake, *Laticauda semifasciata. Proc. 7th Int. Congr. Pharmacol. 1978.* New York: Pergamon. Page 149. (Abstract)

18-45
Harvey, A. L.; Tamiya, N. 1980.
    Role of phospholipase A activity in the neuromuscular paralysis produced by some components isolated from the venom of the seasnake, *Laticauda semifasciata*. *Toxicon* 18(1):65–69.

18-46
Harvey, A. L.; Hider, R. C.; Hodges, S. J.; Joubert, F. J. 1984.
    Structure-activity studies of homologues of short chain neurotoxins from Elapid snake venoms. *Br. J. Pharmacol.* 82:709–716.

18-47
Ibrahim, S. A.; Thompson, R. H. S. 1965.
    Action of phospholipase A on human red cell ghosts and intact erythrocytes. *Biochim. Biophys. Acta* 99:331–341.

18-48
Igata, A.; Kawahira, M.; Fukunaga, H.; Fukuoka, T.; Arimura, K.; Matsuo, T. 1981.
    Induction of experimental autoimmune myasthenia gravis using erabutoxin. *Jpn. Med. Res. Found. Publ.* No. 14:71–80.

18-49
Ishikawa, Y.; Menez, A.; Hori, H.; Yoshida, H.; Tamiya, N. 1977.
    Structure of snake toxins and their affinity to the acetylcholine receptor of fish electric organ. *Toxicon* 15:477–488.

18-50
Ishikawa, Y.; Shimada, Y. 1983.
    The acetylcholine receptors of human muscles are not stained with rhodamine-labeled erabutoxin b. *Brain Res.* 266:159–162.

18-51
Ishikawa, Y.; Shimada, Y. 1983.
    Visualization of acetylcholine receptors at the neuromuscular junction by using erabutoxin b and receptor antibody. *J. Histochem. Cytochem.* 31(1A):255. (Abstract)

18-52
Ishikawa, Y.; Kano, M.; Tamiya, N.; Shimada, Y. 1985.
    Acetylcholine receptors of human skeletal muscle: a species difference detected by snake neurotoxins. *Toxicon* 23(4):578. (Abstract)

18-53
Jeffrey, P. L. 1979.
Uses of snake neurotoxins in the study of post-synaptic surface proteins. Chubb, I. W.; Geffen, L. B., eds. *Neurotoxins: fundamental & clinical advances.* Adelaide: Adelaide University Union Press. Pp. 35–47.

18-54
Jimenez Porras, J. M. 1968.
Pharmacology of peptides and proteins in snake venoms. *Annu. Rev. Pharmacol.* 8:299–318.

18-55
Kato, E.; Kuba, K.; Koketsu, K. 1978.
Effects of erabutoxins on neuromuscular transmission in frog skeletal muscles. *J. Pharmacol. Exp. Ther.* 204(2):446–453.

18-56
Kato, E.; Kuba, K.; Koketsu, K. 1980.
Effects of erabutoxins on the cholinergic receptors of bullfrog sympathetic ganglion cells. *Brain Res.* 191:294–298.

18-57
Kato, E.; Narahashi, T. 1982.
Low sensitivity of the neuroblastoma cell cholinergic receptors to erabutoxins and (alpha)-bungarotoxin. *Brain Res.* 245:159–162.

18-58
Kawaji, K.; Honda, H. 1968.
Ultrastructural changes in rat liver following sea snake venom administration. *J. Electron. Microsc.* 17:92. (Abstract)

18-59
Kellaway, C. H.; Holden, H. F. 1932.
The peripheral action of the Australian snake venoms. 1. The curari-like action on frogs. *Aust. J. Exp. Biol. Med. Sci.* 10:167–179.

18-60
Kellaway, C. H.; Cherry, R. O.; Williams, F. E. 1932.
The peripheral action of the Australian snake venoms. 2. The curari-like action in mammals. *Aust. J. Exp. Biol. Med. Sci.* 10:181–194.

18-60.5
Koketsu, K.; Akasu, T.; Miyagawa, M.; Hirai, K. 1982.

Biogenic antagonists of the nicotinic receptor: their interactions with erabutoxin. *Brain Res.* 250(2):390–393.

18-61
Kuraishi, Y.; Misu, Y.; Takagi, H.; Hayashi, K. 1977.

Neuromuscular blocking actions of neurotoxins isolated from *Laticauda semifasciata*, *Naja naja* and *Naja naja atra*—a comparative assay. *Jpn. J. Pharmacol.* 27:464–467.

18-62
Lamb, G.; Hunter, W. K. 1907.

On the action of venoms of different species of poisonous snakes on the nervous system. *Lancet* 2:1017–1019.

18-63
Lee, C. Y. 1970.

Elapid neurotoxins and their mode of action. *Clin. Toxicol.* 3(3):457–472.

18-64
Lee, C. Y. 1971.

Chemistry and pharmacology of polypeptide toxins in snake venoms. *Annu. Rev. Pharmacol.* 12:265–286.

18-65
Lee, C. Y. 1971.

Mode of action of cobra venom and its purified toxins. Simpson, L. L., ed. *Neuropoisons: their pathophysiological actions.* New York: Plenum. Vol. 1, pp. 21–70.

18-66
Lee, C. Y. 1972.

Classification of polypeptide toxins from elapid and sea snake venoms according to their pharmacological properties and chemical structures. *J. Formosan Med. Assoc.* 71:311–317.

18-67
Lee, C. Y. 1973.

Chemistry and pharmacology of purified toxins from elapid and sea snake venoms. *Pharmacology and the future of man; Proc. 5th Int. Congr. Pharmacol.,* San Francisco, 1972. Basel: Karger. Vol. 2, pp. 210–232.

18-68
Lee, C. Y.; Chang, C. C.; Chen, Y. M. 1972.
Reversibility of neuromuscular blockade by neurotoxins from elapid and sea snake venoms. *J. Formosan Med. Assoc.* 71:344–349.

18-69
Lee, C. Y.; Chen, Y. M. 1975.
Species differences in reversibility of neuromuscular blockade by elapid and sea snake neurotoxins. *Toxicon* 13(2):106. (Abstract)

18-70
Lee, C. Y.; Chen, Y. M. 1976.
Species differences in reversibility of neuromuscular blockade by elapid and sea snake neurotoxins. Ohsaka, A.; Kayashi, K.; Sawai, Y., eds. *Animal, plant and microbial toxins*. New York: Plenum. Vol. 2, pp. 193–204.

18-71
Lee, C. Y.; Lee, S. Y. 1979.
Cardiovascular effects of snake venoms. Lee, C. Y., ed. *Snake venoms*. New York: Springer-Verlag. Pp. 547–590.

18-72
Limpus, C. J. 1975.
A study of the ecology and toxicology of subtropical Queensland sea snakes (Hydrophiinae). Brisbane, Australia: University of Queensland. 197 pp. (Thesis)

18-73
Lin, M. W. 1975.
Venomous snakes of southeast Asia and some physiological and biochemical aspects of their venoms. Chicago, IL: Northeastern Illinois University. 65 pp. (Thesis)

18-73.5
Lin, N. Q.; Tu, A. T. 1989.
Formation of conjugated acetycholine receptor by cross-linking the sulfhydryl group of subunits and its effect on sea snake neurotoxin binding. *Biochem. Arch.* 5:133–118.

18-74
Maeda, R. H.; Puffer, H. W.; Barber, D. W.; Tamiya, N. 1975.
Use of the denervated guinea pig diaphragm to elucidate the site of action of selected sea snake venoms. *Proc. West. Pharmacol. Soc.* 18:335–338.

18-75
Marshall, L. R.; Herrmann, R. P. 1983.
    Coagulant and anticoagulant actions of Australian snake venoms. *Thromb. Haemostasis* 50(3):707–711.

18-76
Martin, C. J.; Lamb, G. 1907.
    Snake-poison and snake-bite. Allbutt, T. C.; Rolleston, H. D., eds. *A system of medicine*. London: Macmillan. Vol. 2, no. 2, pp. 783–821.

18-76.5
Maeume, J. 1966.
    Les venins de serpents agents modificateurs de la coagulation sanguine. *Toxicon* 4:25–58.

18-77
Mebs, D. 1978.
    Pharmacology of reptilian venoms. Gans, C.; Gans, K. A., eds. *Biology of the reptilia*. New York: Academic. Vol. 8, pp. 437–560.

18-78
Meldrum, B. S. 1965.
    The actions of snake venoms on nerve and muscle. The pharmacology of phospholipase A and of polypeptide toxins. *Pharmacol. Rev.* 17(4):393–445.

18-79
Meldrum, B. S.; Thompson, R. H. S. 1962.
    The action of snake venoms on the membrane permeability of brain, muscle and red blood cells. *Guy's Hosp. Rep.* 111:87–97.

18-79.3
Mori, N.; Tu, A. T. 1988.
    Cross-linking of acetylcholine receptor subunits and its effects on sea snake neurotoxin binding. *Biochem. Arch.* 4:85–89.

18-79.5
Mori, N.; Tu, A. T. 1988.
    Isolation and primary structure of the major toxin from sea snake, *Acalyptophis peronii*, venom. *Arch. Biochem. Biophys* 260(1):10–17.

18-80
Murai, S.; Ogura, Y. 1974.
Pharmacological studies on erabutoxin. II. Effects on the circulatory and respiratory system of rabbits. *Folia Pharmacol. Jpn.* 70:177P–178P. (In Japanese) (Abstract)

18-81
Murai, S.; Ogura, Y. 1974.
Pharmacological studies on erabutoxin. II. Effects on the circulatory and respiratory system of rabbits. *Jpn. J. Pharmacol.* 24(Suppl):114. (Abstract)

18-82
Murai, S.; Ogura, Y. 1977.
Effect of erabutoxin on respiration of rabbits. *Jpn. J. Pharmacol.* 27:721–726.

18-83
Ng, R. H.; Howard, B. D. 1980.
Mitochondria and sarcoplasmic reticulum as model targets for neurotoxic and myotoxic phospholipases A_2. *Proc. Nat. Acad. Sci. U.S.A.* 77(3):1346–1350.

18-84
Nistri, A.; Arenson, M. S. 1978.
Effect of erabutoxin b on acetylcholine release and root potentials of the frog spinal cord. *Eur. J. Pharmacol.* 47:245–248.

18-85
Ogura, Y.; Murai, S. 1974.
Pharmacological studies on erabutoxin I: On the isolated phrenic nerve diaphragm preparation of the rat. *Folia Pharmacol. Jpn.* 70:177P. (In Japanese) (Abstract)

18-86
Ogura, Y.; Murai, S. 1974.
Pharmacological studies on erabutoxin I. On the isolated phrenic nerve-diaphragm preparation of the rat. *Jpn. J. Pharmacol.* 24(Suppl):114. (Abstract)

18-87
Ogura, Y.; Kikuchi, T. 1976.
Pharmacological studies on erabutoxin, III. Comparison of effect of erabutoxin and related compounds on neuromuscular transmission. *Jpn. J. Pharmacol.* 26(Suppl):165P. (Abstract)

18-88
Ogura, Y.; Kikuchi, T. 1980.
Pharmacological studies on the sea snake venom, erabutoxin. VI. Effects on the responses of isolated chronically denervated soleus muscles of the mouse to acetylcholine. *Jpn. J. Pharmacol.* 29(Suppl):135P. (Abstract)

18-89
Parmentier, J.; Carpenter, D. 1976.
Blocking action of snake venom neurotoxins at receptor sites to putative central nervous system transmitters. Ohsaka, A.; Hayashi, K.; Sawai, Y., eds. *Animal, plant, and microbial toxins.* New York: Plenum. Vol. 2, pp. 179–191.

18-90
Phillips, S. J. 1972.
The effect of snake and bee venoms on cardiovascular hemodynamics and function. DeVries, A.; Kochva, E., eds. *Toxins of animal and plant origin.* New York: Gordon and Breach. Vol. 2, pp. 683–701.

18-91
Picado Twight, C. 1931.
Serpientes venenosas de Costa Rica sus venenos seroterapia anti-ofidica. Costa Rica: Editorial Universidad de Costa Rica. (Reissued 1976 by Editorial Universidad de Costa Rica)

18-92
Pickwell, G. V.; Vick, J. A.; Shipman, W. H.; Grenan, M. M. 1973.
Production, toxicity and preliminary pharmacology of venom from the sea snake, *Pelamis platurus* with observations on its probable threat to man along middle America. Worthen, L. R., ed. *Food-Drugs from the Sea, Proceedings of the 3rd Conf., 1972.* Washington, DC: Marine Technology Society. Pp. 247–265.

18-93
Puffer, H. W.; Barber, D. W.; Tamiya, N. 1975.
Activity of selected sea snake venoms on the isolated nerve-diaphragm preparation. *Toxicon* 13:115–116. (Abstract)

18-94
Puffer, H. W.; Barber, D. W.; Maeda, R. H.; Tamiya, N. 1976.
Activity of selected sea snake venoms on the isolated nerve-diaphragm preparation. Ohsaka, A.; Hayashi, K.; Sawai, Y., eds. *Animal, plant and microbial toxins.* New York: Plenum. Vol. 2, pp. 161–168.

18-95
Rogers, L. 1903.
  On the physiological action of the poison of the Hydrophidae. *Proc. R. Soc.*
71:481–496.

18-96
Rogers, L. 1903.
  On the physiological action of the poison of the Hydrophidae. Part II. Action
on the circulatory, respiration, and nervous sytems. *Proc. R. Soc.* 72:305–319.

18-96.5
Rogers, L. 1904.
  A lecture on the physiological action and antidotes of snake venoms with a
practical method of treatment of snake bites. *Lancet* 166(4192) (Feb.6):349–355.

18-97
Rogers, L. 1905.
  The physiological action and antidotes of Colubrine and Viperine snake venoms.
Philos. Trans. R. Soc. 197B:123–191.

18-98
Rosenberg, P. 1965.
  Effects of venoms on the squid giant axon. *Toxicon* 3:125–131.

18-98.5
Rosenberg, P. 1966.
  Use of venoms in studies on nerve excitation. *Mem. Inst. Butantan (Sao
Paulo)* 33(2):477–508.

18-99
Rosenberg, P. 1971.
  The use of snake venoms as pharmacological tools in studying nerve activity.
Simpson, L. L., ed. *Neuropoisons: their pathophysiological actions.* New York:
Plenum. Vol. 1, pp. 111–137.

18-100
Rowan, E. G.; Harvey, A. L.; Tamiya, N. 1985.
  Neuromuscular effects of a toxic phospholipase A_2 isolated from venom of the
Solomon Island sea snake *Laticauda colubrina*. *Toxicon* 23(4):606.

18-100.5
Rowan, E. G.; Harvey, A. L.; Takasaki, C.; Tamiya, N. 1989.
  Neuromuscular effects of a toxic phospholipase A_2 and its nontoxic homologue
from the venom of the sea snake, *Laticauda colubrina*. *Toxicon* 27(5):587–591.

18-101
Russell, F. E. 1967.
    Comparative pharmacology of some animal toxins. *Fed. Proc.* 26(4):1206–1224.

18-102
Russell, F. E. 1967.
    Pharmacology of animal venoms. *Clin. Pharmacol. Therap.* 8(6):849–873.

18-103
Russell, F. E.; Puffer, H. W. 1971.
    Pharmacology of snake venoms. Minton, S. A., ed. *Snake venoms and enven-
omation.* New York: Marcel Dekker. Pp. 87–98.

18-104
Sato, S.; Abe, T.; Tamiya, N. 1970.
    Binding of iodinated erabutoxin b, a sea snake toxin, to the endplates of the
mouse diaphragm. *Toxicon* 8:313–314.

18-105
Schmidt, M. E.; Abdelbaki, Y. Z.; Tu, A. T. 1976.
    Nephrotoxic action of rattlesnake and sea snake venoms: an electron-micro-
scopic study. *J. Path.* 118:75–81.

18-106
Schwab, M. S.; Puffer, H. W. 1976.
    A study of the site of blockade of the vertebrate neuromuscular junction by
erabutoxin a. *Fed. Proc.* 35(3):800. (Abstract)

18-107
Schwab, M. C.; Puffer, H. W. 1978.
    Preliminary electrophysiology of erabutoxin-a. *Toxicon* 78 (Suppl.1):393–395.

18-107.3
Singh, Y. N. 1984.
    Effect of freeze-drying and storage on the toxicity of the venom of the common
Fijian sea snake, *Laticauda colubrina* (Schneider). *South Pac. J. Nat. Sci.* 593–98.

18-107.7
Singh, Y. N.; Guinea, M. L. 1984.
    Yield and toxicity studies on the venom of the common Fijian sea snake,
Laticauda colubrina (Schneider). *South Pac. J. Nat. Sci.* 5:71–92.

18-108
Southcott, R. V. 1975.
The neurologic effects of noxious marine creatures. *Contemp. Neurol. Ser.* 12:165–258.

18-109
Southcott, R. V. 1979.
Marine toxins. Vinken, P. J.; Bruyn, G. W., eds. *Handbook of clinical neurology.* Amsterdam: Elsevier/North Holland Biomedical Press. Vol. 37, No. 2, pp. 27–106.

18-110
Sutherland, S. K.; Coulter, A. R.; Harris, R. D.; Lovering, K. E.; Roberts, I. D. 1981.
A study of the major Australian snake venoms in the monkey (*Macaca fascicularis*). I. The movement of injected venom, methods which retard this movement, and the response to antivenoms. *Pathology* 13:13–27.

18-111
Sutherland, S. K.; Campbell, D. G.; Stubbs, A. E. 1981.
A study of the major Australian snake venoms in the monkey (*Macaca fascicularis*). II. Myolytic and haematological effects of venoms. *Pathology* 13:705–715.

18-112
Takamizawa, T. 1970.
Effects of erabutoxin B on the membrane properties of frog sartorius muscle cells. *Tohoku J. Exp. Med.* 101:339–350.

18-113
Tamiya, N.; Arai, H. 1966.
Studies on sea-snake venoms: crystallization of erabutoxins a and b from *Laticauda semifasciata* venom. *Biochem. J.* 99:624–630.

18-114
Taub, A. M.; Elliott, W. B. 1964.
Some effects of snake venoms on mitochondria. *Toxicon* 2:87–92.

18-115
Toh, H. T.; Geh, S. L.; Chan, K. E. 1975.
Ultrastructural changes in neuromuscular junction of the guinea-pig caused by chronic injection of crude *Enhydrina schistosa* venom. *Malays. J. Sci.* 3A:39–47.

18-116

Tu, A. T. 1977.

Venoms: chemistry and molecular biology. New York: Wiley.

18-117

Tu, A. T. 1982.

Venom toxicology and composition. Tu, A. T., ed. *Survey of contemporary toxicology.* New York: Wiley. Vol. 2, pp. 145–201. (Review)

18-117.5

Tu, A. T. 1988.

Pharmacology. Halstead, B. W. *Poisonous and venomous marine animals of the world.* 2nd rev. ed. Princeton: Darwin Press. Pp. 1081–1088.

18-118

Tu, A. T.; Passey, R. B. 1969.

Effects of snake venoms on mammalian cells in tissue culture. *Toxicon* 6:277–280.

18-119

Tu, A. T.; Giltner, J. B. 1974.

Cytotoxic effects of snake venoms on KB and Yoshida sarcoma cells. *Res. Commun. Chem. Pathol. Pharmacol.* 9(4):783–786.

18-120

Tu, T. 1959.

Toxicological studies on the venom of a sea snake, *Laticauda semifasciata* (Reinwardt) of Formosan waters. *J. Formosan Med. Assoc.* 58(4):182–203.

18-121

Tu, T. 1967.

Toxicological studies on the venom of the sea snake *Laticauda laticaudata affinis.* Russell, F. E.; Saunders, P. R., eds. *Animal toxins.* Oxford: Pergamon. Pp. 245–248.

18-122

Tu, T. 1974.

Effects on the venom of a yellow-bellied sea snake, *Pelamis platurus,* on respiration and the circulatory system. *J. Col.-Wyo. Acad. Sci.* 7(5):63. (Abstract)

18-123
Tu, T.; Tu, A. T.; Lin, T. S. 1976.
Some pharmacological properties of the venom, venom fractions and pure toxin of the yellow-bellied sea snake *Pelamis platurus*. *J. Pharm. Pharmacol.* 28:139–145.

18-123.5
Tu, T. C.; Lin, M. J.; Yang, H. M.; Lin, H. J.; Chen, C. N. 1962.
Toxicological studies on the venom of a sea snake, *Laticauda colubrina* (Schneider) (First report.) *J. Formosan Med. Assoc.* 61:1296. (also cited as 61:122) (Abstract)

18-124
Uwatoko Setoguchi, Y. 1970.
Studies on sea snake venom (VI). Pharmacological properties of *Laticauda semifasciata* venom and purification of toxic components, acid phosphomonoesterase and phospholipase A in the venom. *Acta Med. Univ. Kagoshima.* 12(1):73–96.

18-125
Vick, J. A. 1970
Venomous sea snakes. Halstead, B. W. *Poisonous and venomous marine animals of the world*. Washington, DC: Government Printing Office. Vol. 3, pp. 974–975.

18-126
Vick, J. A. 1971.
Symptomatology of experimental and clinical crotalid envenomation. Simpson, L. L., ed. *Neuropoisons: their pathophysiological actions*. New York: Plenum. Vol. 1, pp. 71–86.

18-127
Walker, M. J. A.; Yeoh, P. N. 1974.
The in vitro neuromuscular blocking properties of sea snake (*Enhydrina schistosa*) venom. *Eur. J. Pharmacol.* 28:199–208.

18-127.5
Walkinshaw, M. D.; Saenger, W.; Maelicke, A. 1981.
Multiple binding sites of snake neurotoxins. Balaban, M., ed. *Structural aspects of recognition and assembly in biological macromolecules*. Philadelphia: Balaban International Science Services. Pp. 313–324.

18-128
Werner, R. M.; Vick, J. A. 1977.
   Resistance of the opossum (*Didelphis virginiana*) to envenomation by snakes
of the family Crotalidae. *Toxicon* 15(1):29–33.

18-129
Yang, T. Y.; Lee, C. Y. 1978.
   Pharmacological studies on the venom of a sea snake, *Hydrophis cyanocinctus.*
Toxicon 78 (Suppl.1):415–428.

18-130
Yeoh, P. N. 1977.
   Sea snake *Enhydrina schistosa* venom and d-tubocurarine effects on ganglionic
action potentials recorded from the monkey superior cervical ganglion. *Abst. Int.
Union Pharmacol. Proc. Int. Congr. Pharmacol, 6th, Helsinki, Finland, 1975.* New
York: Pergamon. Page 558. (Abstract)

18-131
Yeoh, P. N.; Walker, M. J. A. 1974.
   Effect of a sea snake (*Enhydrina schistosa*) venom on the ganglionic nicotinic
actions of acetylcholine. *J. Pharm. Pharmacol.* 26:441–447.

18-132
Ziegler, F. D.; Vazquez Colon, L.; Elliott, W. B.; Gans, C.; Taub, A. 1967.
   Studies on energy metabolism following treatment of mitochondria with several
snake venoms. Russell, F. E.; Saunders, P. R., eds. *Animal toxins.* New York:
Pergamon. Pp. 236–243.

18-133
Zimmerman, K. D. 1989.
   The effects of the venom of Aipysurus laevis on its prey species. Armidale,
N.S.W.: University of New England. 211 pp. (Dissertation)

18-134
Zimmerman, S. E.; Heatwole, H. 1987.
   Olive sea snake venom. Covacevich, J.; Davie, P.; Pearn, J., eds. *Toxic plants
and animals: a guide to Australia.* Brisbane: Queensland Museum Press. Pp. 205–
213.

Genus-Species Index
for Chapter 18

Aipysurus duboisii 18-72
Aipysurus eydouxii 18-72, 18-73, 16-116
Aipysurus foliosquama 18-73
Aipysurus fuscus 18-73
Aipysurus laevis 18-3, 18-27, 18-27.3, 18-27.7, 18-46, 18-49, 18-72, 18-74, 18-77, 18-79.5, 18-87, 18-94, 18-134
Astrotia stokesii 18-3, 18-27, 18-27.3, 18-27.7, 18-73, 18-74, 18-79.5, 18-87, 18-94, 18-116, 18-133, 18-134
Disteria 18-27.3
Disteira cyanocincta 18-95
Disteira melanocephala 18-120
Disteira viperina 18-95
Emydocephalus 18-27.3
Emydocephalus annulatus 18-72, 18-73, 18-134
Enhydrina 18-25, 18-34, 18-79.5, 18-96
Enhydrina bengalensis 18-95, 18-96.5, 18-97, 19-120
Enhydrina schistosa 18-5, 18-5.5, 18-6, 18-12, 18-13, 18-14, 18-15, 18-16, 18-17, 18-18, 18-20, 18-23, 18-24, 18-26, 18-27.3, 18-28, 18-29.5, 18-35, 18-35.5, 18-36, 18-37, 18-39, 18-40, 18-41, 18-47, 18-59, 18-60, 18-64, 18-66, 18-67, 18-72, 18-73, 18-74, 18-76.5, 18-77, 18-78, 18-79, 18-83, 18-98, 18-98.5, 18-99, 18-102, 18-103, 18-108, 18-109, 18-110, 18-111, 18-113, 18-114, 18-115, 18-116, 18-117.5, 18-118, 18-119, 18-120, 18-127, 18-130, 18-131, 18-132, 18-134
Enhydrina valakadien 18-30, 18-33, 18-62, 18-76, 18-78, 18-120
Enhydris curtus 18-30, 18-33
Erabutoxin (see Sea snake)
Hydrelaps 18-27.3
Hydropheidae 18-54
Hydrophidae 18-29
Hydrophiidae 18-53, 18-101
Hydrophis belcheri 18-3, 18-77, 18-116
Hydrophis beleheri 18-94, 18-95
Hydrophis brookii 18-73
Hydrophis coerulescens 18-73
Hydrophis cyanocinctus 18-4, 18-6, 18-16, 18-17, 18-18, 18-20, 18-27.3, 18-49, 18-71, 18-73, 18-76.5, 18-77, 18-79.5, 18-94, 18-108, 18-109, 18-112, 18-113, 18-116, 18-117.5, 18-129
Hydrophis elegans 18-3, 18-72, 18-74, 18-77, 18-94, 18-116
Hydrophis fasciatus 18-6, 18-77
Hydrophis fasciatus atriceps 18-73
Hydrophis fasciatus fasciatus 18-73
Hydrophis gracilis 18-33
Hydrophis kingi 18-72
Hydrophis klossi 18-6, 18-73, 18-77, 18-116
Hydrophis lapemoides 18-27.3
Hydrophis major 18-72
Hydrophis melanosoma 18-6, 18-73, 18-77, 18-116
Hydrophis nigrocinctus 18-73
Hydrophis obscurus 18-73
Hydrophis ornatus 18-73, 18-77, 18-79.5, 18-116
Hydrophis spiralis 18-6, 18-73, 18-77, 18-116
Hydrophis stricticollis 18-73
Hydrophis torquatus 18-73
Hydrophis torquatus aagaard 18-73
Hydrophis torquatus diadema 18-73, 18-116
Hydrus platurus 18-95, 18-120
Kerilia 18-27.3

Chapter 19

VENOM—IMMUNOLOGY, ANTIVENINS, DETOXIFICATION

19-1

Abe, T.; Tamiya, N. 1979.

Immunological studies on erabutoxin b, a sea snake toxin. Attempts to locate the amino acid residues determining antigenicity. *Toxicon* 17(6):571–582.

19-1.3

Anonymous. 1958.

Serpents marins venimeux (Hydrophiidae) et leurs venins. *Inst. Pasteur Viet-Nam Rapp. Annu. Fonct. Tech.* 1958:53–56.

19-1.6

Anonymous. 1960.

Serpents marins venimeux (Hydrophiidae) et leurs venins. *Inst. Pasteur Viet-Nam Rapp. Annu. Fonct. Tech.* 1960:52–61.

19-2

Anonymous. 1976.

Sea snake antidote sought. *Tico Times (San Jose, Costa Rica)* Feb.13:3.

19-3

Barme, M.; Huard, M.; Mai, N. X. 1962.

Preparation d'un serum anti-venin d'hydrophiides. *Ann. Inst. Pasteur (Paris)* 102(4):497–500.

19-4

Baxter, E. H.; Gallichio, H. A.. 1974.

Cross-neutralization by tiger snake (*Notechis scutatus*) antivenene and sea snake (*Enhydrina schistosa*) antivenene against several sea snake venoms. *Toxicon* 12:273–278.

19-5

Baxter, E. H.; Gallichio, H. A. 1976.

Protection against sea snake envenomation: comparative potency of four anti-venenes. *Toxicon* 14:347–355.

19-6

Boquet, P. 1979.

Immunological properties of snake venoms. Lee, C. Y., ed. *Snake venoms*. New York: Springer-Verlag. Pp. 751–824.

19-7

Boquet, P.; Izard, Y.; Ronsseray, A. M. 1970.

Essai de classification des proteines toxiques extraites des venins de serpents. *C.R. Acad. Sci. Paris* 271D:1456–1459.

19-8

Boquet, P.; Izard, Y.; Ronsseray, A. M. 1972.

An attempt to classify by serological techniques the toxic proteins of low molecular weight extracted from Elapidae and Hydrophiidae venoms. *J. Formosan Med. Assoc.* 71(6):307–310.

19-9

Boquet, P.; Izard, Y.; Dumarey, C.; Detrait, J. 1973.

Antigenicite et pouvoir immunogene des toxines extraites des venins d'Elapidae et d'Hydrophiidae. *Experientia (Basel)* 29(12):1467–1471.

19-10

Boquet, P.; Poilleux, G.; Dumarey, C.; Izard, Y.; Ronsseray, A. M. 1973.

An attempt to classify the toxic proteins of Elapidae and Hydrophiidae venoms. *Toxicon* 11(4):333–340.

19-11

Boquet, P.; Dumarey, C.; Joseph, D. 1977.

Nouvelles recherches sur l'antigenicite des proteines toxiques des venins d'Elapidae et d'Hydrophiidae. *C.R. Acad. Sci. Paris* 284D(23):2439–2442. (In French with English abstract)

19-12

Boquet, P.; Dumarey, C.; Joseph, D. 1978.

The problem of the antigenicity of some short toxins of Elapidae and Hydrophiidae venoms. Rosenberg, P., ed. *Toxins: animal, plant and microbial*. New York: Pergamon. Pp. 71–76. (Also listed as *Toxicon* 78 (Suppl.1):71–76.)

19-13
Carey, J. E.; Wright, E. A. 1960.
The toxicity and immunological properties of some sea-snake venoms with a particular reference to that of *Enhydrina schistosa*. *Trans. R. Soc. Trop. Med. Hyg.* 54(1):50–67.

19-14
Chandor, S.; Puffer, H.; Tamiya, N. 1975.
Immunological aspects of venom of sea snakes from the Indo-Pacific. *Toxicon* 13:89. (Abstract)

19-15
Chang, C. C.; Yang, C. C. 1973.
Immunochemical studies on the tryptophan-modified cobrotoxin. *Biochim. Biophys. Acta* 295:595–604.

19-16
Chippaux, J. P.; Goyffon, M. 1983.
Producers of antivenomous sera. *Toxicon* 21(6):739–752.

19-17
Coulter, A. R.; Broad, A. J.; Sutherland, S. K.; Cox, J. C. 1979.
Australian snake venoms. A quantitative study of the presynaptic neurotoxin, notexin, and related substances by radioimmunoassay. Chubb, I. W.; Geffen, L. B., eds. *Neurotoxins: fundamental & clinical advances*. Adelaide: Adelaide University Union Press. Page 259.

19-18
Coulter, A. R.; Harris, R. D.; Sutherland, S. K. 1981.
Enzyme immunoassay and radioimmunoassay: their use in the study of Australian and exotic snake venoms. Banks, C. B.; Martin, A. A., eds. *Proceedings of the Melbourne Herpetological Symposium*. Melbourne: Zoological Board of Victoria. Pp. 39–43.

19-19
Elliott, W. B. 1978.
Chemistry and immunology of reptilian venoms. Gans, C.; Gans, K. A., eds. *Biology of the reptilia*. New York: Academic. Vol. 8, pp. 163–436.

19-20
Fukami, M.; Hattori, Z.; Inoue, T.; Sato, M.; Ajiki, Y.; Kogure, Y.; Okonogi, T. 1978.
Studies of persimmon tannin. II. Detoxifying action of P-Ex against snake venoms and bacterial toxins. *Annu. Rep. Sankyo Res. Lab.* 30:104–111. (In Japanese with English abstract)

19-21
Garnet, J. R. 1970.
    Snake venoms and antivenenes. *Australas. J. Pharm. Sci. Suppl.* 91:S65–S68.

19-22
Gawade, S. P.; Gaitonde, B. B. 1979.
    Immunological studies on monovalent *Enhydrina schistosa* (common sea snake) antivenin. *Toxicon* 17 (Suppl.1):54. (Abstract)

19-23
Gawade, S. P.; Gaitonde, B. B. 1980.
    Immunological studies on monovalent *Enhydrina schistosa* antivenin. *Indian J. Med. Res.* 72:895–900.

19-24
Gawade, S. P.; Buduk, D. P.; Gaitonde, B. B. 1980.
    Preparation of monovalent common Indian sea snake (*Enhydrina schistosa*) antivenin. *Indian J. Med. Res.* 72:747–752.

19-25
Gopalakrishnakone, P.; Hawgood, B. J.; Theakston, R. D. G. 1981.
    Specificity of antibodies to the reconstituted crotoxin complex, from the venom of South American rattlesnake (*Crotalus durissus terrificus*), using enzyme-linked immunosorbent assay (ELISA) and double immunodiffusion. *Toxicon* 19:131–139.

19-26
Kaire, G. H. 1964.
    Sea-snake anti-venom. *Med. J. Aust.* 2:729 (Letter)

19-27
Kankonkar, R. C.; Rao, S. S.; Vad, N. E.; Sant, M. V. 1972.
    Efficacy of Haffkine Institute polyvalent antivenin against Indian snake venoms. *Indian J. Med. Res.* 60:512–516.

19-28
Kawamura, Y.; Sawai, Y.; Kaneta, H. 1981.
    Comparative potency of two seasnake antivenoms. *Snake* 13:151–153.

19-29
Lamb, G. 1905.
    The specificity of antivenomous sera, with special reference to a serum prepared with the venom of *Daboia russellii*. *Sci. Mem. Med. Offrs. Army India* Ser. 2, No. 16:1–18.

19-30
Lamb, G. 1906.
Snake venoms and their antidotes: an account of recent research. *J. Bombay Nat. Hist. Soc.* 17:13–22.

19-31
Lin, M. W. 1975.
Venomous snakes of southeast Asia and some physiological and biochemical aspects of their venoms. Chicago, IL: Northeastern Illinois University. 65 pp. (Thesis) (Review)

19-32
Madsen, T.; Lundstrom, H.; Fohlman, J. 1979.
Purification of monospecific antisera against the venom of the cape cobra (*Naja nivea*). *Toxicon* 17(3):326–330.

19-33
Maegraith, B. G. 1958.
Serpents marins venimeux (Hydrophiidae) et leurs venins. *Inst. Pasteur Viet-Nam Rapp. Annu. Fonct. Tech.* 1958:53–56.

19-34
Minton, S. A. 1967.
Paraspecific protection by elapid and sea snake antivenins. *Toxicon* 5:47–55.

19-35
Minton, S. A. 1973.
Common antigens in snake sera and venoms. DeVries, A.; Kochva, E., eds. *Toxins of animal and plant origins.* New York: Gordon and Breach. Volume 3, pp. 905–917.

19-36
Minton, S. A. 1979.
Common antigens in snake venoms. Lee, C. Y., ed. *Snake venoms.* New York: Springer-Verlag. Pp. 847–862.

19-37
Minton, S. A. 1980.
Paraspecific neutralization by antivenoms. *Snake* 12(1–2):165. (Abstract)

19-38
Okonogi, T.; Hattori, Z.; Igarashi, I. 1967.
Experimental studies on immunization against sea-snake venom. *Jpn. J. Bacteriol.* 22(3):173–177. (In Japanese with English abstract)

19-39
Okonogi, T.; Hattori, Z.; Watanabe, M.; Amagai, E. 1970.
Neutralization of seasnake venom with goat antivenin. *Snake* 2:18–21. (In Japanese with English abstract)

19-40
Okonogi, T.; Hattori, Z.; Amagai, E.; Sawai, Y.; Kawamura, Y. 1972.
Studies of immunity against the venom of *Lapemis hardwicki* with a special reference to a pilot production of therapeutic antivenin horse serum. *Snake* 4:84–88.

19-41
Okonogi, T.; Hattori, Z. 1978.
Detoxifying effect of persimmon-tannin against snake venoms. *Snake* 9:91–96. (In Japanese with English abstract)

19-42
Okonogi, T.; Hattori, Z.; Ogiso, A.; Mitsui, S. 1979.
Detoxification by persimmon tannin of snake venoms and bacterial toxins. *Toxicon* 17:524–527.

19-43
Okonogi, T.; Hattori, Z.; Fukami, M.; Inoue, T.; Sato, M.; Ajiki, Y.; Kogure, Y. 1980.
A miscellany report of sea snakes. *Snake* 12(1–2):19–24. (In Japanese with English abstract)

19-43.5
Okonogi, T.; Morishima, K. 1987.
Detoxification by persimmon tannin of snake venoms. Gopalakrishnakone, P.; Tan, C. K., eds. *Progress in venom and toxin research*. Singapore: National University of Singapore. Pp. 248–250.

19-44
Pickwell, G. V. 1978.
Comparative immunology of sea snake venoms. Rosenberg, P., ed. *Toxins: animal, plant and microbial*. New York: Pergamon. Page 105. (Also listed as *Toxicon* 78 (Suppl.1):105.) (Abstract)

19-45
Pickwell, G. V. 1979.
Experiences with sea snake antivenins and antitoxins. *Toxicon* 17(Suppl 1):140. (Abstract)

19-46
Puffer, H. W.; Chandor, S.; Tamiya, N. 1976.
Immunological aspects of venom of sea snakes from the Indo-Pacific. Ohsaka, A.; Hayashi, K.; Sawai, Y., eds. *Animal, plant and microbial toxins*. New York: Plenum. Vol. 2, pp. 403–406.

19-47
Reid, H. A. 1962.
Sea-snake antivenene: successful trial. *Br. Med. J.* 1(5304):576–579.

19-48
Reid, H. A. 1975.
Antivenom in sea-snake bite poisoning. *Lancet* 1975:622–627.

19-49
Russell, F. E.; Lauritzen, L. 1966.
Antivenins. *Trans. R. Soc. Trop. Med. Hyg.* 60(6):797–810.

19-50
Saint Girons, H.; Detrait, J. 1980.
Communautes antigeniques des venins et systematique des Elapinae. *Bijdr. Dierkd.* 50(1):96–104. (In French with English abstract)

19-51
Sato, S.; Ogahara, H.; Tamiya, N. 1972.
Immunochemistry of erabutoxins. *Toxicon* 10:239–243.

19-52
Sawai, Y.; Kawamura, Y.; Okonogi, T.; Hattori, Z.; Kato, I.; Saito, M.; Muto, S. 1978.
Study on the production of polyvalent antivenin against seasnake venoms. *Snake* 9:63–66.

19-53
Tamiya, N.; Abe, T. 1979.
Antigenicity determining amino acid residues of erabutoxin b. *Toxicon* 17 (Suppl.1):186. (Abstract)

19-54
Tamiya, N.; Abe, T. 1980.
Antigenicity-determining amino acid residues of erabutoxin b. Eaker, D.; Wadstroem, F., eds. *Natural toxins*. New York: Pergamon. Pp. 91–98.

19-55

Tan, N. H. 1983.

   Improvement of Malayan cobra (*Naja naja sputatrix*) antivenin. *Toxicon* 21(1):75–79.

19-56

Taub, A. M. 1964.

   Antivenins available for the treatment of snake bite. *Toxicon* 2:71–77.

19-57

Theakston, R. D. G.; Lloyd Jones, M. J.; Reid, H. A. 1977.

   Micro-ELISA for detecting and assaying snake venom and venom antibody. *Lancet* 1(8039):639–641.

19-58

Tidswell, F. 1906.

   Researches on Australian venoms: snake-bite, snake-venom and antivenine; the poison of the platypus, the poison of the red-spotted spider. Sydney: William Applegate Gullick, Government Printer.

19-59

Tu, A. T. 1977.

   Venoms: chemistry and molecular biology. New York: Wiley.

19-60

Tu, A. T.; Ganthavorn, S. 1969.

   Immunological properties and neutralization of sea-snake venoms from Southeast Asia. *Am. J. Trop. Med. Hyg.* 18(1):151–154.

19-61

Tu, A. T.; Salafranca, E. S. 1974.

   Immunological properties and neutralization of sea snake venoms (II). *Am. J. Trop. Med. Hyg.* 23(1):135–138.

19-62

Vick, J. A.; Von Bredow, J.; Grenan, M. M.; Pickwell, G. V. 1975.

   Sea snake antivenin and experimental envenomation therapy. *Maryland, Edgewood Arsenal Technical Report EB-TR-75043.* 26 pp. (Also published in Dunson, W. A., ed. 1975. *The biology of sea snakes.* Baltimore: University Park Press. Pp 463–485.)

Genus-Species Index
for Chapter 19

Chapter 20

VENOM—SEA SNAKE BITE IN HUMANS, SYMPTOMS, BITE TREATMENT, EPIDEMIOLOGY

20-.05

Adams, A. R. D.; Maegraith, B. G. 1966.

Clinical tropical diseases. 4th ed. Philadelphia: F. A. Davis.

20-1

Ambrose, M. S. 1956.

Snakebite in Central America. Buckley, E. E.; Porges, N., eds. *Venoms.* Washington, DC: American Association for the Advancement of Science. Pp. 323–329.

20-1.5

Angel M., R. 1982.

Serpientes de Colombia; guia practica para su clasificacion y tratamiento del envenenamiento causado por sus mordeduras. Medellin, Colombia: Facultad Nacional de Agronomia, Universidad Nacional de Colombia. (Also cited as *Rev. Fac. Nac. Agron. Medellin* 36:(1):1–171)

20-2

Anonymous. 1962.

Sea snake kills man. *Sea Front.* 8(5):309.

20-3

Anonymous. 1974.

Sea snakes kill swimmers. *Herpetol. Rev.* 5(2):41.

20-4

Anonymous. 1974.

Yield and lethality of snake venoms. *Yearb. Herpetol.* 8(1):224.

20-5
Anonymous. 1981 (?).
   Sea snake. *West. Aust. Shell Collect.* No. 28:4p.

20-6
Ariff, A. W. 1957.
   Sea-snake bites. *Br. Med. J.* No. 5043:525. (Letter)

20-7
Arnold, R. E. 1977.
   Poison! sea snakes. *Fld Stream* 82(Nov.):130.

20-8
Audley, I. 1985.
   A case of sea-snake envenomation. *Med. J. Aust.* 143(11):532. (Letter)

20-9
Auerbach, P. S. 1984.
   Hazardous marine animals. *Emerg. Med. Clin. North Am.* 2(3):531–544.

20-9.5
Auerbach, P. S. 1987.
   A medical guide to hazardous marine life. San Pedro, CA: Best Publishing
Co.

20-10
Aung Khin, M. 1980.
   The problem of snakebites in Burma. *Snake* 12(1–2):125–127.

20-11
Bassett Smith, P. W. 1903.
   Snake-bites and poisonous fishes. *J. Bombay Nat. Hist. Soc.* 15(1):112–130.

20-11.5
Bates, L. B. 1928.
   Snakes and snake-bite accidents of the Panama Canal Zone. *Bull. Antivenin
Inst. Am.* 2(2):21–32.

20-12
Bokma, H. 1941.
   Doodelijke vergiftiging door den beet van een zeeslang *Enhydrina schistosa*
(Daudin.). *Geneesk. Tijdschr. Ned.-Indie* 81:1926–1931.

20-13
Bokma, H. 1942.
Nog eens een beet van een zeeslang. *Geneesk. Tijdschr. Ned.-Indie* 82:87.

20-14
Bolanos-Herrera, R. 1971.
Nuevos recursos contra el ofidismo en centroamerica. 2nd ed. Costa Rica: Instituto Clodomiro Picado.

20-15
Bougainville, L. de. 1772
A voyage round the world. Performed by order of his most Christian Majesty in the years 1766, 1767, 1768, and 1769. London: J. Nourse and T. Davies. New York: DaCapo Press; 1967.

20-16
Brown, J. H. 1973.
Toxicology and pharmacology of venoms from poisonous snakes. Springfield, IL: Charles C. Thomas.

20-17
Campbell, C. H. 1964.
Venomous snake bite and its treatment in the territory of Papua and New Guinea. *Papua New Guinea Med. J.* 7(1):1–11.

20-18
Campbell, C. H. 1975.
The effects of snake venoms and their neurotoxins on the nervous system of man and animals. Hornabrook, R. W., ed. *Topics on tropical neurology.* Philadelphia: F. A. Davis. Pp. 259–293.

20-19
Campbell, C. H. 1979.
Snake bite and snake venoms: their effects on the nervous system. Vinken, P. J.; Bruyn, G. W., eds. *Handbook of clinical neurology.* New York: Elsevier/North Holland Biomedical Press. Vol. 37, no. 2, pp. 1–25.

20-19.5
Castellani, A.; Chalmers, A. J. 1901.
Manual of tropical medicine. New York: William Wood.

20-20
Castellani, A.; Chalmers, A. J. 1913.
Manual of tropical medicine. 2nd ed. London: Balliere, Tindall and Cox.

20-21

Chapman, D. S. 1968.

The symptomatology, pathology, and treatment of the bites of venomous snakes of Central and Southern Africa. Buecherl, W.; Buckley, E. E.; Deulofeu, V., eds. *Venomous animals and their venoms.* New York: Academic. Vol. 1, pp. 463–527.

20-22

Chevers, N. 1870.

A manual of medical jurisprudence for India including the outline of a history of crime against the person in India. Calcutta: Thacker, Spink & Co.

20-23

Clark, H. C. 1942.

Venomous snakes. Some Central American records: incidence of snake-bite accidents. *Am. J. Trop. Med.* 22(1):37–49.

20-25

Corkill, N. L. 1932.

An inquiry into snake-bite in Iraq. *Indian J. Med. Res.* 20(2):599–625.

20-26

Corkill, N. L. 1932.

Snakes and snake bite in Iraq: a handbook for medical officers. London: Balliere, Tindall and Cox.

20-27

Corkill, N. L. 1933.

An inquiry into snake-bite in Iraq—concld. *Indian J. Med. Res.* 20(3):679–696.

20-28

Day, F. 1869.

On the bite of the sea-snake. *Indian Med. Gaz.* 4:92.

20-29

Dobb, G. J. 1986.

Sea-snake envenomation. *Med. J. Aust.* 144(2):112. (Letter).

20-30

Dunn, E. R. 1951.

Venomous reptiles of the tropics. Shattuck, G. C. *Diseases of the tropics.* New York: Appleton-Century-Crofts. Pp. 741–755.

20-31
Edmonds, C. 1975.
Dangerous marine animals of the Indo-Pacific region. Newport, Victoria: Wedneil Publications.

20-32
Edmonds, C. 1976.
Dangerous marine animals. *Aust. Fam. Physician* 5:381–407.

20-33
Edmonds, C. 1981.
Marine animal injuries to man. 2nd ed. Newport, Australia: Wedneil Publications.

20-33.5
Edmonds, C.; Lowry, C.; Pennefather, J. 1976.
Diving and subaquatic medicine. Mosman, N.S.W.: Diving Medical Centre.

20-33.7
Edmonds, C.; Lowry, C.; Pennefather, J. 1983.
Diving and subaquatic medicine. 2nd rev. ed. Seaforth, Australia: Diving Medical Centre.

20-33.8
Fabre, R. 1960.
Toxicologie et oceanographie. *Rev. Gen. Sci. Pures Appl.* 76(1–2):11–38.

20-34
Fairley, N. H. 1934.
Snake bite: its mechanism and modern treatment. *Proc. R. Soc. Med.* 27:1083–1094.

20-35
Faust, E. S. 1924.
Die tropischen intoxikationskrankheiten. II. Vergiftungen durch tierische gifte. Mense, C., ed. *Handbuch der tropenkrankheiten.* Leipzig: Johann Ambrosius Barth. Ser. 3, Vol. 2, pp. 865–1004.

20-36
Fayrer, J. 1870.
The thanatophidia of India. Deaths by snake-bite in the Bengal presidency during 1869. *Indian Med. Gaz.* 5:217–219.

20-37
Fayrer, J. 1872.
   The thanatophidia of India, being a description of the venomous snakes of the Indian Peninsula, with an account of the influence of their poison on life; and a series of experiments. London: Churchill.

20-38
Fayrer, J. 1874.
   The thanatophidia of India, being a description of the venomous snakes of the Indian Peninsula with an account of the influence of their poison on life; and a series of experiements. 2nd ed. rev'd and enl'd. London: Churchill.

20-39
Fayrer, J. 1893.
   On serpent-worship and on the venomous snakes of India and the mortality caused by them. *J. Trans. Victoria Inst.* 2:85–122.

20-40
Fernando, M.; Gooneratne, W. 1983.
   Sea-snake envenoming. *Ceylon Med. J.* 28(3):131–143.

20-41
Fossen, A. 1940.
   Vergiftiging door den beet van zeeslangen. *Geneesk. Tijdschr. Ned.-Indie* 80:1164–1166. (In Dutch with English abstract)

20-42
Freyvogel, T. A.; Hofmann, E. 1965.
   Schlangenbisse und ihre behandlung. *Acta Trop.* 22(1):11–36.

20-43
Fulde, G. W. O.; Smith, F. 1984.
   Sea snake envenomation at Bondi. *Med. J. Aust.* 141:44–45.

20-44
Garnet, J. R. 1968.
   Venomous Australian animals dangerous to man. Melbourne: Commonwealth Serum Laboratories.

20-44.5
Gopalakrishnakone, P. 1984.
   Histopathological changes induced by the sea snake *Enhydrina schistosa* venom on murine muscle and neuromuscular junction. Meier, J.; Stocker, K.; Freyvogel,

T. A., eds. *Proceedings of the 6th European symposium on animal, plant and microbial toxins*. Basel. Page 89. (Abstract)

20-45
Halstead, B. W. 1969.
Toxicology of marine animals. Firth, F. E., ed. *The encyclopedia of marine resources*. New York: Van Nostrand Reinhold. Pp. 675–687.

20-46
Halstead, B. W. 1976.
Hazardous marine life. Strauss, R. H., ed. *Diving medicine*. San Francisco: Grune & Stratton. Pp. 227–256.

20-47
Halstead, B. W. 1978.
Poisonous and venomous marine animals of the world. Rev. ed. Princeton: Darwin Press.

20-48
Heatwole, H. 1975.
Attacks by sea snakes on divers. Dunson, W. A., ed. *Biology of sea snakes*. Baltimore: University Park Press. Pp. 503–516.

20-49
Heatwole, H. 1978.
Sea snake attack: myth or menace? *Skindiving Aust.* 8:40–45.

20-49.5
Heatwole, H. 1987.
Sea snakes. Kensington, N.S.W.: New South Wales University Press.

20-50
Hentsch, H. F. G. 1955.
Ueber die behandlung von giftschlangenbissen. *Z. Tropenmed. Parasit.* 6:252–255.

20-51
Institut Pasteur du Viet-Nam. 1960.
Serpents marins venimeux (Hydrophiidae) et leurs venins. *Inst. Pasteur Viet-Nam Rapp. Annu. Fonct. Tech.* 1960:52–61.

20-52
Jenkins, M. S.; Russell, F. E. 1973.
Physikalische therapie der Klapperschlangenbissverletzung; Physical therapy for injuries produced by rattlesnakes. Kaiser, E., ed. *Tier- und pflanzengifte; Animal and plant toxins*. Muenchen: Wilhelm Goldmann. Pp. 195–199.

20-53
Karunaratne, K. E. de S.; Panabokke, R. G. 1972.
Sea snake poisoning—case report. *J. Trop. Med. Hyg.* 75(5):91–94.

20-54
Keele, C. A. 1963.
Venoms and the causes of pain. *New Sci.* 17(327):396–399.

20-55
Kermorgant, A. 1902.
Les serpents de mer et leur venin. *Ann. Hyg. Med. Colon.* 5:431–435.

20-56
Knowles, R. 1921.
The mechanism and treatment of snake-bite in India. *Trans. R. Soc. Trop. Med. Hyg.* 15:71–103.

20-57
Kopstein, F. 1930.
Die giftschlangen Javas und ihre bedeutung fuer den menschen. *Z. Morphol. Oekol. Tiere* 19:339–363.

20-57.5
Kopstein, F. 1932.
Die giftiere Javas und ihre bedeutung fuer den menschen. *Meded. Dienst Volksgezond. Ned.-Indie* 21:222–256.

20-58
Lim, B. L.; Ibrahim, A. B. b. 1970.
Bites and stings by venomous animals with special reference to snake bites in West Malaysia. *Med. J. Malays.* 25(2):128–141.

20-59
Lim, T. W. 1980.
Epidemiology of snake-bites in Malaysia. *Snake* 12:119–124.

20-60
Lin, M. W. 1975.
Venomous snakes of southeast Asia and some physiological and biochemical aspects of their venoms. Chicago, IL: Northeastern Illinois University. 65 pp. (Thesis)

20-61
Linaweaver, P. G. 1967.
Toxic marine life. *Mil. Med.* 132:347–442.

20-62
Mackie, T. T.; Hunter, G. W.; Worth, C. B. 1954.
A manual of tropical Medicine. 2nd ed. Philadelphia: W. B. Saunders.

20-62.5
Manson Bahr, P. H. 1954.
Manson's tropical diseases: a manual of the diseases of warm climates. 15th ed. London: Cassell. (16th ed. published in 1971)

20-63
Marinkelle, C. J. 1966.
Accidents by venomous animals in Colombia. *Ind. Med. Surg.* 1966:988–992.

20-64
Marriott, M. L. 1982.
Dangers of the sea. *Of Sea and Shore* 12(3):168.

20-65
Marsden, A. T. H.; Reid, H. A. 1961.
Pathology of sea-snake poisoning. *Br. Med. J.* 1(5235):1290–1293.

20-66
Medem, F. 1968.
El desarrollo de la herpetologia en Colombia. *Rev. Acad. Colomb. Cienc. Exactas Fis. Nat.* 13(50):149–199.

20-67
Mercer, H. P.; McGill, J. J.; Ibrahim, R. A. 1981.
Envenomation by sea snake in Queensland. *Med. J. Aust.* 1(3):130–132.

20-68
Minton, S. A. 1967.
Snakebite. Beeson, P. B.; McDermott, W., eds. *Cecil-Loeb Textbook of medicine.* 12th ed. Philadelphia: W. B. Saunders. Pp. 420–425.

20-68.5
Minton, S. A. 1971.
   Venom diseases. Beeson, P. B.; McDermott, W., eds. *Cecil-Loeb Textbook of medicine*. 13th ed. Philadelphia: W. B. Saunders. Pp. 76–82.

20-69
Minton, S. A. 1974.
   Venom diseases. Springfield, IL: Charles C. Thomas.

20-70
M'Kenzie, A. 1820.
   An account of venomous sea snakes, on the coast of Madras. *Asiat. Res.* 13:329–336.

20-71
Okonogi, T. 1973.
   Venomous sea snake bite. *Snake* 5(1–2):156–161. (In Japanese with English abstract)

20-72
Peal, H. W. 1903.
   Antivenine as an antidote for sea-snake bite. *Indian Med. Gaz.* 38:276–277.

20-73
Peel, J. S. 1967.
   Poisonous New Zealand "animals". *Pharm. J.N.Z.* 1967:19–21.

20-74
Pickwell, G. V.; Vick, J. A.; Shipman, W. H.; Grenan, M. M. 1973.
   Production, toxicity and preliminary pharmacology of venom from the sea snake, *Pelamis platurus* with observations on its probable threat to man along middle America. Worthen, L. R., ed. *Food-Drugs from the Sea, Proc. 3rd Conf., 1972*. Washington, DC: Marine Technology Society. Pp. 247–265.

20-75
Rappolt, R. T.; Quinn, H.; Curtis, L.; Minton, S. A.; Murphy, J. B. 1978.
   Medical toxicologist's notebook: snakebite treatment and international antivenin index. *Clin. Toxicol.* 13(3):409–438.

20-76
Reid, H. A. 1956.
   Sea-snake bite research. *Trans. R. Soc. Trop. Med. Hyg.* 50(6):517–542.

20-77
Reid, H. A. 1956.
Sea-snake bites. *Br. Med. J.* 2(4984):73–78.

20-78
Reid, H. A. 1956.
Three fatal cases of sea snakebite. Buckely, E. E.; Porges, N., eds. *Venoms.* Washington, DC: American Association for the Advancement of Science. Pp. 367–371.

20-79
Reid, H. A. 1957.
Antivenene reaction following accidental sea-snake bite. *Br. Med. J.* 2(5035):26–29.

20-80
Reid, H. A. 1957.
Snake-bites. *Br. Med. J.* 2:942. (Letter)

20-81
Reid, H. A. 1959.
Sea-snake bite and poisoning. *Practitioner* 183:530–534.

20-82
Reid, H. A. 1960.
The natural history of sea-snake bite and poisoning: a clinical study. Edinburgh, Scotland: University of Edinburgh. 55 pp. (Medical Dissertation)

20-83
Reid, H. A. 1961.
Diagnosis, prognosis, and treatment of sea-snake bite. *Lancet* 1961:399–402.

20-84
Reid, H. A. 1961.
Myoglobinuria and sea-snake bite poisoning. *Br. Med. J.* 1(5235):1284–1289.

20-84.5
Reid, H. A. 1962.
Malayan snakebite. *Proc. 9th Pacif. Sci. Congr.* 17:374. (Abstract)

20-85
Reid, H. A. 1963.
Snakebite in Malaya. Keegan, A.; MacFarlane, M. V., eds. *Venomous and poisonous animals and noxious plants of the Pacific region.* New York: Macmillan. Pp. 335–362.

20-85.5
Reid, H. A. 1966.
Clinical notes on snakebite. Manson-Bahr, P. H. *Manson's tropical diseases; a manual of the diseases of warm climates.* 16th ed. Baltimore: Williams and Wilkins. Pp. 757–760.

20-86
Reid, H. A. 1968.
Snakebite in the tropics. *Br. Med. J.* 3:359–362.

20-87
Reid, H. A. 1971.
The principles of snakebite treatment. Minton, S. A. *Snake venoms and envenomation.* New York: Marcel Dekker. Pp. 127–136.

20-88
Reid, H. A. 1973.
Clinical aspects of animal toxins. DeVries, A.; Kochva, E., eds. *Toxins of animal and plant origin.* New York: Gordon and Breach. Vol. 3, pp. 127–136.

20-89
Reid, H. A. 1975.
Epidemiology of sea-snake bites. *J. Trop. Med. Hyg.* 78(5):106–113.

20-90
Reid, H. A. 1975.
Epidemiology and clinical aspects of sea snake bites. Dunson, W. A., ed. *The biology of sea snakes.* Baltimore: University Park Press. Pp. 417–462.

20-91
Reid, H. A. 1979.
Pathophysiology of snake bite. *Toxicon* (Suppl.1), 17(1):149. (Abstract)

20-92
Reid, H. A. 1979.
Symptomatology, pathology, and treatment of the bites of sea snakes. Lee, C. Y., ed. *Snake venoms.* New York: Springer-Verlag. Pp. 922–955.

20-93

Reid, H. A. 1980.

Antivenom treatment of sea snake bites. *Snake* 12(1–2):176. (Abstract)

20-94

Reid, H. A. 1980.

Epidemiology of sea snake bites. *Snake* 12(1–2):140. (Abstract)

20-95

Reid, H. A.; Lim, K. J. 1957.

Sea-snake bite: a survey of fishing villages in north-west Malaya. *Br. Med. J.* 2:1266–1272.

20-96

Reid, H. A.; Theakston, R. D. G. 1983.

The management of snake bite. *Bull. W.H.O.* 61(6):885–895.

20-97

Richards, V. 1873.

Dr. Fayrer's treatment of snake-bite by artificial respiration. *Indian Med. Gaz.* 1873:118–120.

20-98

Romer, J. D. 1959.

Aid to the recognition of venomous snakes in Hong Kong with recommendations for first aid treatment of bites. Hong Kong: Government Printer. 16 pp.

20-99

Rosanelli, J. D. 1961.

Sea-snake poisoning and myoglobinuria. *Br. Med. J.* 2:49. (Letter)

20-100

Sawai, Y. 1976.

Medical treatment of snakebites. 2. Southeast Asia. *Snake* 8(1):1–30.

20-101

Sawai, Y. 1979.

Study on snakebites in the Asian areas. *Toxicon* 17 (Suppl.1):159. (Abstract)

20-102

Sawai, Y. 1980.

Epidemiological study on snakebites in the Asian areas. *Snake* 12(1–2):115–203.

20-103
Sawai, Y.; Koba, K.; Okonogi, T.; Mishima, S.; Kawamura, Y.; Chinzei, H.; Ibrahim, A. B. b.; Devaraj, T.; Phong Aksara, S.; Puranananda, C.; Salafranca, E. S.; Sumpaico, J. S.; Tseng, C. S.; Taylor, J. F.; Wu, C. S.; Kuo, T. P. 1971.
   An epidemiological study of snakebites in the Southeast Asia. *Snake* 3(2):97–128. (In Japanese with English abstract)

20-104
Sawai, Y.; Koba, K.; Okonogi, T.; Mishima, S.; Kawamura, Y.; Chinzei, H.; Ibrahim, A. B. b.; Devaraj, T.; Phong Aksara, S.; Puranananda, C.; Salafranca, E. S.; Sumpaico, J. S.; Tseng, C. S.; Taylor, J. F.; Wu, C. S.; Kuo, T. P. 1972.
   An epidemiological study of snakebites in the Southeast Asia. *Jpn. J. Exp. Med.* 42(3):285–307. (In Japanese with English abstract)

20-105
Sawai, Y.; Koba, K.; Okonogi, T.; Mishima, S.; Kawamura, Y.; Chinzei, H.; Ibrahim, A. B. b.; Devaraj, T.; Phong Aksara, S.; Puranananda, C.; Salafranca, E. S.; Sumpaico, J. S.; Tseng, C. S.; Taylor, J. F.; Wu, C. S.; Kuo, T. P. 1973.
   Epidemiologische studie ueber schlangenbisse in sudostasien; Epidemiological study of snakebites in the Southeast Asia. *Tier-und Pflanzengifte; Animal and plant toxins*. Muenchen: Verlag. Pp. 191–194.

20-106
Sawai, Y.; Honma, M. 1975.
   Snakebites in India. *Snake* 7(1):1–16. (In Japanese with English abstract)

20-107
Sawai, Y.; Mishima, S.; Tseng, C. S.; Toriba, M.; Lim, B. L.; Ramachandran, C. P.; Devaraj, T.; Phong Aksara, S.; Romer, J. D. 1978.
   Study on seasnakes and the bites in Malaysia, Thailand and Hong Kong. *Snake* 9(2):48–62. (In Japanese with English abstract)

20-108
Schultz, K. E. 1983.
   Dangerous marine organisms. Kravis, T. C.; Warner, C. G., eds. *Emergency medicine, a comprehensive review*. Rockville, MD: Aspen. Pp. 465–489.

20-108.5
Scortecci, G. 1936.
   L'ofidismo nel'impereo italiano d'Etiopia. *Ric. Sci.* 1(11–12):1–3.

20-109
Sims, J. K. 1984.
Dangerous marine life. Shilling, C. W.; Carlston, C. B.; Mathias, R. A., eds. *The physician's guide to diving medicine.* New York: Plenum. Pp. 427–440.

20-110
Sitprija, V. 1979.
Renal diseases in snakebite. *Toxicon* 17(Suppl 1):172. (Abstract)

20-111
Sitprija, V.; Sribhibhadh, R.; Benyajati, C. 1971.
Haemodialysis in poisoning by sea-snake venom. *Br. Med. J.* 3:218–219.

20-112
Sitprija, V.; Sribhibhadh, R.; Benyajati, C.; Tangchai, P. 1973.
Acute renal failure in snakebite. DeVries, A.; Kochva, E., eds. *Toxins of animal and plant origin.* New York: Gordon & Breach. Vol. 3, pp. 1013–1018.

20-113
Sitprija, V.; Benyajati, C.; Boonpucknavig, V. 1976.
Renal involvement in snakebite. Ohsaka, A.; Hayashi, K.; Sawai, Y., eds. *Animal, plant and microbial toxins.* New York: Plenum. Vol. 2, pp. 438–495.

20-114
Sitprija, V.; Boonpucknavig, V. 1977.
The kidney in tropical snakebite. *Clin. Nephrol.* 8(3):377–383.

20-115
Sitprija, V.; Boonpucknavig, V. 1979.
Snake venoms and nephrotoxicity. Lee, C. Y., ed. *Snake venoms.* New York: Springer-Verlag. Pp. 997–1018.

20-116
Smith, M. 1955.
Deaths from snake-bite. *Country Life* 117:324–325.

20-117
Southcott, R. V. 1977.
Australian venomous and poisonous fishes. *Clin. Toxicol.* 10(3):291–325.

20-118
Stahel, E.; Freyvogel, T. A. 1982.
Animaux venimeux et veneneux: aspects medicaux. *Ann. Soc. Belg. Med. Trop.* 62(1):3–23.

20-119
Strauss, M. B.; Orris, W. L. 1974.
Injuries to divers by marine animals: a simplified approach to recognition and management. *Mil. Med.* 139:129–130.

20-120
Sutherland, S. K. 1974.
Venomous Australian creatures: the action of their toxins and the care of the envenomated patient. *Anaesthesiol. Intens. Care Med.* 2(4):316–328.

20-121
Sutherland, S. K. 1979.
Australian venoms and care of the envenomed patient. Parkville, Victoria, Australia: University of Melbourne. 323 pp. (Medical Dissertation)

20-122
Sutherland, S. K. 1980.
Epidemiology of snakebite in Australia. Presented at the International Seminar on Eipdemiology and Medical Treatment of Snake Bites, August 25–28, Naha, Japan. *Snake* 12(1–2):138–139.

20-123
Sutherland, S. K. 1983.
Australian animal toxins: the creatures, their toxins and care of the poisoned patient. New York: Oxford University Press.

20-124
Sutherland, S. K.; Coulter, A. R.; Harris, R. D.; Lovering, K. E.; Roberts, I. D. 1981.
A study of the major Australian snake venoms in the monkey (*Macaca fascicularis*). I. The movement of injected venom, methods which retard this movement, and the response to antivenoms. *Pathology* 13:13–27.

20-125
Swaroop, S.; Grab, B. 1954.
Snakebite mortality in the world. *Bull. W.H.O.* 10:35–76.

20-126
Swaroop, S.; Grab, B. 1955.
The snakebite mortality problem in the world. Buckley, E. E.; Porges, N., eds. *Venoms.* Washington, DC: American Association for the Advancement of Science. Pp. 439–446.

20-126.5
Taub, A. M.; Dunson, W. A. 1966.
  A new gland in sea snakes (Squamata, Reptilia). *Am. Zool.* 6(4):565. (Abstract)

20-127
Toriba, M. 1984.
  Notes on sea snakes: a correction. *Snake* 16(1):75–76. (In Japanese with English abstract)

20-128
Trethewie, E. R. 1971.
  The pathology, symptomatology, and treatment of snake bite in Australia. Buecherl, W.; Buckley, E. E., eds. *Venomous animals and their venom.* New York: Academic. Vol. 2, pp. 103–113.

20-128.5
Trinca, J. C. 1979.
  Venomous bites and stings: use of antivenoms. Trinca, J. C., ed. *C.S.L. medical handbook.* Parkville, Victoria: Commonwealth Serum Laboratories. Pp. 183–215. (First printed 1966)

20-129
Tu, A. T. 1982.
  Venom toxicology and composition. Tu, A. T., ed. *Survey of contemporary toxicology.* New York: Wiley. Vol. 2, pp. 145–201.

20-130
U.S. Navy Department. 1970
  U.S. Navy diving manual. Washington, DC: Government Printing Office.

20-131
Vick, J. A.; Von Bredow, J.; Grenan, M. M.; Pickwell, G. V. 1975.
  Sea snake antivenin and experimental envenomation therapy. *Maryland, Edgewood Arsenal Technical Report* EB-TR-75043. 26 pages.

20-132
Vick, J. A.; Von Bredow, J.; Grenan, M. M.; Pickwell, G. V. 1975.
  Sea snake antivenin and experimental envenomation therapy. Dunson, W. A., ed. *The biology of sea snakes.* Baltimore: University Park Press. Pp. 463–485.

20-133
Visser, J. 1966.
  Poisonous snakes of Southern Africa and the treatment of snakebite. Cape Town: Howard Timmins.

20-134
Visser, J.; Chapman, D. S. 1978.
   Snakes and snakebite: venomous snakes and management of snakebite in Southern Africa. Cape Town: Purnell.

20-135
Watt, G. 1984.
   Snakebite in the Philippines. *Toxicon* 23(4):629. (Abstract)

20-136
Watt, G.; Theakston, R. D. G. 1985.
   Seasnake bites in a freshwater lake. *Am. J. Trop. Med. Hyg.* 34(4):770–773.

20-136.5
Williamson, J.; Fenner, P.; Acott, C. 1987.
   Medical aspects of marine envenomation and poisoning. Covacevich, J.; Davie, P.; Pearn, J.; eds. *Toxic plants and animals: a guide for Australia.* South Brisbane: Queensland Museum. Pp. 215–40.

20-137
Zimmerman, K. D. 1988.
   The question of sea snake aggression. *Herpetofauna* 18:11–12.

Genus-Species Index
for Chapter 20

Chapter 21

SEA SNAKES IN CAPTIVITY

21-.03
Annandale, N. 1912.
> The Madras aquarium. *J. Bombay Nat. Hist. Soc.* 21(2):693–694.

21-.05
Anonymous. 1910.
> The marine aquarium, Madras. *Nature (Lond.)* 82(2101):411–412.

21-2
Anonymous. 1927.
> (Report of) "Salamander" ortsgruppe Berlin. *Blaett. Aquar.-u. Terr. (Stuttgart)* 38:88–89.

21-3
Anonymous. 1970.
> New York aquarium exhibits sea snakes. *Anim. Kingdom* 73(5):31.

21-4
Anonymous. 1973.
> Scripps scientist captures venomous sea snakes. *Sentinel (Pacific Beach: CA)*, Feb. 7:A4.

21-5
Anonymous. 1983.
> Rare poisonous snakes on display at Scripps. *San Diego Union*, July 2:B11.

21-6
Bacon, J. P. 1983.
> Yellow-bellied sea snakes. *Zoonooz (San Diego)* 56(6):17.

21-7

Cansdale, G. S. 1952.

Recent arrivals at the London Zoo. *Zoo Life (Lond.)* 7:135–139.

21-8

Chhapgar, B. F.; Kewalramani, H. G. 1967.

Mating and other observations on sea snakes in captivity. *J. Bombay Nat. Hist. Soc.* 64(3):563–564.

21-9

Gans, C. 1979.

Momentarily excessive construction as the basis for protoadaptation. *Evolution* 33(1):227–233.

21-9.5

Hornell, J. 1923.

The Madras marine aquarium. *Madras Fish. Bull.* 14(5):57–96.

21-10

Klemmer, K. 1962.

Seeschlangen und ihre haltung im aquarium. *Natur Mus. (Arhus)* 92(3):99–105.

21-11

Klemmer, K. 1966.

Observations on the biology of sea snakes—Hydrophiidae—with remarks on their systematics. *Mem. Inst. Butantan (Sao Paulo)* 33(1):101–103.

21-12

Klemmer, K. 1967.

Observations on the sea-snake *Laticauda laticaudata* in captivity. *Int. Zoo Yearb.* 7:229–231.

21-13

Mays, C. E.; Nickerson, M. A. 1968.

Notes on shedding in the sea snake, *Laticauda semifasciata* (Reinwardt) in captivity. *Copeia* 1968(3):619.

21-13.5

Millard, W. O. 1911.

A marine aquarium. *J. Bombay Nat. Hist. Soc.* 20(4):1177–1179.

21-14
Mishima, S.; Yamazato, S. 1970.
   Feeding of the snakes kept indoors at the Japan Snake Institute. *Snake* 2:59–68. (In Japanese with English abstract)

21-15
Murthy, T. S. N.; Rao, V. K. R. 1977.
   On "mouth-rot" disease in the captive sea snakes. *Snake* 9:29–30.

21-16
Nakamoto, E.; Toriba, M. 1986.
   Successful artificial incubation of the eggs of erabu sea snake, *Pseudolaticauda semifasciata* (Reinwardt). *Snake* 18:55–56.

21-17
Phillips, C. 1964.
   The captive sea: life behind the scenes of the great modern oceanariums. Philadelphia: Chilton Books.

21-18
Phillips, C. 1968.
   Some unusual aquatic snakes. *Drum Croaker* 68(1):11–15.

21-19
Raj, B. S. 1926.
   Parturitions of electric rays and a sea snake in the Marine Aquarium, Madras. *J. Bombay Nat. Hist. Soc.* 31:828.

21-20
Reid, H. A. 1956.
   Sea snakes in captivity. *Proc. Zool. Soc. Lond.* 127:581.

21-21
Sachs, W. 1938.
   Lebende seeschlangen im Berliner Aquarium. *Blaett. Aquar.-u. Terr. (Stuttgart)* 49:15.

21-22
Shaw, C. E. 1961.
   Snakes of the sea. *Zoonooz (San Diego)* 34(7):3–5.

21-23
Shaw, C. E. 1962.
   Sea snakes at the San Diego Zoo. *Int. Zoo Yearb.* 4:49–52.

21-24
Shockley, C. H. 1949.
     Herpetological notes for Ras Jiunri, Baluchistan. *Herpetologica* 5(6):121–123.

21-25
Tochimoto, T.; Uchida, I. 1973.
     On the artificial incubation and breeding of sea snake (*Laticauda semifasciata*).
Jpn. J. Herpetol. 5(2):27. (In Japanese) (Abstract)

21-25.5
Trejos, J. F. 1937.
     Geografia de Costa Rica, fisica, politica y economica. San Jose, Costa Rica:
Imprenta Universal.

21-26
Wolterstorff, W. 1934.
     Seeschlange in den Strassen Magdeburgs! *Blaett. Aquar.-u. Terr. (Stuttgart)*
45:135.

21-27
Zeiller, W. 1969.
     Maintenance of the yellowbellied seasnake, *Pelamis platurus*, in captivity.
Copeia 1969(2):407–408.

Genus-Species Index
for Chapter 21

Chapter 22

USES BY MAN

22-1
Anonymous. 1982.
Sea "snake oil". *San Diego Union*, Apr.6:B4.

22-2
Bacolod, P. T. 1984.
Notes on sea snake fishery on Gato Islet, Cebu Island, Philippines and a proposal for a conservation and management program. *Philipp. Sci.* 21:155–163.

22-3
Beirn, J. T. 1979.
Sea snakes of Gato Island: an exotic resource of the Philippines. *Explor. J.* 57(2):90–91.

22-4
Cahill, T. 1981.
Coming back from Gato Island: poisonous entanglements in the Philippines. *Outside* 1981:30–34, 80–81.

22-5
Campbell, H. 1905.
The diet of the Precibiculturists. *Br. Med. J.* 1905:979–981.

22-5.3
Dang, H. V.; Pham, K. C.; Nguyen, D. M; Hoang, N. H.; Nguyen, T. V.; Phan, Q. K. 1984.
Etudes preliminaires de la bile et du venin du serpent de mer *Lapemis hardwickii* (Den com). *Rev. Pharm.* 1984:48–52.

22-5.5
Haddon, A. C. 1912.
    Reports of the Cambridge Anthropological Expedition to Torres Straits. Cambridge: Cambridge University Press. Volume 4.

22-6
Haneda, T. 1977.
    Therapeutic lipids from Hydrophiidae. Japan. Kokai 77 70,008. 1977 Jun. 10. 4 p. Cl. A61K35/58; Appl 75/145,121. (In Japanese) (Patent)

22-7
Haneda, T. 1979.
    Sea snake oil for oil injection compositions. Japan. Kokai 79,110,316. 1979 Aug. 29. 4 p. Cl. A61K9/08; Appl 78/16,968. (In Japanese) (Patent)

22-8
Haneda, T. 1983.
    Therapeutic lipids from Laticauda semifasciata and Hydrophis. Japan. Kokai JP 58 29,712 (83 29,712). 1983 Feb. 22. 5 p. Cl. A61K35/58; Appl. 81/127,915. (In Japanese) (Patent)

22-8.5
Heatwole, H. 1987.
    Sea snakes. Kensington, N.S.W.: New South Wales University Press.

22-9
Herre, A. W. C. T.; Rabor, D. S. 1949.
    Notes on Philippine sea snakes of the genus *Laticauda. Copeia* 1949(4):282–284.

22-10
Irvine, F. R. 1954.
    Snakes as food for man. *Br. J. Herpetol.* 1(10):183–189.

22-10.5
Man, E. H. 1883.
    On the aboriginal inhabitants of the Andaman Islands. *J. Anthropol. Inst. London* 12:327–434.

22-11
Punay, E. Y. 1975.
    Commercial sea snake fisheries in the Philippines. Dunson, W. A., ed. *The biology of sea snakes.* Baltimore: University Park Press. Pp. 489–502.

22-12
Punay, E. Y. 1980.
   Schlangenfang auf gato. *Tauchen* 3(6):59–60, 62–63.

22-13
Seale, A. 1911.
   The fishery resources of the Philippine Islands, Part IV. Miscellaneous marine products. *Philipp. J. Sci. Sec. D. Gen. Biol., Ethnol., Anthropol.* 6(6):283–320.

22-14
Wall, F. 1906.
   The snake and its natural foes. *J. Bombay Nat. Hist. Soc.* 17(2):375–395.

22-15
Wallach, L. 1984.
   Sea serpents for burn victims. *Omni* 7(1):54.

22-16
Wolff, T. 1955.
   Rennellese names of animals. *Nat. Hist. Rennell Isl. Br. Solomon Isl.* 1:59–63.

Genus-Species Index
for Chapter 22

Chapter 23

GENERAL REFERENCES—NONTECHNICAL

23-.03
Abe, R. 1907.
  Sea snakes in Province Izumo. *Zool. Mag. (Tokyo)* 19:252–253. (In Japanese)

23-.05
Adams, A. 1848.
  Notes from a journal of research into the natural history of the countries visited during the voyage of H.M.S. Samarang, under the command of Captain Sir E. Belcher. Belcher, E. *Narrative of the voyage of H.M.S. Samarang during the years 1843–46; employed surveying the islands of the Eastern Archipelago; accompanied by a brief vocabulary of the principal languages.* London: Reeve, Benham, and Reeve. Volumes 1 and 2.

23-1
Aflalo, F. G. 1896.
  A sketch of the natural history of Australia with some notes on sport. London: Macmillan.

23-1.5
Aitken, E. H. 1901.
  Common birds of Bombay. Bombay: Thacker.

23-2
Alvarez del Toro, M. 1960.
  Los reptiles de Chiapas. Tuxtla Gutierrez, Mexico: Instituto Zoologico del Estado.

23-3
Alvarez del Toro, M. 1972.
  Los reptiles de Chiapas. 2nd ed. rev. Tuxtla Gutierrez, Chiapas, Mexico: Instituto de Historia Natural.

23-3.5
Alvarez del Toro, M. 1982.
Los reptiles de Chaipas. 3rd ed. rev. Tuxtla Guiterrez, Chaipas, Mexico: Instituto de Historia Natural.

23-4
Amos, A. F.; Langseth, M. G.; Markl, R. G. 1972.
Visible oceanic saline fronts. Gordon, A. L., ed. *Studies in physical oceanography: a tribute to Georg Wuest on his 80th birthday*. New York: Gordon and Breach. Vol. 1, pp. 49–62.

23-4.5
Angel, F. 1942.
Petit atlas des amphibiens et reptiles. II. Lacertiliens Ophidiens. Paris: Editions N. Boubee.

23-4.7
Anonymous. 1868.
Sea serpent. *Chambers's encyclopedia, a dictionary of universal knowledge for the people*. London: W. and R. Chambers. Volume 8.

23-5
Anonymous 1948.
Chased by deadly sea snake. *Argus (Melbourne)* Feb. 2:1.

23-6
Anonymous. 1961.
Snake's alive. *Sunday Advertiser (Honolulu)*. April 2:B6.

23-8
Anonymous. 1970.
Ecology and the canal. *Sci. News* 97(15):364–365.

23-9
Anonymous. 1971.
Channel for sea snakes. *Nature (Lond.)* 232:11–12.

23-10
Anonymous. 1971.
Panama Canal warning: ripe for vipers? *Sr. Sch.* 98:9.

23-12
Anonymous. 1976.
Dangerous sea creatures; based on the television series. (New York): Time-Life Films.

23-12.5
Anonymous. 1985.
How sea snakes avoid the bends. *Biol. Action* No. 17, Sept.:1 page

23-12.7
Anonymous. 1987.
Sea snakes keep their heads in deep water. *New Sci.* 1987(1562):34.

23-13
Arnold, R. E. 1973.
What to do about bites and stings of venomous animals. New York: Macmillan.

23-14
Arnold, R. E. 1977.
Poison! Sea snakes. *Fld Stream* 82(Nov.):130.

23-15
Ashe, J. 1965.
Snakes of East Africa. *Africana* 2(4):6–8.

23-16
Atlantic-Pacific Interoceanic Canal Study Commission. 1971.
Interoceanic canal studies 1970. Washington, DC: Government Printing Office.

23-17
Bailey, D. 1970.
A thermal barrier for snakes. *Biomed. News* 1970(12):14.

23-18
Banerjee, R. P. 1893.
Ophiology or a description of the snakes of India. *Med. Reporter* 2:74–77.

23-20
Bannikov, A. G.; Dareoskii, I. S.; Rustamov, A. K. 1971.
Amphibia and reptiles of the U.S.S.R. Moscow. (In Russian)

23-21
Barbour, T. 1926.
   Reptiles and amphibians; their habits and adaptations. Boston: Houghton Mifflin.

23-22
Barbour, T. 1934.
   Reptiles and amphibians; their habits and adaptations. Rev. ed. Boston: Houghton Mifflin.

23-23
Barrett, C. 1924.
   Reptile life in Australia. *Nat. Hist.* 24:42–59.

23-24
Barrett, C. 1950.
   Reptiles of Australia: crocodiles, snakes and lizards. Toronto: Cassell.

23-24.5
Becke, L. 1925.
   Some sea snakes of James Shervinton, and other tales of the south seas. *Bull. N.Y. Zool. Soc.* 28(5):131–133.

23-25
Beebe, W. 1926.
   The Arcturus *adventure: an account of the New York Zoological Society's first oceanographic expedition.* New York: G. P. Putnam's Sons.

23-25.5
Behler, J. L. 1979.
   Operation sea snakes. *Anim. Kindgom Rept.* 82(6):N2–N3.

23-26
Beirn, J. T. 1979.
   Sea snakes of Gato Island: an exotic resource of the Philippines. *Explor. J.* 57(2):90–91.

23-28
Bellairs, A. d'A. 1957.
   Reptiles: life history, evolution, and structure. New York: Harper & Brothers.

23-29
Bellairs, A.; Carrington, R. 1966.
   The world of reptiles. New York: American Elsevier.

23-30
Bennett, G. 1860.
Gatherings of a naturalist in Australasia: being observations principally on the animal and vegetable productions of New South Wales, New Zealand and some Austral islands. London: John Van Voorst.

23-30.5
Bertin, L. 1950.
La vie des animaux. Paris: Librairie Larousse. Volume 2.

23-31
Bisson, R. 1962.
Skindiving safari to the Kenn Reefs. *Fathom* 1:(6p)

23-32
Bligh, W. 1792.
A voyage to the south sea, undertaken by command of His Majesty, for the purpose of conveying the bread-fruit tree to the West Indies, in His Majesty's Ship, the Bounty. London: George Nicol, 1792. Adelaide: Libraries Board of South Australia; 1969.

23-33
Blumenbach, J. F. 1810.
Abbildungen naturhistorischer gegenstaende. Nr.28: *Anguis platuros.* Goettingen: Heinrich Dieterich.

23-34
Bodeker, P. 1971.
Shark Bay—born of the sea. *Walkabout (Melbourne)* 1971(May):42–45.

23-35
Bogert, C. M. 1954.
Amphibians and reptiles of the world. Drimmer, F., ed. *Animal kingdom.* New York: Greystone. Vol. 2, pp. 1189–1390.

23-35.5
Boulenger, E. G. 1914.
Reptiles and batrachians. New York: E. P. Dutton.

23-36
Boulenger, E. G. 1934.
The aquarium book. New York: D. Appleton-Century.

23-36.3
Boulenger, E. G. 1937.
   World natural history. New York: C. Scribner's Sons.

23-36.7
Branch, W. R. 1986.
   Vagrant voyager, the yellow-bellied sea-snake. *Afr. Wildl.* 40:224–227.

23-37
Braun, M. 1904.
   Ueber seeschlangen. *Schr. Ges. Koenigsb.* 45:68–71.

23-38
Brehm, A. 1913.
   Die lurch und kriechtiere. *Brehms Tierleben*. Leipzig: Bibliographisches Institut. Volume 5, Part 2.

23-39
Brehm, A. E. 1865.
   Les reptiles et les batrachiens. Paris: J.-B. Balliere.

23-40
Brenning, M. 1895.
   Die vergiftungen durch schlangen. Stuttgart: Ferdinand Enke.

23-41
Briggs, J. C. 1972.
   Marine biological effects of a sea-level Panama Canal. *Proc. Int. Congr. Zool.* 17(3):1–5.

23-41.5
Brown, J. N. B. 1987.
   Sea snakes of the Arabian Gulf. *Gazelle Dubai Nat. Hist. Group Newsl.* 1987:4 pages

23-42
Buckland, F. 1879.
   Sea-snake caught in telegraph submarine wire. *Land and Water*, Nov. 15:414.

23-42.5
Burgersdijk, L. A. J. 1873.
   Die dieren, afgebeeld, beschreven en in hunne levenswijze geschetst. Leiden: D. Noothoven van Goor.

23-43

Carr, A.; The Editors of *Life*. 1963.

The reptiles. Sacramento: California State Department of Education.

23-44

Cendrero, L. 1972

Zoologia hispanoamericana: vertebrados. Mexico: Editorial Porrua.

23-45

Cochran, D. M. 1943.

Poisonous reptiles of the world: a wartime handbook. Washington, DC: Smithsonian Institution.

23-46

Cochran, D. M. 1944.

Dangerous reptiles. *Smithson. Inst. Annu. Rep.* 1943(1944):275–323.

23-47

Cochrane, R. 1968.

Dangers of the reef and sea. Brisbane, N.S.W.: Jacaranda.

23-48

Codoceo, M. 1957.

Serpientes marinas. *Mus. Nac. Hist. Nat. Not. Mens. (Santiago)* 1(6):7.

23-49

Cogger, H. G. 1959.

Sea-snakes. *Aust. Mus. Mag.* 13(2):37–41.

23-50

Cogger, H. G. 1960.

Snakes, lizards and chelonians. *Aust. Mus. Mag.* 13(8):250–253.

23-51

Cogger, H. G. 1963.

Sea-snakes. *Bull. Post-Grad. Comm. Med. Univ. Sydney* 18(12)Suppl.:5–7.

23-52

Cogger, H. 1967.

Australian reptiles in colour. Sydney: A. H. & A. W. Reed.

23-53
Cogger, H. G. 1967.
   The snakes of Australia. McMichael, D. F., ed. *A treasury of Australian wildlfe*. Sidney: Ure Smith. Pp. 152–158.

23-54
Cogger, H.; Heatwole, H. 1978.
   Serpents of Australian seas. *Aust. Nat. Hist.* 19(8):261–265.

23-55
Corkill, N. L. 1962.
   The snakes of Aden Territory. *Port of Aden Annual*, 1961–1962:78–81.

23-56
Cromie, W. J. 1966.
   The living world of the sea. Englewood Cliffs, NJ: Prentice-Hall.

23-57
Crompton, J. 1963.
   The snake. London: Faber and Faber.

23-58
Cropp, B. 1962.
   Sea snakes; sea snakes scare spearfisherman at Swains. *Skin Diver* 11(10):30–31,53.

23-59
Cropp, B. 1969.
   Great Barrier Reef adventure. *Skin Diver* 1969:46–49.

23-60
Cropp, B. 1970.
   Sea snake. Burton, M.; Burton, R., eds. *International wildlfe encyclopedia*. New York: Marshall Cavendish Corp. 15:2084.

23-61
Cropp, B. 1970.
   Sea snakes. *Oceans Mag.* 3(2):48–54.

23-62
Curran, C.; Kauffeld, C. 1937.
   Les serpents. Paris: Payot.

23-63

Curran, C. H.; Kauffeld, C. 1937.

Snakes and their ways. New York: Harper.

23-64

D. G. 1981.

On the banded sea snake. *Sea Secrets* 25(4):5.

23-65

Dahl, F. 1898.

Die verbreitung der thiere auf hoher see. II. *Sitz. Deut. Akad. Wiss. Berlin* 1898:102–118.

23-66

D'Aulaire, E.; D'Aulaire, P. O. 1980.

Meet the sea snake—cautiously. *Int. Wildl.* 10(3):40–47.

23-67

Dawson, S. 1985.

Deadly but not dangerous. *Sea Front.* 1985:282–286.

23-68

Day, F. 1879.

The example of sea-snake (*Pelamis bicolor*). *Land and Water*, Nov.15:414.

23-69

Deas, W. 1972.

Sea snakes. *Skindiving Aust.* 2:26–27.

23-70

Deckert, K. 1967.

Urania tierreich. Leipzig: Urania-Verlag. Volume 4.

23-71

DeCouet, H. G. 1979.

Seeschlangen (Hydrophiidae). *Tauchen* 2(5):55–58.

23-72

DeCouet, H. G. 1980.

Seeschlangen: elegant, aber gefaehrlich. *Tauchen* 3(6):55–59.

23-72.3

Delsman, H. C. 1951.

Dierenleven in Indonesie. 's-Gravenhage: W. Van Hoeve.

23-72.7
Devaney, C. 1976.
   Serpents of the deep. Dent, N., ed. *The sea*. London: New Caxton Library Service Ltd. Vol. 4, pp. 945–947.

23-73
Ditmars, R. L. 1910.
   Reptiles of the world: tortoises and turtles, crocodilians, lizards and snakes of the eastern and western hemispheres. New York: Sturgis & Walton.

23-74
Ditmars, R. L. 1931.
   Snakes of the world. New York: Macmillan.

23-75
Ditmars, R. L. 1936.
   The reptiles of North America: a review of the crocodilians, lizards, snakes, turtles and tortoises inhabiting the United States and Northern Mexico. New York: Doubleday.

23-76
Ditmars, R. L. 1936.
   Reptiles of the world: the crocodilians, lizards, snakes, turtles and tortoises of the eastern and western hemispheres. New rev. ed. New York: Macmillan.

23-77
Doughton, R. 1960.
   On safari to remote Marion Reef with Ron and Denyse Doughton. *Skindiving Aust.* 3:(4p).

23-78
Dowling, H. G. 1966.
   Poisonous snakes of Vietnam. *Anim. Kingdom* 69(2):34–43.

23-80
Dowling, H. G. 1981.
   Snake. *Encyclopedia Americana*. Danbury, CT: Grolier. Vol. 25, pp. 85–101.

23-80.3
Dowling, H. G. 1986.
   Snakes. Halliday, T. R.; Adler, K., eds. *The encyclopedia of reptiles and amphibians*. New York: Facts on File. Pp. 112–127.

23-80.7
Dozier, T. A. 1976.
  Dangerous sea creatures: based on the television series Wild, wild world of animals. New York: Time-Life Films.

23-81
Dunson, W. A. 1971.
  The sea snakes are coming. *Nat. Hist.* 80(9):52–61.

23-82
Dunson, W. A. 1975.
  Sea snakes and the sea level canal controversy. Dunson, W. A., ed. *The biology of sea snakes.* Baltimore: University Park Press. Pp. 517–524.

23-83
Fayrer, J. 1877.
  Venomous animals. *Edinb. Med. J.* 23(27):97–117.

23-84
Fayrer, J. 1893.
  On serpent-worship on the venomous snakes of India and the mortality caused by them. *J. Trans. Victoria Inst.* 2:85–122.

23-85
Felix, J. 1968.
  Seeschlangen in Vietnam. *Monatsschr. Ornithol. Vivarienkd. Ausg. B Aquarien Terrarien* 15(8):270–273.

23-86
FitzSimons, F. W. 1912.
  The snakes of South Africa: their venom and the treatment of snake bite. New ed. Cape Town: T. Maskew Miller.

23-87
FitzSimons, F. W. 1921.
  The snakes of South Africa: their venom and the treatment of snake bite. 3rd rev. ed. Cape Town: T. Maskew Miller.

23-88
FitzSimons, V. F. M. 1962.
  The snakes of Southern Africa. London: MacDonald.

23-89
FitzSimons, V. F. M. 1970.
  A field guide to the snakes of Southern Africa. London: Collins.

23-90
Fonseca, F. da. 1949.
  Animais peconhentos. Sao Paulo: Instituto Butantan.

23-91
Frade, F.; Manacas, S. 1955.
  Serpentes do Ultramar Portugues. *Garcia de Orta (Lisb.)* 3(4):547–553.

23-92
Friedrich, H. 1969.
  Marine biology; an introduction to its problems and results. Seattle: University
of Washington.

23-93
Gadow, H. 1920.
  Amphibia and reptiles. London: Macmillan.

23-93.5
Gasperetti, J. 1977.
  Snakes in Arabia. *J. Saudi Arabian Nat. Hist. Soc.* 19:3–16.

23-94
Gawade, S. P.; Bhise, S. B. 1977.
  Sea snakes and their venoms. *Sci. Rep.* 1977:587–588.

23-95
Gertychowa, R. 1972.
  Waz morski a nowy Kanal Panamski. *Chronmy Przyr. Ojczysta* 28(4):60–65.

23-95.3
Gervais, P. 1847.
  Ophidiens. *Dictionnarie universel d'histoire naturelle*. Paris: Renard, Marti-
net. Volume 9.

23-95.6
Gharpurey, K. G. 1935.
  The snakes of India. Bombay: Popular Book Depot.

23-96

Gharpurey, K. G. 1937.

The snakes of India. 2nd ed. Bombay: Popular Book Depot.

23-97

Gharpurey, K. G. 1954.

The snakes of India & Pakistan. 4th ed. Bombay: Popular Book Depot.

23-98

Gharpurey, K. G. 1962.

Snakes of India and Pakistan. 5th ed. Bombay: Popular Prakashan.

23-98.5

Giebel, C. G. 1861.

Die naturgeschichte der thierreichs. Leipzig: D.Wigand. Volume 3

23-99

Gilchrist, J. D. F. 1911.

South African zoology: a text book for the use of students, teachers and others in South Africa. Cape Town: T. Maskew Miller.

23-100

Gillett, K. 1968.

The Australian Great Barrier Reef in colour. Sydney: A. H. & A. W. Reed.

23-101

Gillett, K.; McNeill, F. 1959.

The Great Barrier Reef and adjacent isles. Sydney: Coral Press Pty. Ltd.

23-102

Gilluly, R. H. 1971.

Consequences of a sea-level canal: a new study of sea snakes reinforces ecologists' concerns about the proposed canal. *Sci. News* 99(3):52–53,

23-103

Gistel, J. von. 1848.

Naturgeschichte aus thierreichs. Stuttgart: Hoffman'sche.

23-104

Gistel, J. 1851.

Naturgeschichte des thierreichs fuer hoehere schulen. 2nd ed. Stuttgart: Scheitlin & Krais.

23-105
Goin, C. J.; Goin, O. B. 1971.
   Introduction to herpetology. 2nd ed. San Francisco: W. H. Freeman.

23-106
Goin, C. J.; Goin, O. B.; Zug, G. R. 1978.
   Introduction to herpetology. 3rd ed. San Francisco: W. H. Freeman.

23-107
Gow, G. F. 1976.
   Snakes of Australia. Sydney: Angus and Robertson.

23-108
Gow, G. F. 1977.
   Snakes of the Darwin area. Darwin: The Museums and Art Galleries Board of
the Northern Territory.

23-109
Gow, G. F. 1983.
   Snakes of Australia. Rev. ed. Sydney: Angus & Robertson.

23-109.5
Gray, J. E. 1849.
   Vertebrata. Adams, A. *The zoology of the voyage of H.M.S. Samarang; under
the command of Captain Sir Edward Belcher, during the years 1843–1846.* London:
Reeve, Benham, and Reeve.

23-110
Grocott, R. G.; Sadler, G. G. 1958.
   The poisonous snakes of Panama. Ancon: Panama Canal Zone Printing Plant.

23-111
Guibe, J. 1970.
   La systematique des reptiles actuels. Grasse, P. P., ed. *Traite de zoologie.*
Paris: Masson et Cie. Vol. 14, no. 3, pp. 1054–1160.

23-112
Halstead, B. W. 1959.
   Dangerous marine animals. Cambridge, MD: Cornell Maritime Press.

23-113
Halstead, B. W. 1980.
   Dangerous marine animals that bite, sting, shock, are non-edible. 2nd ed.
Centreville, MD: Cornell Maritime Press.

23-113.5
Hanitsch, R. 1908.
   Guide to the zoological collections of the Raffles Museum, Singapore. Singapore.

23-114
Harding, J. 1962.
   A closer look at sea snakes. *Fathom* 1:20–24.

23-115
Heatwole, H. 1975.
   Attacks by sea snakes on divers. Dunson, W. A., ed. *The biology of sea snakes.* Baltimore: University Park Press. Pp. 503–516.

23-115.5
Heatwole, H. 1977.
   Serpentes (sea snakes). Saenger, P., ed. *The Great Barrier Reef: a diver's guide.* Brisbane: Scientific Committee, Australian Underwater Committee. Pp. 149–154.

23-116
Heatwole, H. 1978.
   Sea snake attack: myth or menace? *Skindiving Aust.* 8:40–45.

23-116.5
Hediger, H. 1948.
   Kleine tropen-zoologie. *Acta Trop.* 1948 (Suppl 1):1–182.

23-116.7
Heidenreich, E.; Wolff, M. 1937.
   Die tiere des meeres und des strandes. Berger, A.; Schmid, J. *Das reich der tiere: das tier in seinem lebensraum.* Berlin: Ullstein. Vol. 1, pp. 373–566.

23-117
Helm, T. 1965.
   A world of snakes. New York: Dodd, Mead & Co.

23-118
Helm, T. 1976.
   Dangerous sea creatures: a complete guide to hazardous marine life. New York: Funk & Wagnalls.

23-119
Heuvelmans, B. 1968.
   In the wake of the sea-serpents. New York: Hill and Wang.

23-120
Hoesel, J. K. P. van. 1959.
   Ophidia javanica. Bangor: Pertjetakan Archipel.

23-121
Honma, Y.; Kitami, T. 1978.
   A stuffed specimen of sea snake, *Pelamis platurus*, presented as a treasure to the Ushio Shrine, Sado Island. *Bull. Sado Mus.* No. 26:1–4.

23-122
Hopley, C. C. 1879.
   Sea snakes. *Land and Water*, Apr.5:278.

23-122.5
Hornell, J. 1923.
   The Madras marine aquarium. *Madras Fish. Bull.* 14(5):57–96.

23-122.7
Jahn, T. 1967.
   Der farbige brehm. Wien: Herder.

23-123
Jones, M. L.; Manning, R. B. 1971.
   A two-ocean bouillabaisse can result if and when sea-level canal is dug. *Smithsonian* 2(9):12–21.

23-124
Jourdan, A. G. L. 1925.
   A propos du serpent de mer. *Nature (Paris)* No. 2697 (Suppl.):185–186.

23-125
Ka-Vai. 1945.
   Sea-snakes of the tropics. *Walkabout (Melbourne)* 11(8):33–34.

23-126
Keegan, H. L. 1958.
   Some venomous animals of the Far East. Camp Zuma, Japan: U.S. Army 406th Medical General Laboratory.

23-127
Keegan, H. L. 1960.
   Some venomous and noxious animals of the Far East. Rev. ed. Camp Zuma, Japan: U.S. Army 406th Medical General Laboratory.

23-128
Keegan, H. L.; Weaver, R. E.; Toshioka, S.; Matsui, T. 1964
Some venomous and noxious animals of East and Southeast Asia. 3rd ed. Washington, DC: U.S. Army 406th Medical General Laboratory.

23-128.5
Keller, C. 1895.
Das leben des meers. Leipzig: C. H. Tauchnitz.

23-129
Kempfer, A. 1980.
Wenn die schlange naeher kommt. *Tauchen* 3(6):61.

23-130
Kent, F. B. 1972.
The biological unknowns of a new Panama Canal. *Wash. Post* January 18:B1.

23-130.5
Keswal, J. A. 1886.
Waters of western India. Part II. Konkan and coast. *J. Bombay Nat. Hist. Soc.* 1(4):153–175.

23-131
Kinghorn, J. R. 1929.
Snakes of Australia. Sydney: Angus & Robertson.

23-132
Kinghorn, J. R.; Kellaway, C. H. 1943.
The dangerous snakes of the south west Pacific area. Melbourne: Victoria Railways Printing Works.

23-133
Kinghorn, J. R. 1943.
Some New Guinea reptiles. *Aust. Mus. Mag.* 8:112–117.

23-134
Kinghorn, J. R. 1956.
Snakes of Australia. 2nd ed. Cape Town: Angus and Robertson.

23-135
Kinghorn, J. R. 1964.
The snakes of Australia. Rev. ed. Sydney: Angus and Robertson.

23-136
Klemmer, K. 1971.
    Giftnattern und seeschlangen. Grzimek, B., ed. *Grzimek Tierleben: Enzyklopedie des tierreichs.* Zurich: Kindler Verlag. Vol. 6, pp. 424–450.

23-137
Klemmer, K. 1975.
    Cobras and sea snakes. Grzimek, B., ed. *Grzimek's animal life encyclopedia.* London: Van Nostrand Reinhold. Vol. 6, pp. 415–438.

23-137.5
Knauer, F. 1882.
    Die kriechthiere und lurch. Martin, P. L. *Illustrirte naturgeschichte der thiere.* Leipzig: F. A. Brockhaus. Volume 2.

23-138
Kneeland, S. 1884.
    Apropos of the "sea serpents". *Proc. Boston Soc. Nat. Hist.* 23:163–164. (Letter)

23-139
Kopstein, F. 1930.
    De javaansche gifslangen en haar beteekenis voor den mensch. Weltevreden: Nederlandsch-Indische Natuurhistorische Vereeniging.

23-140
Kropach, C. 1972.
    Pelamis platurus as a potential colonizer of the Caribbean Sea. *Bull. Biol. Soc. Wash.* 2:267–269.

23-140.5
Kurka, A.; Pfleger, V. 1984.
    Jedovati zivocichove (poisonous snakes). Prague: Academic Praha.

23-140.7
Kuroda, N. 1924.
    Laticauda semifasciata from the northernmost part of its range. *Zool. Mag. (Tokyo)* 36:388–390.

23-140.9
Laboute, P.; Magnier, Y. 1979.
    Underwater guide to New Caledonia. Papeete, Tahiti: Editions du Pacifique.

23-141
LeMare, D. W. 1952.
Poisonous Malayan fish. *Med. J. Malaya* 7(1):1–8.

23-142
Lenz, H. O. 1870.
Schlangen und schlangenfeinde. *Der schlangenkunde zweite sehr veraenderte auflage*. Gotha: E. F. Thienemann.

23-143
Liat, L. B. 1979.
Poisonous snakes of Peninsular Malaysia. Kuala Lumpur: Malayan Nature Society. (Second edition published in 1982.)

23-143.5
Liat, L. B. 1987.
Snakes of public health importance. Gopalakrishnakone, P.; Tan, C. K., eds. *Progress in venom and toxin research*. Singapore: National University of Singapore.

23-144
Longman, H. A. 1926.
Snakes. Jose, A. W.; Carter, H. J., eds. *The Australian encyclopaedia*. Sydney: Angus & Robertson. Vol. 2, pp. 470–475.

23-145
Lord, C. E. 1919.
Tasmanian snakes. Hobart: J. Walch & Sons.

23-145.5
Loveridge, A. 1928.
The snake-like eel. *Bull. Antivenin Inst. Am.* 2(2)51–52.

23-146
Loveridge, A. 1946.
Reptiles of the Pacific world. New York: Macmillan.

23-147
Lowe, C. H. 1968.
Fauna of desert environments with desert disease information. McGinnies, W. G.; Goldman, B. J.; Paylore, P., eds. *Deserts of the world; an appraisal of research into their physical and biological environments*. Tucson: University of Arizona Press. Pp. 567–645.

23-148
Lowe, W. P. 1932.
   The trail that is always new. London: Gurney and Jacobson.

23-149
Lucas, A. H. S.; LeSouef, W. H. D. 1909.
   The animals of Australia: mammals, reptiles and amphibians. Melbourne:
Whitcombe and Tombs.

23-150
Lydekker, R. 1896.
   The royal natural history. New York: Frederick Warne. Volume 5.

23-151
Lydekker, R. 1899.
   The new natural history. New York: Merrill & Baker. Volume 5.

23-151.3
Lydekker, R. 1906.
   *Guide to the gallery of reptilia and amphibia in the Department of Zoology of
the British Museum (Natural History) Cromwell Road, London, S. W.* London:
The Trustees,

23-152
Macinnes, I. G. 1951.
   *Australian fisheries: a handbook prepared for the second meeting of the Indo-
Pacific Council, Sydney, April 1950.* Sydney: Halstead Press.

23-153
MacLeish, K. 1972.
   Diving with sea snakes. *Nat. Geogr. Mag.* 141(4):565–578.

23-154
MacPherson, J. 1924.
   Snakes and snake bite. *Sydney Univ. Med. J.* 18:175–187.

23-155
McCann, C. 1966.
   Key to the marine turtles and snakes occurring in New Zealand. *Tuatara*
14(2):73–81.

23-156
McClane, A. J. 1965.
   Cobras of the sea. *Fld Stream* 69:74, 150, 153–154.

23-156.3
McCoy, M. 1981.
   Tugihono. *Paradise* 31:35–38.

23-156.5
McCoy, M. 1985.
   I bought 200 poisonous snakes from children. *Geo* 7(1):90–101.

23-157
McKeown, S. 1978.
   Hawaiian reptiles and amphibians. Honolulu: Oriental Publishing Co.

23-157.2
Mertens, R. 1951.
   Zoologische wanderungen in El Salvador. *Nat. Volk (Frankf.)* 81(7):162–167.

23-157.7
Mertens, R. 1952.
   El Salvador: biologische reisen im lande der vulkane. Frankfurt am Main:
Waldemar Kramer.

23-158
Mertens, R. 1960.
   The world of amphibians and reptiles. New York: McGraw-Hill.

23-159
Miller, J. W., ed. 1979.
   NOAA diving manual: diving for science and technology. 2nd ed. Washington,
DC: Government Printing Office.

23-161
Minton, S. A. 1970.
   Identification of poisonous snakes. *Clin. Toxicol.* 3(3):347–362.

23-162
Minton, S. A. 1971.
   Identification of poisonous snakes. Minton, S. A., ed. *Snake venoms and
envenomation.* New York: Marcel Dekker. Pp. 1–16.

23-163
Minton, S. A. 1977.
   Spines and stings and sea snakes. *Consultant (Phila.)* 17:45–63.

23-163.2
Minton, S. A. 1986.
    The 11 families of snakes. Halliday, T. R.; Adler, K., eds. *The encyclopedia of reptiles and amphibians*. New York: Facts on File. Pp. 125–131.

23-163.7
Minton, S. A. 1989.
    The life and times of sea snakes. *Newsl. Tucson Herpetol. Soc.* 2(4):28–33.

23-163.8
Minton, S. A.; Minton, M. R. 1969.
    Venomous reptiles. New York: Charles Scribner's Sons.

23-164
Minton, S. A.; Heatwole, H. 1978.
    Snakes & the sea. *Oceans Mag.* 11(2):53–56.

23-165
Minton, S. A.; Minton, M. R. 1980.
    Venomous reptiles. Rev. ed. New York: Charles Scribner's Sons.

23-166
Mocquard, F. 1907.
    Sur les reptiles aquatiques de l'Indo-Chine. *Paris. Bull. Soc. Centr. Aquiculture* 19:209–214.

23-167
Monestel, Y. 1976.
    Here, the snake. *Tico Times (San Jose, Costa Rica)*, May 7:2.

23-167.5
Murthy, T. S. N. 1986.
    The snake book of India. Dehra Dun, India: R. P. Singh Gahlot.

23-168
Murthy, T. S. N.; Rao, K. V. R. 1975.
    Snakes in the sea. *Illus. Weekly India* 96(43):17,19.

23-169
Murthy, T. S. N.; Rao, K. V. R. 1977.
    Sea-snakes. *Everyday Sci.* 22:27–30.

23-170

Myers, A. 1985.

Getting a grip on sea snakes. *West. Boatman* 3(5):60–61.

23-171

Myers, C. W. 1972.

The status of herpetology in Panama. *Bull. Biol. Soc. Wash.* 2:199–209.

23-171.5

Nakamura, K. 1957.

Reptilia. *Encyclopaedia zoologica illustrated in colours.* Tokyo: Hokuryu-kan. Vol. 1, pp. 283–324. (In Japanese)

23-172

Nakamura, K.; Ueno, S. I. 1963.

Japanese reptiles and amphibians in colour. Osaka: Hoikusha. (Reprinted in 1971)

23-173

Neill, W. T. 1958.

The occurrence of amphibians and reptiles in saltwater areas, and a bibliography. *Bull. Mar. Sci. Gulf Caribb.* 8(1):1–97.

23-173.5

Okada, Y. 1927.

Reptilia. Uchida, K. *Figuraro de Japanaj bestoj.* Tokyo: Hokuryukwan. Pp. 191–234. (In Japanese)

23-174

Owner, F. 1970.

Reptiles of Australia and New Zealand. Sydney: Angus & Robertson.

23-174.5

Parker, H. W. 1931.

Class Reptilia. Pycraft, W. P., ed. *The standard natural history from amoeba to man.* New York: Frederick Warne. Pp. 511–563.

23-175

Parker, H. W. 1965.

Natural history of snakes. London: British Museum (Natural History).

23-175.5
Parker, H. W. 1967.
   Snake. *Encyclopedia Britannica*. Chicago: Encyclopedia Britannica. Vol. 20, pp. 717–721.

23-176
Parker, H. W. 1977.
   Snakes: a natural history. 2nd ed. rev. and enl'd. by A. G. C. Grandison. Ithaca, NY: Cornell University Press. (Original title: *Natural history of snakes*)

23-176.5
Patzelt, E. 1978.
   Fauna del Ecuador. Quito: Editorial las Casas. (2nd ed. published in 1979)

23-177
Paulson, D. R. 1967.
   Searching for sea serpents. *Sea Front*. 13(4):244–250.

23-178
Pelletier, G. 1956.
   Sea snakes. *Skin Diver* 5(5):12.

23-179
Pickwell, G. V. 1972.
   Sea snakes of Viet Nam and Southeast Asia. Pickwell, G. V.; Evans, W. E., eds. *Handbook of dangerous animals for field personnel*. San Diego, CA: Naval Undersea Center.

23-180
Pickwell, G. V. 1972.
   The venomous sea snakes. *Fauna* (4):17–32.

23-181
Pigulevsky, S. V. 1966.
   Poisonous animals: toxicology of vertebrates. Leningrad: Print House "Medicina."

23-182
Pinner, E. 1974.
   Seeschlangen im Pazifischen Ozean. *Naturw. Rundsch.* 27(1):31.

23-183
Pinney, R. 1981.
   The snake book. Garden City, NY: Doubleday.

23-184
Pope, C. H. 1949.
   Snakes alive and how they live. New York: Viking.

23-185
Pope, C. H. 1955.
   The reptile world. New York: Knopf.

23-186
Popovici, Z.; Angelescu, V. 1954.
   La economia del mar y sus relaciones con la alimentacion de la humanidad.
Buenos Aires: Imprenta y Casa Editora "Coni" (cited also as *Publ. Ext. Cult. Mus.
Argent. Cienc. Nat.* No. 8) Volume 1.

23-186.5
Pringle, J. A. 1954.
   Common snakes. New York: Longmans, Green.

23-187
Punay, E. Y. 1980.
   Schlangenfang auf gato. *Tauchen* 3(6):59–60,62–63.

23-188
Rafinesque, C. S. 1817.
   Dissertation on water snakes, sea snakes and sea serpents. *Am. Month. Mag.
Crit. Rev.* 1:431–435. (Reprinted in *Philos. Mag.* 54:361–367)

23-189
Rice, T. 1979.
   Dangerous marine animals. *Of Sea and Shore* 1979:131–133,141.

23-190
Richardson, M. 1972.
   The fascination of reptiles. New York: Hill and Wang.

23-190.5
Robin, A. 1923.
   Les vertebrates. Joubin, L. *Les invertebres.* Paris: Librairie Larousse. Pp.
157–332.

23-191
Romer, J. D. 1965.
   *Illustrated guide to the venomous snakes of Hong Kong with recommendations
for first aid treatment of bites.* 2nd ed. rev. Hong Kong: Government Printer.

23-192
Rose, W. 1929.
   Veld & vlei: an account of South African frogs, toads, lizards, snakes, & tortoises. Cape Town: Speciality Press of South Africa.

23-193
Rose, W. 1955.
   Snakes—mainly South African. Cape Town: Maskew Miller.

23-194
Rose, W. 1962.
   Reptiles and amphibians of Southern Africa. Rev. and enl. Cape Town: Maskew Miller.

23-195
Rosenberg, M. J. 1976.
   Sea snakes. *J. No. Ohio Assoc. Herpetol.* 3(1):25–29.

23-196
Rosenberg, P. 1980.
   Common names index: poisonous animals, plants and bacteria. *Toxicon* 18(1):11–54.

23-197
Ruiter, L. C. de. 1958.
   Zeeslangen I. *Lacerta* 16:86–87.

23-198
Ruiter, L. C. de. 1958.
   Zeeslangen II. *Lacerta* 17:1–2.

23-199
Ruiter, L. C. de. 1959.
   Zeeslangen III. *Lacerta* 17:19–21.

23-200
Russell, F. E.; Minton, S. 1978.
   Poisonous snakes. *Clin. Med.* 85(11):13–22.

23-201
Satyamurti, S. T. 1960.
   Guide to the snakes exhibited in the reptile gallery. Madras: Government Museum.

23-202
Schall, J. 1969.
Sea snakes—mysterious mariners. *Frontiers* 32(3):12–15.

23-203
Schmidt, K. P.; Davis, D. D. 1941.
Field book of snakes of the United States and Canada. New York: G. P. Putnam's
Sons.

23-204
Schmidt, K. P.; Inger, R. F. 1967.
Living reptiles of the world. Garden City, NY: Doubleday.

23-204.5
Schnee, S. 1906.
Die schlangeninsel bei herbertshoehe. *Aus der Natur (Berlin)* 1(21):669–670.

23-205
Schnoo, P. 1848.
Einiges ueber seeschlangen. *Naturw. Beob.* 39(3):90–96.

23-205.5
Schoenhals, L. C. 1988.
A Spanish-English glossary of Mexican flora and fauna. Mexico, D. F.: Instituto
Linguistico de Verano.

23-206
Scholey, K. 1985.
Sea serpents. *BBC Wildl.* 3(10):462–466.

23-207
Scholey, K. 1986.
Die giftigsten tiere der welt. *Tier (Stutt.)* 1986(10):31–33.

23-208
Serventy, V.; Raymond, R. 1973.
Marine reptiles. *Aust. Wildl. Heritage* 1:91–96.

23-209
Sharland, M. 1962.
*Tasmanian wild life: a popular account of the furred land mammals, snakes
and introduced mammals of Tasmania.* Melbourne: Melbourne University Press.

23-210
Shaw, C. E. 1961.
   Snakes of the sea. *Zoonooz (San Diego)* 34(7):3–5.

23-210.5
Shelford, R. W. C. 1917.
   A naturalist in Borneo. New York: E. P. Dutton.

23-211
Shihab, M. 1982.
   Poisonous snakes. *Newsl. Ahmadi Nat. Hist. Field Study Group.* No. 21:21–23.

23-212
Shuntov, V. P. 1963.
   Morskie zmei. *Priroda (Mosc.)* 3:103–104. (In Russian)

23-213
Shuntov, V. P. 1965.
   Sea snakes. Washington, DC: U.S. Naval Oceanographic Office. 3 pages. (Translation 278-1-3 of "Morskie zmei", *Priroda (Mosc.)* 3:103–104, by the U.S. Naval Oceanographic Office)

23-214
Slater, K. R. 1961.
   Reptiles in New Guinea. *Aust. Terr.* 1:27–35.

23-215
Slevin, J. R. 1935.
   An account of the reptiles inhabiting the Galapagos Islands. *Bull. N.Y. Zool. Soc.* 38(1):2–24.

23-216
Smedley, N. 1930.
   Poisonous snakes and their venom. *Hongkong Nat.* 1(4):155–160.

23-217
Smith, D.; Westlake, M. 1982.
   The diver's guide to the Philippines. Hongkong: Unicorn Books.

23-218
Smith, F. G. W. 1957.
   Sea snakes are real. *Sea Front.* 3(2):87–93.

23-219
Smith, J. L. B. 1958.
   The deadly sea-snakes. *Veld & Vlei* 3(10):20–21,26.

23-220
Smith, J. L. B. 1968.
   High tide. Cape Town: Books of Africa.

23-220.5
Smith, M. A. 1967.
   Sea snake. *Chambers's encyclopaedia*. Rev. ed. New York: Pergamon. Vol. 12,
p.382.

23-221
Stackhouse, J. 1970.
   Australia's venomous wildlfe. New York: Paul Hamlyn.

23-221.5
Stebbins, R. C. 1985.
   A field guide to western reptiles and amphibians. Boston: Houghton Mifflin.

23-222
Stidworthy, J. 1971.
   Snakes of the world. New York: Bantam Books.

23-223
Storr, G. M. 1979.
   Dangerous snakes of western Australia. Perth: Western Australian Museum.

23-223.5
Suckling, H. J. 1876.
   *Ceylon, a general description of the island, historical, physical, statistical.
Containing the most recent information*. London: Chapman & Hall. Volume 2.

23-224
Sullivan, W. 1970.
   Panama Canal: what if snakes and starfish change oceans? *New York Times,
The Week in Review*, Dec.13, IV:12:1–3.

23-225
Sweeney, J. B. 1972.
   A pictorial history of sea monsters and other dangerous marine life. New York:
Crown.

23-226
Talizin, F. F. 1970.
   Poisonous animals of the land and sea. Moscow. (In Russian)

23-227
Taschenberg, O. 1909
   Die giftigen tiere. Stuttgart: Ferdinand Enke.

23-228
Tayless, J. 1968.
   Serpents in paradise. *Pac. Discovery* 21(3):24–26.

23-229
Taylor, R. 1965.
   Sea snakes. *Diver Int.* 2(6):4–8.

23-230
Taylor, V. 1975.
   Sea snakes. *Skin Diver* 1975(Oct.):47.

23-231
Taylor, V. 1978.
   Gentle but deadly: Australia's surprising sea snakes. *Wildlife* (*Lond.*) 20(10):448–451.

23-232
Tennent, J. E. 1861.
   Sketches of the natural history of Ceylon with narratives and anecdotes illustrative of the habits and instincts of the mammalia, birds, reptiles, fishes, insects, &c. including a monograph of the elephant and a description of the modes of capturing and training it. London: Longman, Green, Longman, and Roberts.

23-232.5
Trejos, J. F. 1937.
   Geografia de Costa Rica, fisica, politica y economica. San Jose, Costa Rica: Imprenta Universal.

23-233
Troschel, F. H.; Ruthe, J. F. 1853.
   Handbuch der zoologie. 4th ed. Berlin: Luderitz.

23-233.5
Troschel, F. H.; Ruthe, J. F. 1859.
   Handbuch der zoologie. 5th ed. Berlin: Luderitz.

23-234.
Tweedie, M. W. F. 1941.
Poisonous animals of Malaya. Singapore: Malaya Publishing House.

23-235
Tweedie, M. W. F. 1941.
The poisonous snakes of Malaya. Part III: The vipers and sea snakes. *Malay. Nat. J.* 1(4):128–134.

23-236
Tweedie, M. W. F. 1956.
Poisonous snakes in Malaya. Singapore: Government Printing.

23-236.5
Tweedie, M. W. F.; Harrison, J. L. 1954.
Malayan animal life. New York: Longmans, Green.

23-236.7
Ueno, S. 1965.
Reptilia. Okada, Y.; Uchida, K.; Uchida, T. *New illustrated encyclopedia of the fauna of Japan*. Tokyo: Hokuryu-kan. Pp. 531–547.

23-237
U.S. Department of the Air Force. 1969.
Air Force manual: survival training edition. Washington, DC: Government Printing Office.

23-239.2
U.S. Department of the Army. 1986.
Survival. Washington, DC: Government Printing Office. (*Field Manual 21–76*.)

23-239.3
U.S. Navy. Bureau of Medicine and Surgery. 1962.
Poisonous snakes of the world; a manual for use by U.S. amphibious forces. Washington, DC: Government Printing Office.

23-239.7
U.S. Navy. Bureau of Medicine and Surgery. 1970.
Poisonous snakes of the world; a manual for use by U.S. amphibious forces. Rev. ed. Washington, DC: Government Printing Office.

23-239.8
Van der Meer Mohr, J. C. 1928.
   Peoleo berhala. *Trop. Natuur (Weltevreden)* 17(6):85–97.

23-239.9
Van der Sleen, W. G. N. 1932.
   Zeeslangen. *Trop. Natuur (Weltevreden)* 21(5):84.

23-240
Verrill, A. H. 1943.
   Moeurs etranges des reptiles. Payot: Paris.

23-241
Visser, J. 1966.
   Poisonous snakes of Southern Africa and the treatment of snakebite. Cape Town: Howard Timmins.

23-242
Volsoe, H. 1956.
   Sea snakes. *The Galathea deep sea expedition 1950–1952.* New York: Macmillan. Pp. 87–75.

23-243
Voris, H. K. 1983.
   Pelamis platurus (culebra del mar, pelagic sea snake). Janzen, D. H., ed. *Costa Rican natural history.* Chicago: University of Chicago Press. Pp. 411–412.

23-244
Voris, H. K.; Voris, H.; Jeffries, W. B. 1983.
   Sea snakes: mark-release-recapture. *Bull. Field Mus. Nat. Hist.* 54(9):5–10.

23-245
Voss, G. L. 1972.
   Zoogeographical considerations of the impact of a sea-level canal across Panama. *Proc. Int. Congr. Zool.* 17(3):1–20.

23-246
Wagner, P. 1984.
   Poisonous sea snake comes ashore at Lanikai, then dies. *Honolulu Star Bull.*, Sept. 20:A-12.

23-247
Waite, E. R. 1898.
   A popular account of Australian snakes with a complete list of the species and an introduction to their habits and organisation. Sydney: Thomas Shine.

23-247.2
Wall, F. 1906.
   The poisonous snakes of India and how to recognise them. *J. Bombay Nat. Hist. Soc.* 17(1):51–71.

23-247.7
Wall, F. 1908.
   The poisonous terrestrial snakes of our British Indian dominions and how to recognise them. 2nd ed. Bombay: Bombay Natural History Society.

23-248
Wall, F. 1917.
   The poisonous terrestrial snakes of our British Indian dominions (including Ceylon) and how to recognize them. With symptoms of snake poisoning and treatment. 3rd ed. Rangoon: Smart & Mookerdum. (Reprint of *J. Bombay Nat. Hist. Soc.* 26:430–437)

23-249
Weathersbee, C. 1968.
   Linking the oceans: a sea-level Panama canal would open the gates to completely unforeseeable ecological results. *Sci. News* 94(23):578–581.

23-250
Whitaker, R. 1978.
   Common Indian snakes: a field guide. New Delhi: Macmillan.

23-251
Whitehead, H. 1957.
   Sea snakes: Indian coastal waters. *Mar. Observer* 27(176):77.

23-252
Williston, S. W. 1914.
   Water reptiles of the past and present. Chicago: University of Chicago Press.

23-253
Wilson, A. 1879.
   Leisure-time studies, chiefly biological. London: Chatto and Windus.

23-253.5
Wood, J. G. 1863.
   The illustrated natural history. London: Routledge, Warne, and Routledge.
Volume 3.

23-254
Worrell, E. 1952.
   Dangerous snakes of Australia. London: Angus and Robertson.

23-255
Worrell, E. 1961.
   Dangerous snakes of Australia and New Guinea. 4th ed. Sydney: Angus &
Robertson.

23-256
Worrell, E. 1963.
   Dangerous snakes of Australia and New Guinea. 5th ed. Sydney: Angus and
Robertson.

23-257
Worrell, E. 1963.
   *Australian snakes, crocodiles, tortoises, turtles, lizards: describing all Aus-
tralian species, their appearance, their haunts, their habits, with over 300 illustrations, many in full colour.* Sydney: Angus and Robertson.

23-258
Worrell, E. 1966.
   Australian snakes, crocodiles, tortoises, turtles, lizards. Sydney: Angus and
Robertson.

23-258.5
Worrell, E. 1966.
   *Australian wildlfe; best known birds, mammals, reptiles, plants of Australia
and New Guinea.* Sydney: Angus and Robertson.

23-259
Wright, R. 1969.
   With hook, line and snorkel in the South Pacific. Sydney: Pacific Publications.

Genus-Species Index
for Chapter 23

Chapter 24

ADDITIONAL BIBLIOGRAPHIC SOURCES

24-.01

Bauer, A. M. 1985.

An annotated bibliography of New Caledonian herpetology. *Smithson. Herpetol. Inf. Serv.* 66:1–15.

24-1

Campbell, C. H. 1976.

Snake bite, snake venoms and venomous snakes of Australia and New Guinea: an annotated bibliography. Canberra: School of Public Health and Tropical Medicine, University of Sydney.

24-2

Culotta, W. 1973.

Critical and evaluative bibliography of the literature on sea snakes, 1960–. Completed as partial fulfillment for the Masters of Library and Information Science, University of California, Los Angeles. Pp. 1–74.

24-3

DeSilva, A. 1980.

An annotated bibliography of snakes of Sri Lanka. *Snake* 12:61–108.

24-4

Gans, C. 1949.

A bibliography of the herpetology of Japan. *Bull. Am. Mus. Nat. Hist.* 93(6):389–496.

24-5

Gans, C.; Parsons, T. S. 1970.

Taxonomic literature on reptiles. Gans, C., ed. *Biology of the reptilia.* New York: Academic. Vol. 2, pp. 315–333.

24-6
Harmon, R. W.; Pollard, C. B. 1948.
Bibliography of animal venoms. Gainesville, FL: University of Florida Press.

24-7
Murthy, T. S. N. 1977.
Systematic index, distribution and bibliography of the sea-snakes of India. *Indian J. Zootomy* 18(2):131–134.

24-8
Murthy, T. S. N. 1983.
A historical resume and bibliography of the snakes of India. *Snake* 15:113–135.

24-9
Neill, W. T. 1958.
The occurrence of amphibians and reptiles in saltwater areas, and a bibliography. *Bull. Mar. Sci. Gulf Caribb.* 8(1):1–97.

24-10
Rosenfeld, G.; Kelen, E. M. A. 1969.
Bibliography of animal venoms, envenomations, and treatments (Period 1500–1968). Sao Paulo: Butantan Institute.

24-11
Russell, F. E.; Scharffenberg, R. S. 1964.
Bibliography of snake venoms and venomous snakes. West Covina, CA: Bibliographic Associates.

24-12
Russell, F. E.; Gonzalez, H.; Dobson, S. B.; Coats, J. A. 1984.
Bibliography of venomous and poisonous marine animals and their toxins. Los Angeles, CA: Attending Staff Association, Los Angeles County-University of Southern California Medical Center; Tucson, AZ: College of Pharmacy, University of Arizona.

24-13
Schmidt, K. P. 1951.
Annotated bibliography of marine ecological relations of living reptiles (except turtles). *Mar. Life; Occas. Pap.* 1(9):47–54.

24-14
Schmidt, K. P. 1957.
Reptiles (except turtles). *Geol. Soc. Am. Mem.* 67(1):1213–1216.

24-15
Smith, H. M.; Smith, R. B. 1973.
Synopsis of the herpetofauna of Mexico. Augusta, WV: Eric Lundberg. Volume 2.

24-16
Smith, H. M.; Smith, R. B. 1976.
Synopsis of the herpetofauna of Mexico. Source analysis and index to Mexican reptiles. North Bennington, VT: John Johnson. Volume 3.

24-17
Vigle, G. O.; Heatwole, H. 1978.
A bibliography of the Hydrophiidae. Washington, DC: National Museum of Natural History.

24-18
Zangerl, R. 1957.
Reptiles. *Geol. Soc. Am. Mem.* 76:1013–1017.

Author Index

Brattstrom, B. H. 13-6
Braun, M. 5-68, 23-37
Brazenor, C. W. 5-69
Breder, C. M. 8-.05
Brehm, A. 23-38
Brehm, A. E. 3-7.5, 3-8, 3-9, 23-39
Brenning, M. 23-40
Bridgers, W. F. 17-139
Briggs, J. C. 23-41
Broad, A. J. 16-17, 19-17
Brodie, A. F. 16-90
Brongersma, L. D. 3-10, 4-49, 4-50, 4-51, 4-52, 4-53, 5-70, 5-71, 5-72, 5-73, 5-73.2, 5-73.4
Brook, G. A. 18-5, 18-5.5
Brooks, D. R. 10-9, 10-10
Brown, A. E. 5-73.6, 5-73.8, 6-4.8
Brown, J. H. 16-18, 18-6, 20-16
Brown, J. N. B. 5-73.9, 5-73.95, 13-6.5, 23-41.5
Bruggren, A. C. van 9-4
Bruyant, L. 2-9.5
Bryan, W. A. 5-74
Dryckaert, M. C. 17-212
Buckland, F. 23-42
Bucknill, J. A. S. 9-4.4
Buduk, D. P. 19-24
Buffa, P. 12-20
Bullock, T. H. 13-7
Burger, W. L. 4-54
Burgersdijk, L. A. J. 23-42.5
Burggren, W. W. 14-5, 14-6
Burne, J. A. 18-7, 18-8, 18-9, 18-10, 18-11
Burns, B. 12-21, 12-22, 12-23
Burns, G. W. 3-11, 7-1, 11-5
Burt, C. E. 4-55, 5-75
Burt, M. D. 4-55, 5-75
Bush, A. O. 10-11, 10-12
Buth, D. G. 6-4.9
Butler, G. W. 12-24, 12-25
Butler, W. H. 5-75.5
Bystrov, V. F. 17-5, 17-10, 17-120, 17-187

Cadle, J. E. 6-5, 6-6
Cahill, M. 5-76, 7-2
Cahill, T. 22-4
Caira, J. N. 10-10
Caldwell, G. S. 9-5
Calmctte, A. 2-9.5, 16-18.5, 16-19, 16-20
Campbell, A. J. 9-6
Campbell, C. H. 20-17, 20-18, 20-19, 24-1
Campbell, C. W. B. 18-12

Campbell, D. G. 18-111
Campbell, H. 9-7, 22-5
Campbell, I. D. 17-61
Campbell, J. A. 3-11.5, 5-76.5
Campo, R. M. del 5-77, 5-78, 5-79
Cansdale, G. S. 21-7
Cantor, T. 3-12, 3-13, 4-56, 4-57, 5-81, 16-21
Cantor, T. E. 5-80
Caras, R. 3-14
Cardew, A. G. 4-58
Carey, J. E. 16-22, 17-11, 17-12, 18-13, 19-13
Carlsson, F. H. H. 16-16
Carpenter, D. 18-89
Carr, A. 5-82, 23-43
Carrillo de Espinoza, N. 5-82.5
Carrington, R. 3-15, 23-29
Carus, C. G. 12-26
Castellani, A. 20-19.5, 20-20
Cayley, N. W. 9-8
Cendrero, L. 23-44
Cerdas, L. 12-16, 16-13
Chabanaud, P. 12-27
Chacko, P. I. 9-9
Chalmers, A. J. 20-19.5, 20-20
Chan, K. E. 18-14, 18-15, 18-36, 18-115
Chandor, S. 19-14, 19-46
Chang, C. C. 17-37, 17-53, 17-217, 17-218, 18-16, 18-68, 19-15
Chang, H. 17-52
Chang, H. M. 6-21
Chang, L. C. 17-219
Chang, M. L. Y. 5-83
Chang, P. 18-15
Chang, T. W. 16-23
Chang, W. Y. 16-24
Chapman, D. S. 20-21, 20-134
Chasen, F. N. 5-84, 9-4.4
Chattopadhyaya, D. R. 10-13
Chau, W. C. 18-4.5
Chaudhury, L. 17-125
Cheah, L. S. 18-16.3, 18-16.5, 18-40.5
Cheeseman, T. F. 5-85, 5-86
Cheesman, R. E. 5-87
Chen, B. Y. 2-37, 4-169, 4-170, 4-171, 4-172, 5-292, 5-293, 6-21, 6-22, 10-33, 12-75, 15-13
Chen, C. N. 16-116.5, 18-123.5
Chen, R. 5-87.5
Chen, Y. C. 15-9.5, 16-53.5
Chen, Y. H. 17-13
Chen, Y. M. 17-97, 17-99, 18-68, 18-69, 18-70
Chengdu Institute of Biology, Academia Sinica 5-534

Khomenko, L. P. 5-339
Kiewiet de Jong, G. W. 4-140.5
Kikuchi, T. 18-87, 18-88
Kim, H. S. 17-25, 17-74, 17-75, 17-76, 17-77, 17-118, 17-168, 17-169, 17-172
Kimball, M. R. 17-78
Kimura, S. 17-144.7
Kinghorn, J. R. 2-31.2, 4-141, 5-244, 5-245, 5-246, 5-247, 23-131, 23-132, 23-133, 23-134, 23-135
Kini, R. M. 17-79
Kiran, U. 6-18, 12-64
Kitami, T. 5-211, 5-212, 5-213, 5-214, 5-248, 5-249, 23-121
Klauber, L. M. 8-6
Klawe, W. L. 8-7, 8-8
Klein, J. T 1-14
Klemmer, K. 3-26, 4-142, 5-250, 21-10, 21-11, 21-12, 23-136, 23-137
Kloss, C. B. 4-235, 5-430
Knauer, F. 2-31.5, 23-137.5
Kneeland, S. 23-138
Knowles, R. 4-1, 20-56
Ko, R. C. 10-28
Koba, K. 5-251, 5-252, 5-392, 20-103, 20-104, 20-105
Kobayashi, M. 17-117
Kochva, E. 12-65, 16-57, 16-58
Kogure, Y. 19-20, 19-43
Koh, K. H. 16-62
Kojima, K. 17-207, 17-208
Koketsu, K. 18-2, 18-55, 18-56, 18-60.5
Kokubun, Y. 17-119, 17-144.7
Kopstein, F. 2-31.7, 4-142.5, 4-143, 5-253, 5-254, 5-255, 5-255.5, 5-256, 5-256.5, 12-65.5, 20-57, 20-57.5, 23-139
Kosuge, T. 16-49
Kraus, R. 2-32
Krefft, G. 2-33, 2-33.5, 4-144, 4-145, 5-257, 5-258
Kropach, C. 2-34, 3-27, 5-202, 5-259, 5-498, 7-10, 7-12, 7-13, 8-9, 8-10, 9-20, 10-29, 10-66, 12-66, 23-140
Kruger, P. 10-29.4, 10-29.6
Kuba, K. 18-55, 18-56
Kudo, T. 17-220
Kuekenthal, W. 4-228.5
Kuhn, O. 4-145.5, 4-146, 4-147
Kukihara, T. 12-67
Kuntz, R. E. 2-35, 5-260, 5-261, 10-15, 10-18, 10-19, 10-45
Kuo, T. P. 20-103, 20-104, 20-105
Kuraishi, Y. 18-61

Kuramochi, H. 17-144.5
Kurka, A. 23-140.5
Kuroda, N. 23-140.7
Kwiecinski, G. 12-102

Laboute, P. 23-140.9
LaCepede, B. 4-148, 4-149, 4-150
Laidlaw, F. F. 4-151, 5-262, 5-263
Lakjer, T. 12-68
Lamar, W. W. 3-11.5, 5-76.5
Lamarck, J. B. 4-152, 4-153
Lamb, G. 16-68, 18-62, 18-76, 19-29, 19-30
Lamouroux, A. 17-173
Lampe, E. 4-153.2, 4-153.4, 4-153.6
Lance, V. 10-28, 11-16
Lanchester, W. F. 10-30
Langebartel, D. A. 12-69
Langlet, G. 17-108
Langseth, M. G. 23-4
Lanza, B. 5-264
Latifi, M. 5-265
Latreille, P. A. 4-153.8, 4-250
Laurent, R. 5-266
Laurenti, J. N. 4-154
Lauritzen, L. 19-49
Lauterwein, J. 17-80, 17-81
Lavergne, J. 12-53
Lavkumar, K. S. 9-10
Lazdunski, M. 17-81
Lee, C. Y. 16-59, 16-60, 16-60.5, 17-82, 17-83, 17-84, 17-85, 17-97, 17-99, 18-63, 18-64, 18-65, 18-66, 18-67, 18-68, 18-69, 18-70, 18-71, 18-129
Lee, S. Y. 18-71
Lee, T. J. 17-13.5
LeMare, D. W. 23-141
Lemen, C. A. 11-17
Lenfant, C. J. M. 14-38
Lenz, H. O. 23-142
Leong, L. T. 16-62
Leske, N. G. 1-14.5
LeSouef, W. H. D. 5-287, 23-149
Lesson, R. P. 4-154.5, 4-155, 5-266.5
Levey, H. A. 16-61
Levey, H. H. 16-62
Leviton, A. E. 4-234, 5-267
Liang, X. 16-118.5, 17-213.5
Liat, L. B. (see also Lim, B. L.) 4-156, 5-268, 5-269, 7-14, 8-23, 23-143, 23-143.5
Licht, P. 11-10
Lichtenstein, H. 2-35.5, 4-157
Lidth de Jeude, T. W. van 4-157.3, 4-157.5, 5-269.5, 5-270, 5-270.3, 5-270.4, 5-270.5

1

2

3

4

5

1. *Pelamis platurus.* Pelagic or yellow-bellied sea snake. Photo by J. Sneed.
2. *Hydrophis ornatus.* Okinawa. Photo by G. V. Pickwell.
3. *Aipysurus laevis.* Olive sea snake, Great Barrier Reef. Photo by Bio-Ocoanics.
4. *Laticauda colubrina.* Yellow-lipped sea krait. Photo by W. Farmer.
5. *Pelamis* coiling. Pelagic sea snake. Photo by W. Farmer.